Advances in Lubricated Bearings

Advances in Lubricated Bearings

Editors

Hubert Schwarze
Thomas Hagemann

MDPI • Basel • Beijing • Wuhan • Barcelona • Belgrade • Manchester • Tokyo • Cluj • Tianjin

Editors
Hubert Schwarze
Clausthal University of
Technology
Germany

Thomas Hagemann
Clausthal University of
Technology
Germany

Editorial Office
MDPI
St. Alban-Anlage 66
4052 Basel, Switzerland

This is a reprint of articles from the Special Issue published online in the open access journal *Lubricants* (ISSN 2075-4442) (available at: https://www.mdpi.com/journal/lubricants/special_issues/lubricated_bearings).

For citation purposes, cite each article independently as indicated on the article page online and as indicated below:

LastName, A.A.; LastName, B.B.; LastName, C.C. Article Title. *Journal Name* **Year**, *Volume Number*, Page Range.

ISBN 978-3-0365-6289-6 (Hbk)
ISBN 978-3-0365-6290-2 (PDF)

Contents

About the Editors

Hubert Schwarze

Hubert Schwarze (Prof. Dr.-Ing.) is a University Professor and head of the Institute of Tribology and Energy Conversion Machinery at Clausthal University of Technology since 2000. For several decades, the Institute of Tribology and Energy Conversion Machines has been conducting experimental and theoretical research in the field of static and dynamic behavior of sliding radial- and thrust bearings and their interactions with surrounding structures. Through numerous individual and cooperative investigations, practical case studies as well as theoretical studies, and model developments, the research institute has significant expertise in the field. Professor Schwarze's main research interests include tribology with a focus on rotor-slide bearing systems, tribo-life spans, rotor dynamics, and rheology.

Thomas Hagemann

Thomas Hagemann (Dr.-Ing. habil.) is a Senior Researcher at the Institute of Tribology and Energy Conversion Machinery at Clausthal University of Technology. His research focuses on improving the performance, efficiency, and robustness of thrust and journal bearings in challenging machinery applications ranging from high-speed turbomachinery to heavy-duty industrial ones. Besides the development of novel algorithms, he puts a special emphasis on validation with identification data of academic test rigs and practical applications. The results of his research contribute to the TEHL bearing codes of the Institute of Tribology and Energy Conversion Machinery have become valuable tools in practical industrial development and design processes. In his field, he is a key researcher in the German Research Associations for Combustion Engines (FVV) and Drive Technology (FVA).

lubricants

Editorial

Advances in Lubricated Bearings

Hubert Schwarze * and Thomas Hagemann

Institute of Tribology and Energy Conversion Machinery, Clausthal University of Technology, 38678 Clausthal-Zellerfeld, Germany; hagemann@itr.tu-clausthal.de
* Correspondence: schwarze@itr.tu-clausthal.de; Tel.: +49-5323-72-2464

Citation: Schwarze, H.; Hagemann, T. Advances in Lubricated Bearings. *Lubricants* **2022**, *10*, 156. https://doi.org/10.3390/lubricants10070156

Received: 12 July 2022
Accepted: 12 July 2022
Published: 14 July 2022

Publisher's Note: MDPI stays neutral with regard to jurisdictional claims in published maps and institutional affiliations.

Advances in the design and development of lubricated bearings have been a goal of tribology engineers over decades, as the requirements on efficiency, power density, and robustness continuously increase in the history of rotating machinery. Today, numerous applications of bearings exist operating under different boundary conditions with high variety in bearing size, speed, and loads. Each of these aspects involves specific challenges ranging from the manufacturing process to the demands in operation. The latter ones typically include a sufficiently low wear level and acceptable vibrations ensuring safe and stable operation. Besides the general task of optimizing well-known bearing solutions, novel fields of application frequently occur. Solving the issues appearing during operation of all of these bearings in complex drive trains requires innovative approaches based on the experience and the physical understanding of the particular phenomena by tribology engineers. This Special Issue (SI) contributes to the latest steps in understanding bearing operating behavior [1–6], its interaction with lubricants [7,8], and its role as a component in the drive train [9–12].

Stottrop et al. [1] and Buchhorn et al. [2] investigate the operating behavior of original size large turbine tilting-pad bearings on a special test rig. The first study [1] compares the performance of the same test bearing with different lubrication designs provoking flooded conditions in the first case and non-flooded ones in the second case. The non-flooded design exhibits higher film thickness and lower power loss while the temperature level does not change significantly. In their second investigation [2], the authors show that the precise modification of convective heat transfer outside the lubricant gap by systematic modification of oil flow is an appropriate measure to optimize the operating behavior of tilting-pad bearings. In the investigated case, a trailing edge lube oil pocket designed to improve heat transfer at the pads' trailing edge free surface provides significant lower maximum pad metal temperatures with only slightly increasing power loss for the test bearing with evacuated housing. Schüler and Berner [3] propose a different approach to reduce temperatures in high-speed journal bearings. They apply so-called eddy grooves in the highly loaded pad region in order to generate radial flow components to reduce temperature gradients in the film. Experimental results for a five-pad tilting-pad bearing provide a reduction of maximum pad temperature of over 20 K for high surface speeds and simultaneously high mechanical loads. Tauviqirrahman et al. [4] investigate a different aspect of surface structure. In their theoretical study, the authors apply three types of heterogeneous/smooth sliding surface arrangements to a journal bearing in a CFD analysis and gain an improvement of load-carrying capacity and average acoustic power level compared to the conventional plain surface design. Colombo et al. [5] derive a design procedure for passively compensated bearings controlled by diaphragm valves applied in aerostatic bearings. Numerical investigations for a single pad show that significant improvements of stiffness can be achieved by the valve design. Moreover, sensitivity analysis of the design procedure proves its suitability for a wide range of parameters. Chernets et al. [6] present a calculation procedure for metal-polymer bearings operating at the lower end of Stribeck curve at boundary or dry friction. The authors put an emphasis on the impact of Young's modulus reduction with increasing temperature on the predicted results and establish an improvement of the operating behavior due to the decrease in the rigidity of the polymer composites of the bearing bushing.

The tribological interaction of bearing surface and the lubricant is a central question in bearing applications. Khaskhoussi et al. [7] study the hydrophobicity and oleophilic behavior of lead/lead-free bronze coatings for varying dimple diameters and density of laser textured surfaces. The experimental results highlight that the porous textured surface serves as an oil reservoir due to its good oleophilic behavior ensuring better lubrication and higher wear resistance. In addition, relevant hydrophobicity is observed, suggesting that surface texturing promotes the water-repellant barrier effect on the surface. Nassef et al. [8] investigate the dynamic behavior of ball bearings lubricated by lithium grease with different fractions of reduced graphene oxide (rGO) as a nano-additive. In their experiments, a significant reduction of the temperature level and improvement of damping is obtained by the application of rGO compared to base lithium grease showing the significance of this additive in enhancing bearing performance.

In high-speed turbomachinery applications, hydrodynamic journal bearings are the main damping element to ensure limited vibrational amplitudes in the resonance frequencies of the rotor-bearing system and rotordynamic stability within the entire operating range of the machine. In this domain, turbochargers used, for example, in automotive applications represent one of the most complex rotor-bearing system since they run at extremely high rotating frequencies and simultaneously low static loads. Generally, this combination encourages non-linear system behavior since self-excited sub-synchronous vibrations induced by the oil films arise. To enable safe operation, the bearing design commonly features two oil films connected by a full- or semi-floating bush to provide damping by the second film if the other one becomes unstable. Adiletta [9] theoretically studies the impact of non-circular bore profiles on the stability of a rigid rotor supported by full-floating bush bearings. The non-circular profile is applied on the stator side of the outer film. The investigations show an influence of the geometrical shape of the sliding surface as well as of its angular orientation in the bore. Results indicate that a suitable design is able to increase stability threshold speed. Ziese et al. [10] analyze the impact of bearing model complexity on predicted operating behavior of a turbocharger with semi-floating bush bearings. Validation with test data shows that increasing model complexity improves correspondence with experimentally determined vibrations. In particular, the cavitation in the journal bearing and the consideration of the commonly neglected thrust bearing contribute to the enhancements. Completely different issues are involved in the bearing arrangements of slow running planetary gear units for wind turbines. Hagemann et al. [11] investigate the impact of the special load situation in these type of bearings that exist due to the helical gears applied here. Consequently, the planet bearing has to restore high force as well as moment loads generated by the two mesh contacts between sun and planet gear and planet and ring gear, respectively. Different design measures and load situations are analyzed considering operation in the mixed friction regime and potentially occurring wear mechanisms. Moreover, the high flexibility of the structure due to the lightweight requirements and the simultaneously high mechanical loads accompany structure deformations that lead to modifications of the shape of the lubricant gap. Hagemann et al. [12] focus on this phenomenon in the second part of their study and show that its consideration is essential for the simulation results. Generally, the results indicate an improvement of predicted operating conditions by the consideration of structure deformation in the bearing analysis for this application. The peak load in the bearing decreases since the loaded proportion of the sliding surface increases.

Although lubricated bearings have been applied for centuries and studied with continuously improving methods for decades, continuous improvements of bearings and new challenges accompanying with their application are still part of current research. This SI contributes to the progress in this field. The guest editors would like to express their sincere thanks to the authors, reviewers and the editorial staff of MDPI Lubricants that helped to develop and finalize this SI.

Funding: This research received no external funding.

Conflicts of Interest: The authors declare no conflict of interest.

References

1. Stottrop, M.; Buchhorn, N.; Bender, B. Experimental Investigation of a Large Tilting-Pad Journal Bearing—Comparison of a Flooded and Non-Flooded Design. *Lubricants* **2022**, *10*, 83. [CrossRef]
2. Buchhorn, N.; Stottrop, M.; Bender, B. Influence of active cooling at the trailing edge on the thermal behavior of a tilting-pad journal bearing. *Lubricants* **2021**, *9*, 26. [CrossRef]
3. Schüler, E.; Berner, O. Improvement of Tilting-Pad Journal Bearing Operating Characteristics by Application of Eddy Grooves. *Lubricants* **2021**, *9*, 18. [CrossRef]
4. Tauviqirrahman, M.; Jamari, J.; Wicaksono, A.A.; Muchammad, M.; Susilowati, S.; Ngatilah, Y.; Pujiastuti, C. CFD Analysis of Journal Bearing with a Heterogeneous Rough/Smooth Surface. *Lubricants* **2021**, *9*, 88. [CrossRef]
5. Colombo, F.; Lentini, L.; Raparelli, T.; Trivella, A.; Viktorov, V. Design and Analysis of an Aerostatic Pad Controlled by a Diaphragm Valve. *Lubricants* **2021**, *9*, 47. [CrossRef]
6. Chernets, M.; Pashechko, M.; Kornienko, A.; Buketov, A. Study of the Influence of Temperature on Contact Pressures and Resource of Metal-Polymer Plain Bearings with Filled Polyamide PA6 Bushing. *Lubricants* **2022**, *10*, 13. [CrossRef]
7. Khaskhoussi, A.; Risitano, G.; Calabrese, L.; D'Andrea, D. Investigation of the Wettability Properties of Different Textured Lead/Lead-Free Bronze Coatings. *Lubricants* **2022**, *10*, 82. [CrossRef]
8. Nassef, M.G.; Soliman, M.; Nassef, B.G.; Daha, M.A.; Nassef, G.A. Impact of Graphene Nano-Additives to Lithium Grease on the Dynamic and Tribological Behavior of Rolling Bearings. *Lubricants* **2022**, *10*, 29. [CrossRef]
9. Adiletta, G. Stability Effects of Non-Circular Geometry in Floating Ring Bearings. *Lubricants* **2020**, *8*, 99. [CrossRef]
10. Ziese, C.; Irmscher, C.; Nitzschke, S.; Daniel, C.; Woschke, E. Run-Up Simulation of a Semi-Floating Ring Supported Turbocharger Rotor Considering Thrust Bearing and Mass-Conserving Cavitation. *Lubricants* **2021**, *9*, 44. [CrossRef]
11. Hagemann, T.; Ding, H.; Radtke, E. Schwarze, H. Operating behavior of sliding planet gear bearings in wind turbine gearbox applications—Part I: Basic relations. *Lubricants* **2021**, *9*, 97. [CrossRef]
12. Hagemann, T.; Ding, H.; Radtke, E.; Schwarze, H. Operating behavior of sliding planet gear bearings in wind turbine gearbox applications—Part II: Impact of structure deformation. *Lubricants* **2021**, *9*, 98. [CrossRef]

Article

Experimental Investigation of a Large Tilting-Pad Journal Bearing—Comparison of a Flooded and Non-Flooded Design

Michael Stottrop *, Nico Buchhorn and Beate Bender

Product Development, Ruhr-University Bochum, 44801 Bochum, Germany; nico.buchhorn@rub.de (N.B.); beate.bender@rub.de (B.B.)
* Correspondence: michael.stottrop@rub.de

Abstract: In tilting-pad journal bearings (TPJB), power loss corresponds to the internal friction in the shearing of the oil. Besides the lubrication gap, intermediate spaces between the pads account for a notable amount of frictional losses. Against the background of increasing demands for efficiency and sustainable use of resources, the reduction of power loss takes a key position in the further development of bearings. In our research, we compare two bearing lubrication concepts of a five-pad TPJB. Our objective is to work out the influence of different lubrication methods and bearing housing designs on the bearing operation characteristics. We conduct experimental testing of a 500 mm TPJB in two different bearing configurations with respect to the lubrication concept: an oil-flooded and non-flooded bearing design. In the flooded bearing design, oil is supplied via spray-bars and axial seals ensure the inter-pad spaces to be completely filled with oil. The non-flooded design comes without axial seals but oil drain channels to avoid oil accumulation in the bearing. In the latter design, oil is fed in via leading edge grooves (LEG). For the non-flooded bearing design, the experimental data show that the unloaded pads are not completely filled with oil and therefore, no pressure build-up occurs. The absence of additional load on the lower pads compared to the flooded design results in an increase of minimum film thickness. With the non-flooded design, power loss at high speeds is reduced to almost half. As a result, the efficiency of the entire turbomachinery application can be considerably improved.

Keywords: tilting-pad journal bearing; experimental investigation; power loss; lubrication method; flooded bearing design; non-flooded bearing design; spray-bar; leading edge groove

Citation: Stottrop, M.; Buchhorn, N.; Bender, B. Experimental Investigation of a Large Tilting-Pad Journal Bearing—Comparison of a Flooded and Non-Flooded Design. *Lubricants* **2022**, *10*, 83. https://doi.org/10.3390/lubricants10050083

Received: 28 February 2022
Accepted: 28 April 2022
Published: 3 May 2022

Publisher's Note: MDPI stays neutral with regard to jurisdictional claims in published maps and institutional affiliations.

1. Introduction

Hydrodynamic bearings are used in a wide range of applications in rotating machinery. The wear-free operation under full lubrication and the easy mounting are the main advantages. In turbomachinery, tilting pad journal bearings (TPJB) play an important role as they show good stiffness and damping characteristics and do not tend to cause self-excited vibrations. With an increasing demand for better efficiency of high-speed machinery, the aim is to keep the power loss of the bearings as low as possible.

In hydrodynamic bearings, the power loss is equivalent to the heat resulting from friction. The heat is dissipated as an unavoidable consequence of the shearing of the oil, which is vital for the pressure build-up in the lubrication gap between shaft and bearing. Besides the lubrication gap, notable friction power occurs in other areas of the bearing that do not contribute to pressure build-up. In TPJB, the friction in the intermediates space between two pads account for a significant proportion of the power loss. Depending on the mixture of oil and air, the amount of friction power in these pad intermediate spaces can be up to 50% of the total bearing power loss [1]. For a fixed-pad bearing, Hagemann and Schwarze [2] find that up to 35% of the power loss relates to dissipation in the interpad spaces.

Regarding the lubrication concept of TPJB, an oil-flooded and non-flooded design can be assumed as two limiting cases. While in a flooded bearing design the intermediate spaces between the pads are fully filled with oil, in a non-flooded design these spaces are theoretically oil-free. In both cases the intermediate spaces do not contribute to the hydrodynamic load carrying capacity of the bearing, but have an effect on the converted frictional power.

Regardless of the bearing design, a relatively simple approach to reduce friction power loss is to adjust oil supply. Simmons and Dixon [3] investigate experimentally a considerable reduction of lubricant supply for a 200 mm five-pad TPJB in conventional design without compromising the reliability of the bearing. By reducing the lubricant flow rate to half, they lower power loss by 20% at a circumferential speed of 105 m/s. Studies of San Andres et al. [4,5] show similar results for two 102 mm TPJB with flooded seal ends. For a five-pad TPJB [4], a 15% decrease in power loss at shaft speed 74 m/s is realized by a flow reduction of 50%. With the same flow reduction for a four-pad TPJB [5], power loss is reduced by 19% at 64 m/s.

However, the oil supply cannot be reduced at arbitrary rates, since starvation in the lubrication gap and rising temperatures are limiting factors [6]. Therefore, alternative lubrication methods have been developed to supply oil directed towards the leading edge region of the pad or directly into the lubrication gap [7,8]. Harangozo et al. [9] compare the effects of a directed lubrication, where oil is fed in via spray-bars with several inlet holes, a direct lubrication via leading-edge-grooves (LEG) and a conventional flooded lubrication on the performance of a 127 mm four-pad TPJB. With the directed and direct lubrication, the bearing is designed without axial seals, so that bearing is not flooded with oil. In the experimental results, the flooded bearing shows the greatest power loss, while for the directed lubricated bearing the lowest power loss is observed. The authors attribute the differences in the spray-bar and LEG lubricated bearing to the generally lower temperatures and thus, the higher viscosity with the LEG lubricated bearing as well as to the higher shear stress losses within the LEG. Experimental analyses of a 120 mm four-pad TPJB in a spray-bar and direct LEG lubricated configuration at a broad range of rotational speed show similar results [10]. A comparison of both lubrication methods shows a considerably lower frictional power for spray-bar lubrication than for the LEG lubrication at higher speeds. Hagemann and Schwarze relate these findings to the fact, that with the LEG nearly the entire oil is supplied to the lubrication gap, while with the spray-bars a bypass flow emerges.

A proven method for an evacuation of the intermediate spaces between the pads to reduce maximum bearing temperatures is a TPJB design with open end seals and large drain channels [7,11]. Nicholas [12] investigates a 102 mm five-pad TPJB in a flooded pressurized housing design in comparison with an evacuated housing design featuring directed lubrication on the temperatures without considering power loss. Dmochowski and Blair [13] examine a 99 mm five-pad TPJB in an evacuated housing design, realized by enlarging the clearance of the end seals. Compared to a flooded bearing, oil evacuation leads to a reduction in power loss of 25% at highest shaft speeds 77.5 m/s. With an additional reduction in oil flow by approximately 30%, a decrease in power loss of 12% is observed. Bang et al. [14] evaluate a 301 mm six-pad TPJB with conventional and LEG lubrication method, each with and without end seals. Detecting a reduction of power loss in the bearings without end seals, they find the lowest power loss for the conventional lubricated bearing without seals (39.2% lower than for the conventional bearing with seals). As in other works, a decrease in power loss is achieved by reducing oil flow rate. Sano et al. [15] obtain comparable results for a large-scale TPJB with nominal diameter 890 mm. By leaving out the upper two pads of a four-pad TPJB in a non-flooded design with spray-bar lubrication, power loss is decreased to less than half.

In this paper, we experimentally investigate a five-pad TPJB with nominal diameter 500 mm in a flooded and non-flooded design. In the flooded design, the bearing is axially sealed and assumed to be fully filled with oil. By leaving out the axial seals in the non-flooded design, intermediate spaces between pads are only partially filled with oil. The aim of our research is to analyse how the bearing design and lubrication method influence the load carrying capacity and efficiency of the bearing. The experimental testing takes place at the Bochum test rig for large journal bearings. We measure the main operating parameters oil film thickness, oil film pressure and pad temperatures as well as the friction power. We compare the measurement data of both bearing designs and identify the impact of the lubrication method on the bearing characteristics. The approach presented in this paper allows for a direct comparison of different lubrication concepts for a large-scale TPJB. In industrial applications, bearings are frequently operated at reduced oil supply rates without modifying the bearing design. Therefore, oil accumulation may occur despite the non-full flooding. The test bearing in the non-flooded design in our research is designed in such a way that no oil accumulates in the pad intermediate spaces and non-flooding can be explicitly investigated.

2. Experimental Set-Up

2.1. Test Rig

We carry out the experimental testing on the Bochum test rig for large journal bearings. The test rig was built in the 1980s to examine original-sized turbine bearings with a nominal diameter of 500 mm. The maximum length of the test bearing is 500 mm. The shaft is powered by a 1.2 MW DC drive (Brown, Boveri & Cie (BBC, Baden, Switzerland)) and can be accelerated up to a rotational speed of 4000 rpm depending on the friction loss what corresponds to a maximum circumferential velocity of 104.72 m/s. The lubricant used is a turbine oil ISO VG 32 (Esso Teresso 32). Figure 1 shows a top view (left) and a technical drawing (right) of the test rig.

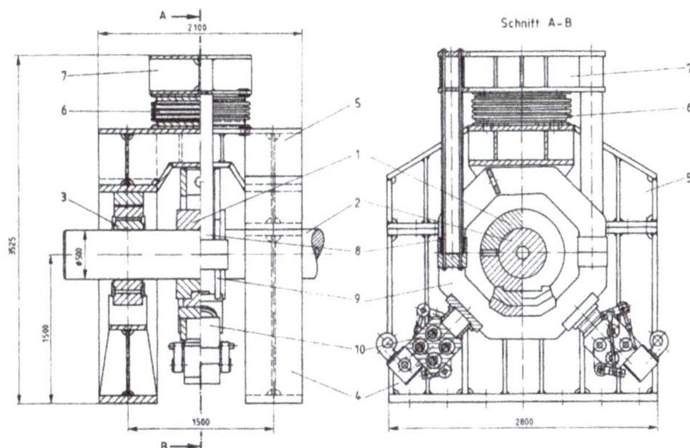

Figure 1. Top view (**left**) and technical drawing (**right**) of the test rig for large journal bearings; Test rig parts: Test bearing (1), shaft (2), support bearings (3), lower part of the test rig frame (4), upper part of the test rig frame (5), pneumatic bellow (6), traverse (7), draw bars (8), rigid frame (9), vibration generators (10).

In the test rig, the shaft (2) is supported by two symmetrically aligned support bearings (3), which are designed as three-pad TPJB. The test bearing (1) is arranged centrally between the support bearings in a rigid frame (9). The rigid frame is connected to the upper part of the test rig frame (5) via two draw bars (8), a traverse (8) and a pneumatic bellow (6). The upper part and the two support bearings are supported by the lower part of the test rig (4). By pressurizing the bellow, the test bearing is pulled against the shaft from below with a maximum bearing force of 1 MN.

The shaft is designed as a hollow shaft and equipped with a film thickness and pressure measuring system. Two capacitive distance sensors and two pressure probes are mounted in the mid-plane of the shaft at 90° to each other with same sensor types being 180° apart. A shifting device at the drive end of the shaft allows to shift the rotating shaft in axial direction and thus the sensor plane across the whole test bearing width during measurement. With this equipment, both fluid film thickness and pressure distribution can be captured in a high-resolution two-dimensional (2D) data field. In circumferential direction 240 data points are measured, while in axial direction up to 4000 data points can be recorded, depending on the rotational speed. Due to the capacitive measurement principle, the fluid film thickness can only be determined reliably in areas where the space between shaft and bearing is completely filled with oil. By contrast, areas of cavitation and oil-air-mixture can be detected by this characteristic.

Beyond the stationary force, the test bearing can be loaded by sinusoidal dynamic forces. Two vibration generators (10) are mounted to the rigid frame (9) at 90° to each other and at 45° to the vertical load direction. The additional dynamic forces created by the vibration generators are detected with the pressure probes. Relative shaft movement is measured by four eddy current distance probes to determine the dynamic spring and damping coefficients of the test bearing. A detailed description of the vibration generators, the measurement system and post-processing is presented in Kukla et al. [16].

2.2. Test Bearing Designs

The test bearing in our investigations is a five-pad TPJB with nominal diameter 500 mm. The bearing is loaded in a load-between-pad configuration with double-tilt supported pads. The tilting in both axial and circumferential direction for a better compensation of misalignment between shaft and bearing is enabled by an elliptical pivot geometry. This pivot geometry results from a manufactured radius in axial direction of both, the pads back and bearing ring. The two lower, highly loaded pads are each equipped with a hydrostatic jacking groove above the pivot area. Both pads are manufactured with an axial concave profile on the running surface, following Kukla et al. [17] to compensate for thermal crowning at high specific loads. The bearing key parameters are summarized in Table 1.

Table 1. Bearing key parameters.

Parameter	Symbol	Unit	Value
Nominal diameter	D	mm	500
Bearing length	B	mm	350
Number of tilting-pads	-	-	5
Angular pad length	Ω	°	56
Pivot offset	-	-	0.6
Relative bearing clearance	Ψ	‰	1.2
Radial clearance	C_R	µm	300
Pad preload	m_p	-	0.538
Pad back radius	r_p	mm	287
Pad back radius (axial)	$r_{p,ax}$	mm	60,000
Bearing ring radius	r_r	mm	322.5
Bearing ring radius (axial)	$r_{r,ax}$	mm	1×10^{20}
Pad thickness	t_{pad}	mm	72.5
White-metal layer thickness	t_{WM}	mm	2.265

We investigate the test bearing in a flooded and non-flooded design. In the flooded design, oil is supplied via spray-bars with each 19 holes in the intermediate space between two pads (Figure 2). In circumferential direction, the pads are kept in place by the spray-bars. Axial sealings on both sides ensure that the bearing is permanently filled with oil and therefore in a flooded condition (Figure 3).

In contrast, no axial sealings are inserted in the non-flooded design and the intermediate space between the pads is enlarged in radial direction. Additionally, oil drain channels and large spacing between pads and axial bearing covers are added to avoid oil accumulation in the bearing except for the lubrication gap (Figure 3). For a sufficient feeding of the lubrication gap with oil, the spray-bars are replaced with a 'leading edge groove' (LEG) in the non-flooded design directly attached at the pads leading edge (Figure 2). This directed lubrication allows to supply oil only in the dedicated areas. The two lower pad are each equipped with additional 'trailing edge grooves' (TEG) for active cooling of the pads trailing edge, since the heat transfer in this region is much lower compared to the flooded bearing design ([18]). A safety mechanism mounted in the pivot area to keep the pads in place hinders the oil feed through the pivot. Hence, oil is fed in via the pads axial edges. In both bearing designs, flooded and non-flooded, the spray-bars respectively the LEG and TEG are supplied separately via several internal galleries and a ring channel between bearing ring and housing. For the investigations in this paper, no oil is supplied in the TEG.

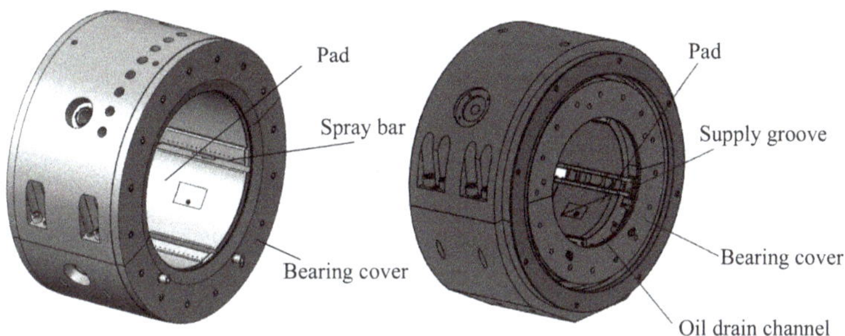

Figure 2. 3D-model of the test bearing in flooded (**left**) and non-flooded design (**right**).

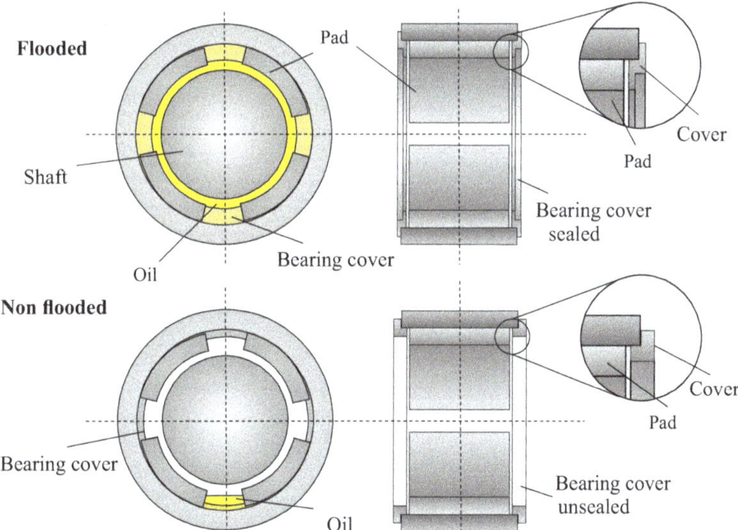

Figure 3. Schematic drawing of the flooded and non-flooded bearing design.

2.3. Calorimetric Determination of Frictional Power

In hydrodynamic bearings, the frictional power is the power converted into heat in the gap between oil-wetted surfaces. The friction power is defined as:

$$P_{fr} = M_{shaft} \cdot \omega_{shaft} \qquad (1)$$

For the test bearing, we assume that the frictional power is mainly dissipated via the lubricating oil. During testing, the oil volume flow rate $\dot{V}$, the oil inlet temperature T_{in} and outlet temperature T_{out} are measured. The out-flowing oil is first led into a mixing tank before it is returned to the oil tank. At the outlet of the mixing tank, the temperature is captured at ten measuring points, which are recorded by a data logger. By sorting out the minima and maxima and subsequent averaging, reliable results can be obtained with regard to the oil outlet temperature. Based on these measurement data, the converted frictional power at constant density ρ and constant specific heat capacity c_p can be determined calorimetrically using the following expression:

$$P_{fr} = \dot{V} \cdot \rho \cdot c_p \cdot (T_{out} - T_{in}) \tag{2}$$

Considering the mass flow $\dot{m}$, which remains constant due to mass conservation, instead of the volume flow $\dot{V}$ and the temperature dependence of specific heat capacity c_p, leads to the following expression:

$$P_{fr} = \dot{m} \cdot \int_{T_{in}}^{T_{out}} c_p(T)\, dT = \dot{V}_{in} \cdot \rho(T_{in}) \cdot \int_{T_{in}}^{T_{out}} c_p(T)\, dT \tag{3}$$

According to [19], the temperature-dependent density for lubricating oil can be calculated using the density ρ_{15} at $15\,°\mathrm{C}$ as follows:

$$\rho(T) = \rho_{15} \cdot e^{-\alpha_{15} \cdot \Delta T \cdot (1 + \alpha_{15} \cdot 0.8 \cdot \Delta T)} \tag{4}$$

with $\alpha_{15} = 0.6278/\rho_{15}$ and $\Delta T = T[°\mathrm{C}] - 15$. For the relevant temperature range, there is a linear relationship between the specific heat capacity c_p and the temperature T, resulting in the following expression for the above integral:

$$\int_{T_{in}}^{T_{out}} c_p(T)\, dT = (T_{out} - T_{in}) \cdot c_p(T_m) \tag{5}$$

using the mean oil temperature $T_m = \frac{T_{in} + T_{out}}{2}$. Taking the correlations shown into account, the equation for the calorimetric determination of the frictional power is:

$$P_{fr} = \dot{V}_{in} \cdot \rho(T_{in}) \cdot c_p(T_m) \cdot (T_{out} - T_{in}). \tag{6}$$

3. Experimental Results

In the following section, the operating characteristics of the non-flooded bearing design are compared with those of the flooded one [20]. For both bearing designs, flooded and non-flooded, the nominal oil flow rate is 7.1 L/s.

As an example, Figure 4 shows the lubricant film thickness and Figure 5 the lubricant film pressure of both bearing designs plotted against the bearing circumference in the bearing center. The peaks in lubricant film thickness for pads 3 and 4 ($\approx 135°$–$150°$ and $207°$–$222°$) and the corresponding disruption in pressure are due to the hydrostatic jacking grooves. In the flooded version, all five pads show complete gap filling, while in the non-flooded version only the lower loaded pads 3 and 4 are fully filled. Here, a significant difference between the two designs becomes apparent. Due to the sealed bearing configuration on the sides, in the flooded design the complete bearing is almost completely filled with oil. Each pad is provided with a sufficient amount of lubricant and a hydrodynamic pressure build-up is generated. By redesigning the spaces between the pads (elimination of the spray-bars) and increasing the axial distance between the pads and the bearing cover in the non-flooded design, pressure build-up of the lubricant is not possible. Only the oil provided at the leading edge is supplied to each pad. To completely fill the gaps of the non-loaded pads, the total volume flow supplied would have to be significantly increased.

The differences described can also be found in the measured pressure distribution. While in the non-flooded bearing design only the loaded pads generate a hydrodynamic pressure build-up, in the flooded design a pressure build-up is observed for all five pads, as shown in Figure 5. As a result, the lower pads 3 and 4 are subject to additional loads from upper pads 1, 2 and 5 in addition to the external load, resulting in higher maximum pressure and smaller film thickness.

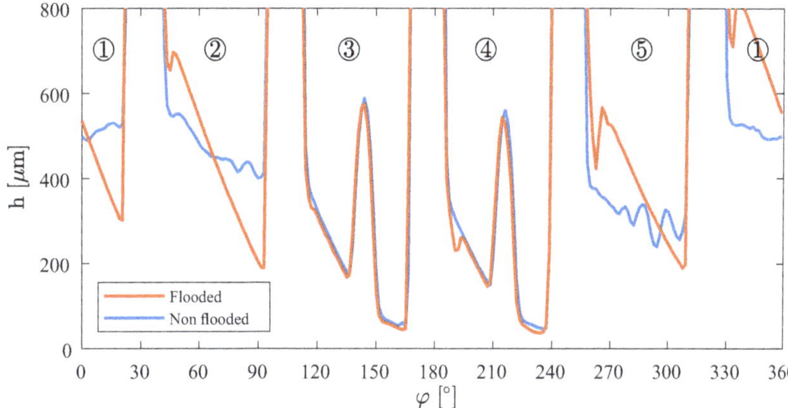

Figure 4. Experimentally determined lubricant film thickness over the bearing circumference in the bearing center (pads 1–5) of the flooded and non-flooded bearing design. Operating point: $n = 3000\,\mathrm{rpm}$ and $\bar{p} = 3.00\,\mathrm{MPa}$.

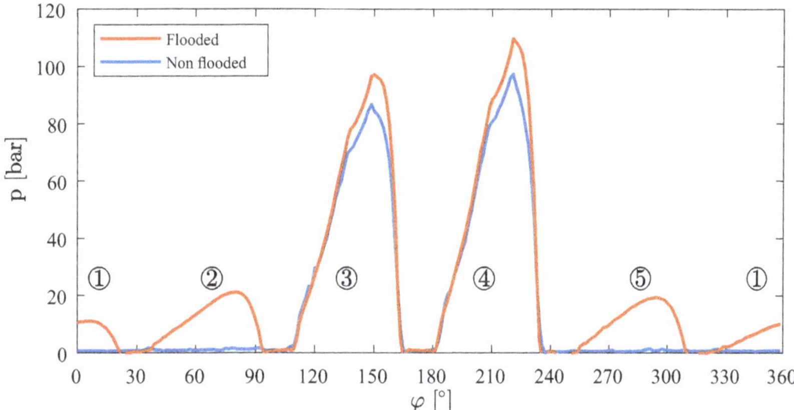

Figure 5. Experimentally determined lubricant film pressure over the bearing circumference in the bearing center (pads 1–5) of the flooded and non-flooded bearing design. Operating point: $n = 3000\,\mathrm{rpm}$ and $\bar{p} = 3.00\,\mathrm{MPa}$.

The temperature curves of a pad of the respective bearing designs are in qualitative agreement, see Figure 6. A comparison of the measured temperatures at the leading edge of the pad shows differences in the gap entry temperatures between the flooded and non-flooded versions of the bearing. For the flooded design, no particular difference across the bearing width can be detected at the pad beginning. A clear difference in the initial temperatures can be seen across the pad width for the non-flooded bearing configuration (ax. edge–center). The temperatures measured laterally are approx. $10\,\mathrm{K}$ below the temperature in the center of the pad. The reason for the temperature differences at the beginning of the pad is the different lubricant supply systems of the two bearing designs. In the flooded version, the fresh oil is supplied via spray-bars in the inter-pad spaces, resulting in a more homogeneous mixing temperature across the bearing width. In the non-flooded design, the lubricant is supplied via a groove at the leading edge of the pad. Due to the groove design, the temperature of the fresh oil supplied dominates in the peripheral areas. In the center of the bearing, the lubricant is mixed with the hot oil carried over from the previous pad, so that the temperatures in this area are higher than those in the peripheral areas and the axial temperature gradient is ultimately established at leading edge.

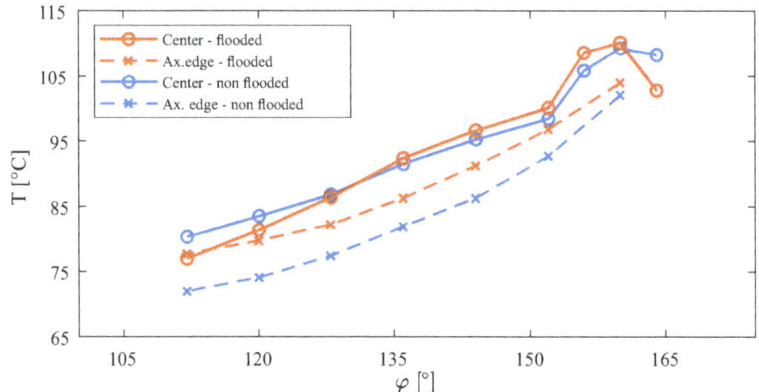

Figure 6. Temperatures of pad 3 in the axial center of the bearing (z = 0 mm) and near the edge (z = −120 mm) for the different bearing designs. Operating point: n = 3000 rpm and $\bar{p}$ = 3.00 MPa.

Figure 7 shows the experimentally determined maximum bearing temperatures (left) and the minimum lubricant film thicknesses (right) of the different bearing designs above the specific bearing load at speed of 3000 rpm. It can be seen that the curves of the maximum bearing temperatures of the flooded and non-flooded variants are consistent between 2.0 MPa and 3.5 MPa. Only in the lower bearing load range and at 4.00 MPa, the non-flooded bearing operates at a lower maximum temperature. The direct comparison of the measured minimum film thickness shows that the bearing in the flooded design has a significantly lower film thickness over the complete load range than in the non-flooded design.

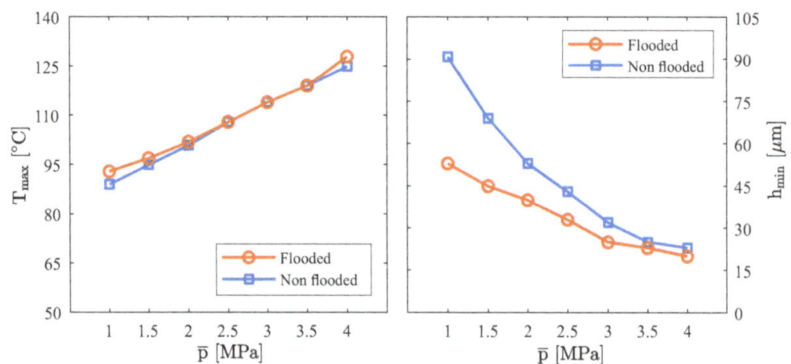

Figure 7. Experimentally determined maximum bearing temperature and frictional power of the different bearing designs as a function of the specific bearing load. Speed: n = 3000 rpm.

A comparison of the experimentally determined power losses of the flooded and non-flooded bearing configurations shows that the non-flooded design allows for a significant reduction of the power loss. For this purpose, the left part of the Figure 8 plots the measured power losses of both variants at a specific bearing load of 2.00 MPa as a function of speed. The corresponding measured shaft temperatures are plotted in the right part of the Figure.

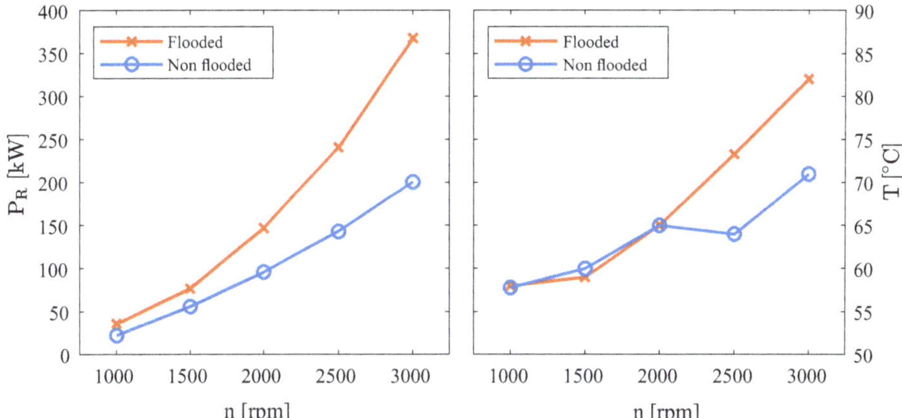

Figure 8. Experimentally determined frictional power and shaft temperature of the flooded and non-flooded bearing design as a function of speed at a specific bearing load of 2.00 MPa.

On the one hand, the reduction in frictional power is due to the non-flooded gaps and, on the other hand, to the incipient deficient lubrication of the unloaded pads. The oil-free gaps and the oil-air mixture in the incompletely filled gaps result in a lower shear stress between the surfaces compared with a fully filled bearing. When looking at the associated measured shaft temperatures (Figure 8 on the right), the occurring deficient lubrication of the unloaded pads in the non-flooded bearing design becomes apparent. While in the flooded lubrication the shaft temperature rises steadily above speed, indicating complete gap filling of all pads above speed, the course of the non-flooded design shows a slight drop in shaft temperature between speeds 2000 rpm and 2500 rpm. This temperature drop is due to the onset of insufficient lubrication of the unloaded pads. In these areas, the shaft is cooled or, in direct comparison with flooded lubrication, the heat input is reduced by the energy dissipated in the lubrication gap. This effect can be observed particularly at comparatively low loads, while at high loads partial filling of the unloaded pads can be observed over the entire speed spectrum (no drop in the measured shaft temperature over the speed). For this purpose, Figure 9 shows the measured shaft temperatures of both bearing variants versus speed for different bearing loads.

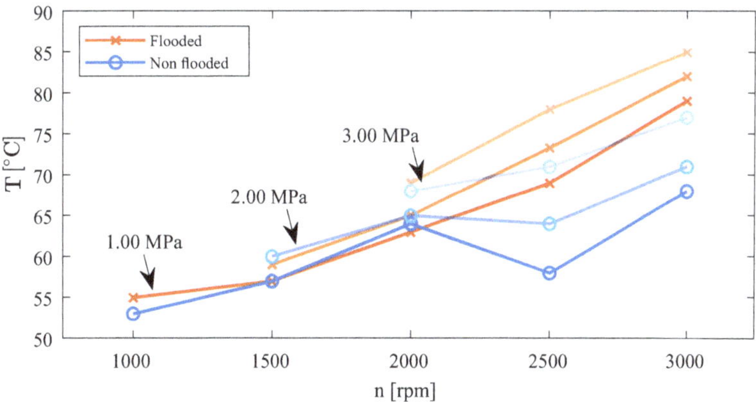

Figure 9. Experimentally determined shaft temperature of the flooded and non-flooded bearing design as a function of speed for different bearing loads.

4. Discussion and Conclusions

In our investigations, we have experimentally examined a bearing design in two different configurations. The two bearings differ only in the oil feed and the axial sealing. All other parameters are identical.

By comparing the experimentally determined operating characteristics of the flooded and non-flooded design, it can be shown that the non-flooded design has a positive influence on the load carrying capacity (higher minimum lubricant film thickness at the same load level). Mainly this can be related to the non-existing hydrodynamic pressure build-up of the unloaded pads. Due to the pressure build-up in the upper pads in the flooded version, there is an additional load on the lower loaded pads. The measured pad temperatures are at an identical level, but differ at the pad edges due to the different oil regimes in the intermediate spaces.

However, the major difference between the two configurations is evident in the measured frictional powers. The power loss at a speed of 3000 rpm can be reduced by almost half by designing the bearing in a non-flooded configuration. This reduction in power loss is attributed to the unloaded pads and the spaces between them not being filled with oil. In these areas, there is less shear stress on the shaft surface and the frictional power converted in the bearing is reduced.

In conclusion, we demonstrate that the concept of lubrication can positively influence the frictional performance of a TPJB without reducing or increasing the operational safety parameters such as minimum lubricant film thickness or maximum bearing temperature.

Author Contributions: Conceptualization, M.S. and N.B.; methodology, M.S. and N.B.; experimental investigation, M.S. and N.B.; writing—original draft preparation, M.S. and N.B.; writing—review and editing, M.S., N.B. and B.B.; supervision, B.B.; project administration, N.B.; funding acquisition, N.B. All authors have read and agreed to the published version of the manuscript.

Funding: This Project is supported by the Federal Ministry for Economic Affairs and Climate Action (BMWK) on the basis of a decision by the German Bundestag. The financial support was assigned by the Industrial Research Association (AiF e. V.) in the project "Verbesserte Effizienz großer, schnelllaufender Radialkippsegmentgleitlager für verlustleistungskritische Anwendungen".

Data Availability Statement: The data presented in this study are available on request from the corresponding author.

Acknowledgments: We acknowledge support by the Open Access Publication Funds of the Ruhr-Universität Bochum. The authors would like to thank D. Schüler for his valuable advice and GTW Alpen, especially C. Weißbacher, for the support of the experimental investigation and project administration.

Conflicts of Interest: The authors declare no conflict of interest.

Abbreviations

B	Bearing length
c_p	Specific heat capacity
C_R	Radial clearance
D	Nominal diameter
h	Film thickness
m_p	Pad preload
$\dot{m}$	Mass flow
M	Torque
n	Rotor speed
p	Film pressure
$\bar{p}$	Specific bearing load
P_{fr}	Frictional power
P_R	Power loss
r	Radius

t	Thickness
T	Temperature
ΔT	Temperature difference
$\dot{V}$	Volume flow
ρ	Density
φ	Angular coordinate
Ψ	Relative bearing clearance
ω	Rotational speed
LEG	Leading edge groove
TEG	Trailing edge groove
TPJB	Tilting-pad journal bearing

References

1. Buchhorn, N. Einfluss einer Niederdrucktasche auf die Wärmeabfuhr an der Segmenthinterkante eines großen Radialkippsegmentlagers. Ph.D. Thesis, Ruhr-Universität Bochum, Bochum, Germany, 2020.
2. Hagemann, T.; Schwarze, H. A Model for Oil Flow and Fluid Temperature Inlet Mixing in Hydrodynamic Journal Bearings. *J. Tribol.* **2019**, *141*, 021701. [CrossRef]
3. Simmons, J.E.L.; Dixon, S.J. Effect of Load Direction, Preload, Clearance Ratio, and Oil Flow on the Perfonmance of a 200 mm Journal Pad Bearing. *Tribol. Trans.* **1994**, *37*, 227–236. [CrossRef]
4. San Andrés, L.; Jani, H.; Kaizar, H.; Thorat, M. On the Effect of Supplied Flow Rate to the Performance of a Tilting-Pad Journal Bearing—Static Load and Dynamic Force Measurements. *J. Eng. Gas Turbines Power* **2020**, *142*, 121006. [CrossRef]
5. San Andrés, L.; Toner, J.; Alcantar, A. Measurements to Quantify the Effect of a Reduced Flow Rate on the Performance of a Tilting Pad Journal Bearing with Flooded Ends. *J. Eng. Gas Turbines Power* **2021**, *143*, 111012. [CrossRef]
6. Nicholas, J.C.; Elliott, G.; Shoup, T.P.; Martin, E. Tilting Pad Journal Bearing Starvation Effects. In Proceedings of the 37th Turbomachinery Symposium, Houston, TX, USA, 8–11 September 2008; pp. 1–10. [CrossRef]
7. Nicholas, J.C. Tilting Pad Bearing Design. In Proceedings of the 23rd Turbomachinery Symposium, Dallas, TX, USA, 13–15 September 1994; pp. 179–194. [CrossRef]
8. Dmochowski, W.; Brockwell, K.; DeCamillo, S.; Mikula, A. A Study of the Thermal Characteristics of the Leading Edge Groove and Conventional Tilting Pad Journal Bearings. *J. Tribol.* **1993**, *115*, 219–226. [CrossRef]
9. Harangozo, A.V.; Stolarski, T.A.; Gozdawa, R.J. The Effect of Different Lubrication Methods on the Performance of a Tilting-Pad Journal Bearing. *Tribol. Trans.* **1991**, *34*, 529–536. [CrossRef]
10. Hagemann, T.; Schwarze, H. Theoretical and Experimental Analyses of Directly Lubricated Tilting-Pad Journal Bearings with Leading Edge Groove. *J. Eng. Gas Turbines Power* **2019**, *141*, 051010. [CrossRef]
11. Tanaka, M. Thermohydrodynamic Performance of a Tilting Pad Journal Bearing with Spot Lubrication. *J. Tribol.* **1991**, *113*, 615–619. [CrossRef]
12. Nicholas, J. Tilting Pad Journal Bearings with Spray-Bar Blockers And By-Pass Cooling For High Speed, High Load Applications. In Proceedings of the 32nd Turbomachinery Symposium, Houston, TX, USA, 8–11 September 2003; pp. 27–38. [CrossRef]
13. Dmochowski, W.M.; Blair, B. Effect of Oil Evacuation on the Static and Dynamic Properties of Tilting Pad Journal Bearings. *Tribol. Trans.* **2006**, *49*, 536–544. [CrossRef]
14. Bang, K.B.; Kim, J.H.; Cho, Y.J. Comparison of power loss and pad temperature for leading edge groove tilting pad journal bearings and conventional tilting pad journal bearings. *Tribol. Int.* **2010**, *43*, 1287–1293. [CrossRef]
15. Sano, T.; Magoshi, R.; Shinohara, T.; Yoshimine, C.; Nishioka, T.; Tochitani, N.; Sumi, Y. Confirmation of performance and reliability of direct lubricated tilting two pads bearing. *Proc. Inst. Mech. Eng. Part J. Eng. Tribol.* **2015**, *229*, 1011–1021. [CrossRef]
16. Kukla, S.; Hagemann, T.; Schwarze, H. Measurement and Prediction of the Dynamic Characteristics of a Large Turbine Tilting-Pad Bearing Under High Circumferential Speeds. In Proceedings of the ASME Turbo Expo: Turbine Technical Conference and Exposition, San Antonio, TX, USA, 3–7 June 2013. [CrossRef]
17. Kukla, S.; Buchhorn, N.; Bender, B. Design of an axially concave pad profile for a large turbine tilting-pad bearing. *Proc. Inst. Mech. Eng. Part J. Eng. Tribol.* **2017**, *231*, 479–488. [CrossRef]
18. Buchhorn, N.; Stottrop, M.; Bender, B. Influence of Active Cooling at the Trailing Edge on the Thermal Behavior of a Tilting-Pad Journal Bearing. *Lubricants* **2021**, *9*, 26. [CrossRef]
19. DIN 51757:2011-01. *Prüfung von Mineralölen und Verwandten Stoffen—Bestimmung der Dichte*; Deutsches Institut für Normung e. V, Beuth Verlag GmbH: Berlin, Germany, 2011. [CrossRef]
20. Kukla, S. Erhöhung der Tragfähigkeit großer Radialkippsegmentlager durch axiale Profilierung der Segmentlauffläche. Ph.D. Thesis, Ruhr-Universität Bochum, Bochum, Germany, 2017.

 *lubricants*

Article

Influence of Active Cooling at the Trailing Edge on the Thermal Behavior of a Tilting-Pad Journal Bearing

Nico Buchhorn, Michael Stottrop * and Beate Bender

Product Development, Ruhr-University Bochum, 44801 Bochum, Germany; nico.buchhorn@rub.de (N.B.); beate.bender@rub.de (B.B.)
* Correspondence: michael.stottrop@rub.de

Abstract: In tilting-pad journal bearings (TPJB) with a non-flooded lubrication concept, higher maximum pad temperatures occur than with a flooded bearing design due to the lower convective heat transfer at the pad edges. In this paper, we present an approach to influence the thermal behavior of a five-pad TPJB by active cooling. The aim of this research is to investigate the influence of additional oil supply grooves at the trailing edge of the two loaded pads on the maximum pad temperature of a large TPJB in non-flooded design. We carry out experimental and numerical investigations for a redesigned test bearing. Within the experimental analysis, the reduction in pad temperature is quantified. A simulation model of the bearing is synthesized with respect to the additional oil supply grooves. The simulation results are compared with the experimental data to derive heat transfer coefficients for the pad surfaces. The experimental results indicate a considerable reduction of the maximum pad temperatures. An overall lower temperature level is observed for the rear pad in circumferential direction (pad 4). The authors attribute this effect by a cooling oil carry-over from the previous pad (3). Within the model limits, a good agreement of the simulation and experimental results can be found.

Keywords: tilting-pad journal bearing; pad deformation; heat transfer; trailing edge groove; CFD; power loss

Citation: Buchhorn, N.; Stottrop, M.; Bender, B. Influence of Active Cooling at the Trailing Edge on the Thermal Behavior of a Tilting-Pad Journal Bearing. *Lubricants* **2021**, *9*, 26. https://doi.org/10.3390/lubricants9030026

Received: 31 January 2021
Accepted: 22 February 2021
Published: 2 March 2021

Publisher's Note: MDPI stays neutral with regard to jurisdictional claims in published maps and institutional affiliations.

1. Introduction

In turbomachinery, hydrodynamic bearings are mainly applied to support rotor shafts. Due to the continuous development of fast running turbomachinery, the requirements for the bearings with regard to load carrying capacity, operational safety and efficiency also increase. To ensure safe operation, wear due to mixed friction is prevented, the permissible maximum temperature is not exceeded and mechanical overstressing of the materials is avoided. In addition, an economic aim is to achieve the lowest possible power loss of the bearings.

The power loss of a bearing corresponds to the power converted into heat due to friction. This friction is caused by shearing of the oil in the lubrication gap and is inevitably linked to the hydrodynamic principle. However, friction does not only occur in the lubricating gap between the shaft and the running surface, where the bearing's load capacity is generated. Also in the areas that do not contribute to the bearing's load carrying capacity, a notable proportion of frictional power arises. In tilting-pad journal bearings (TPJB), the power loss occurs mainly in the intermediate spaces between two consecutive pads. The frictional power in these areas can, depending on the oil-air mixture, amount up to 50% of the total power loss of the bearing.

Basically, there are two lubrication concepts for TPJB: flooded and non-flooded design. With a flooded lubrication concept, the intermediate spaces between the pads are fully filled with oil. In a non-flooded bearing design, there is theoretically no oil in the intermediate space and therefore less friction compared to the flooded design. While power loss can be reduced considerably with a non-flooded lubrication concept, maximum temperatures at

---

the pads trailing edge increase. In a flooded bearing design, high flow velocities and the mixing of fresh and draining oil effect a high convective heat transfer at the leading and trailing edge of the pads. The convective heat transfer at the trailing edge is assumed to have a major influence on the maximum pad temperature, as the intermediate spaces in a non-flooded bearing are almost oil-free. Therefore, the temperature rise with a non-flooded lubrication concept is due to the lower convective heat transfer at the pads edges.

Improving the heat transfer at the trailing edge seems to be a reasonable approach to reduce maximum pad temperatures in non-flooded bearing design. In practice, suitable design changes have to be applied to the corresponding bearings. In this context, we investigate the following research question: How does active cooling at the trailing edge influence the thermal behavior of a large TPJB?

1.1. State of the Art

1.1.1. Bearing Calculation Models

Theoretical models are developed for a better understanding of the physical processes in hydrodynamic bearings. Validating these models with experimental data increases model complexity and enables phenomena such as heat transfer to be considered. For the development of hydrodynamic bearings, such models are vital in order to make accurate predictions of the bearing performance. Currently, there are several theoretical approaches to the calculation of hydrodynamic bearings.

The investigations of Mittwollen [1], Gerdes et al. [2] and Fuchs [3] form the basis for the calculation tool Allgeimes LagerProgramm mit 3-dimensionalem Temperature-influss, English: General Bearing Program with 3-dimensional Temperature Influence (ALP3T). Waltermann [4] determines bearing characteristics considering two-dimensional deformation of the running surface. The journal bearing calculation tool COMputation of Bearings for ROtor Systems (COMBROS-R) represents the current state of the art and offers a wide range of calculation and functions with regard to static and dynamic bearing characteristics, bearing geometry and model depth [5–7]. Simulation results of both tools, ALP3T and COMBROS-R, show good agreement with experimental data [7,8].

According to the current state of the art, the oil flow in the bearing is not modeled except for the lubrication gap. Thus, in TPJB oil flow in oil supply grooves and the intermediate space between the pads is not taken into account. A consideration of coupled heat flows at the interfaces of the individual regions (bearing ring-lubricating film-shaft) is therefore not possible. To describe the heat dissipation at the interfaces of the bearing, heat transfer coefficients (HTC) are required at the free pad surfaces (leading edge, trailing edge, axial edges, pads back) or the outer system boundaries.

1.1.2. Heat Transfer in Thrust Bearings

Various approaches to determine convective HTC theoretically or experimentally can be found in literature. Most of these publications focus on thrust bearings. Ettles et al. [9] develop a method to model heat transfer at the free pad surfaces considering rotational speed, oil viscosity and pad length. The calculation for this method is derived from temperature measurement data of seven different thrust bearings. Based on the empiric approach of Ettles et al. [9] Ettles [10] investigates a large tilting-pad thrust bearing. Heinrichson et al. [11] also use the calculation method proposed by Ettles et al. [9] to determine the HTC at the free pad surfaces and pads back of a thrust bearing. Heinrichson et al. [12] contrast the calculation results with experimental data and obtain a good agreement between simulation and measurement.

A detailed overview of the published HTCs of different authors with regard to thrust bearings is given by Wodtke et al. [13]. A total of 15 different bearings of different diameters (including the bearings investigated by Ettles et al. [9]) are considered. The outer diameters of the bearings vary between 149 mm and 3100 mm. The HTC of all pad surfaces used are between $100\,\mathrm{W/(m^2\,K)}$ and $3000\,\mathrm{W/(m^2\,K)}$. Some authors additionally describe the convective coefficients as a function of the rotational speed in order to model an influence of the flow change in the pads intermediate spaces. Wodtke et al. [13] compare two different approaches to describe the heat transfer in their study. On the one hand, a constant heat transfer at all surfaces of $750\,W/(m^2\,K)$ is assumed and, on the other hand, a HTC determined with the help of computational fluid dynamics (CFD) analysis is considered. The validation based on experimental data shows a considerable improvement of the theoretical results with the variable, but lower HTC.

Papadopoulos et al. [14] define HTC between $20\,W/(m^2\,K)$ and $1000\,W/(m^2\,K)$ when studying a thrust bearing. The ambient temperature is assumed to be equal to the feed temperature of $40\,°C$. A comparison to experimental data is not given.

1.1.3. Heat Transfer in Journal Bearings

Taniguchi et al. [15] present an approach to calculate the bearing properties of a TPJB, also taking into account the heat conduction through the pads. For the HTC they assume $115\,W/(m^2\,K)$ up to $350\,W/(m^2\,K)$ and set a calculated mixing temperature as the ambient temperature. The calculation results show good agreement with the experimental data of a $\varnothing$-479 mm bearing. Within the FVA project no. 577 *Improved Journal Bearing Calculation* Hagemann [5] examines a $\varnothing$-500 mm five-pad TPJB for validation of COMBROS-R. A HTC of $250\,W/(m^2\,K)$ is assumed at all free pad surfaces with an ambient temperature of $50\,°C$. The calculation procedure and the results are also discussed in the publications of Hagemann et al. [16] and Kukla et al. [17] and show good agreement between calculation results and experimental data. Kukla et al. [18] apply an optimization procedure to the same $\varnothing$-500 mm five-pad TPJB already used by Hagemann [5]. They determine the HTC at the free pad surfaces by minimizing the deviations between measured and calculated measuring point temperatures. Depending on the pad surface, the HTC are in a range of $490\,W/(m^2\,K)$ up to $2664\,W/(m^2\,K)$. The experimentally measured oil drain temperature is used as the ambient temperature.

Sano et al. [19] investigate a $\varnothing$-890 mm two-pad TPJB. Based on CFD analyses, which are not described in detail, a range of HTC of $100\,W/(m^2\,K)$ up to $1000\,W/(m^2\,K)$ is given depending on the operating conditions of the bearing. The ambient temperature is set approximately equal to the oil drain temperature.

Hagemann et al. [20] examine both theoretically and experimentally different methods of oil feed of a $\varnothing$-120 mm test bearing. With the help of CFD investigations, HTC at the leading and trailing edge are derived and classified on the basis of experimentally recorded data. Depending on the feed, the values range between $250\,W/(m^2\,K)$ and $1000\,W/(m^2\,K)$ for the leading edge and $1000\,W/(m^2\,K)$ and $1900\,W/(m^2\,K)$ for the trailing edge with an assumed ambient temperature of $50\,°C$. Arihara et al. [21] observe a $\varnothing$-101.6 mm four-pad TPJB at an ambient temperature of $40\,°C$. Due to turbulent flow and oil mixing in the pads intermediate spaces, a notably higher HTC of $1750\,W/(m^2\,K)$ is assumed at the leading and trailing edge than of $500\,W/(m^2\,K)$ for the axial edges. Similar values are given by Hagemann et al. [22] for a high-speed four-pad TPJB with a diameter of 120 mm. The HTC determined by CFD analyses are in a range of $1500\,W/(m^2\,K)$ up to $1900\,W/(m^2\,K)$ at the trailing edge and at the leading edge between $300\,W/(m^2\,K)$ and $400\,W/(m^2\,K)$.

1.1.4. Design Changes for Improvement of Bearing Characteristics

The variety of different HTC indicates insufficient knowledge of the phenomena and the associated heat transfer in the intermediate spaces between the pads. All publications mentioned are concerned with a better prediction or description of the heat transfer without aiming to influence the heat transfer by design modification on a theoretical level. However, several researchers carry out design changes to improve the bearing characteristics of various journal bearings.

Chen et al. [23] and Chen et al. [24] develop an isothermal circular-cylindrical journal bearing, which is equipped with methanol-filled heat transfer chambers below the running surface in the bearing ring. In the thermally loaded areas of the bearing, the methanol evaporates, transports the heat away and heats the unloaded areas of the bearing.

Compared to a conventional bearing, the isothermal bearing has notably lower operating temperatures, as the test data prove. Experimental investigations by Nicholas [25] show that so-called spraybar blockers in combination with bypass cooling have a positive effect on the thermal behavior of the bearing. The spraybar blockers are a kind of scraper edge that reduces the excess hot oil. By means of bypass cooling, fresh oil is fed directly to the pads back, which is designed with cooling grooves. Martsinkovsky et al. [26] arrange oil wipers between the pads in a three-pad TPJB to minimize the effect of hot oil carry-over. In addition, several cooling channels in axial direction are manufactured at the trailing edge below the surface. The authors report a temperature reduction compared to the bearing design without the channels.

Mermertas et al. [27] and Mermertas et al. [28] publish an improved design of a $\varnothing$-900 mm TPJB that previously exhibited high bearing temperatures and a noticeable feed temperature dependence during operation. Based on extensive thermomechanical analyses, the support position, pad dimensions and oil feed are modified. With these modifications, the bearing shows a more robust and reproducible characteristic. Hermes [29] presents a strong chamfering of the pads trailing edge to improve the operating characteristics. Compared to the initial situation, the bearing can be operated at considerably reduced temperatures with the same load capacity.

Kukla [30] improves the steady-state operating characteristics of a $\varnothing$-500 mm five-pad TPJB by means of an axially concave profiling of the running surface. The profiling ensures that the running surface deformations occurring under high thermal loads can be almost completely compensated. Compared to the characteristic values of the critical operating point of the non-profiled initial form, the profiled bearing can be operated at the same operating point with a notably reduced temperature and increased oil film thickness. Kukla et al. [31] and Buchhorn et al. [32] present two theoretical approaches to influence and improve the steady-state operating characteristics of a $\varnothing$-500 mm TPJB without experimental validation. Both approaches focus on a single pad in their simulation model, which considers additional volume flows through the pad into or out of the lubrication gap. Kukla et al. [31] use the oil film pressure at the trailing edge to transfer the heated oil to the pads back. Due to the heating and the more homogeneous temperature distribution, the thermomechanical deformations can be reduced. Buchhorn et al. [32] inject a small volume flow of cold oil into the lubrication gap. The additional cold oil volume flow effects a kind of flushing away of the warm oil from the running surface and therefore reduces the maximum oil film temperatures and the pad deformation.

Most design changes aim at improving the bearing characteristics by geometrical variations of the bearing parameter. Only Nicholas [25] investigates enhanced heat transfer for free pad surfaces. A design modification to improve heat transfer at the trailing edge of a pad is currently not known.

1.2. Scope and Aims

In this paper, we present an approach to influence the thermal behavior of a five-pad TPJB by active cooling. The cooling is realized by adding oil supply grooves at the trailing edge of the two highly loaded pads. The aim of this research is to investigate the influence of these additional oil supply grooves on the maximum pad temperature of a large TPJB in non-flooded design. We carry out experimental and numerical investigations for the redesigned test bearing from Kukla's studies [30]. Within the experimental analysis, the reduction of pad temperature is quantified. A simulation model of the bearing is synthesized with respect to the additional oil supply grooves. We compare the simulation results with the experimental data to derive HTC for the pad surfaces. With these coefficients, the simulation model allows us to make better predictions for the operating characteristics of large TPJB.

2. Experimental Set-Up

2.1. Test Rig

The experiments for this paper are carried out on a test rig for large journal bearings. The test rig was designed in the 1980s to examine original sized turbine bearings under practical operating conditions. The nominal bearing diameter is 500 mm with a maximum length of 500 mm. By means of a 1.2 MW DC drive (Brown, Boveri & Cie. (BBC)), the shaft can be run up to a rotational speed of 4000 rpm depending on friction loss. Thus, circumferential velocities up to 104.7 m/s can be achieved. The lubricant used is turbine oil ISO VG 32 (Esso Teresso 32, Hamburg, Germany). Figure 1 shows the top view of the test rig and a technical drawing.

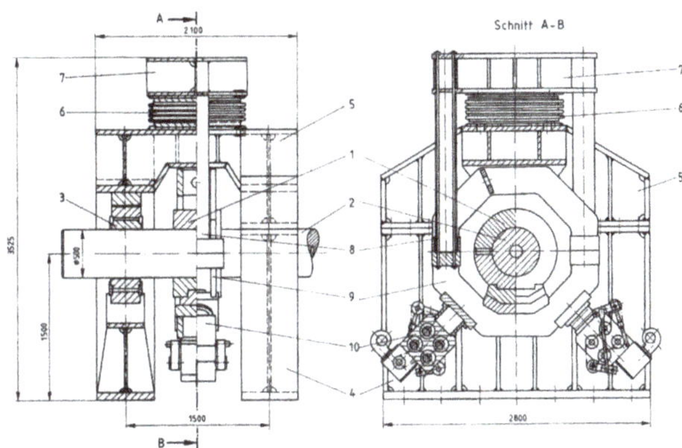

Figure 1. Top view and technical drawing of the test rig; Test rig parts: test bearing (1), shaft (2), support bearings (3), lower part of the test rig frame (4), upper part of the test rig frame (5), pneumatic bellow (6), traverse (7), drawbars (8), rigid frame (9), vibration generators (10).

In the test rig, the shaft (2) is supported by two symmetrically aligned support bearings, which are designed as tilting-pad bearings with three pads (3). The test bearing (1) is attached to a rigid frame (9) centrally between the support bearings. The rigid frame is connected to a pneumatic bellow (6) via two drawbars (8) and a traverse (7). The pneumatic bellow is mounted on the upper part of the test rig frame (5). This upper part and the two support bearings are supported by the lower part of the test rig frame (4). By pressurizing the bellow, the bearing is pulled against the shaft from below. A maximum bearing force of 1 MN can be applied.

The shaft, designed as a hollow shaft, is equipped with to capacitive distance sensors and two piezoelectric pressure sensors. The sensors are arranged in the mid plane of the shaft with a circumferential distance of 90° to each other. Sensors of the same type are 180° apart. The sensor plane can be moved across the whole test bearing width by axially shifting the rotating shaft during measurement. Thus, both the fluid film thickness and pressure distribution can be measured in a high-resolution two-dimensional (2D) data field. In circumferential direction 240 data values are captured. Depending on the rotational speed, up to 4000 data values can be recorded in axial direction. Due to the use of capacitive distance sensors, the film thickness can only be measured reliably in areas where the gap is completely filled with oil. However, areas of cavitation and deficient lubrication can be detected by this characteristic.

In addition to stationary force, the test bearing can be loaded by sinusoidal dynamic forces. Two vibration generators (10) are attached to the rigid frame at 90° to each other and each at 45° to the stationary force in vertical direction. To determine the dynamic coefficients of the test bearing, relative movement between shaft and test bearing is measured by four eddy current distance sensors. Kukla et al. [17] give a detailed description of the vibration generators, the measurement system and post-processing.

2.2. Test Bearing

The test bearing is a five-pad TPJB in a load between pad configuration. The nominal bearing diameter is 500 mm. Both, the pads back and bearing ring are manufactured with a radius in axial direction leading to an elliptical pivot geometry. This pivot geometry enables a double-tilt support of the pads, hence a tilting in both axial and circumferential direction to better compensate for misalignment between shaft and bearing. The two lower, highly loaded pads each feature hydrostatic jacking grooves on the running surface above the pivot area. In both pads, the running surface is manufactured with an axial concave profile following Kukla [30] to compensate for thermal crowning at high specific loads. Table 1 shows the bearing key parameter.

Table 1. Bearing key parameters.

Parameter	Symbol	Unit	Value
Nominal diameter	D	mm	500
Bearing length	B	mm	350
Number of tilting-pads	-	-	5
Angular pad length	Ω	°	56
Pivot offset	-	-	0.6
Relative bearing clearance	Ψ	‰	1.2
Radial clearance	C_R	µm	0.3
Pad preload	m_p	-	0.538
Pad back radius	r_p	mm	287
Pad back radius (axial)	$r_{p,ax}$	mm	60,000
Bearing ring radius	r_r	mm	322.5
Bearing ring radius (axial)	$r_{r,ax}$	mm	1×10^{20}
Pad thickness	t_{pad}	mm	72.5
White-metal layer thickness	t_{WM}	mm	2.265

Oil is supplied via a 'leading edge groove' (LEG) mounted directly on the pads leading edge (Figure 2). This directed lubrication allows a non-flooded lubrication concept for the bearing to reduce friction loss. Therefore, no axial seals are inserted and the intermediate space between the pads is enlarged in radial direction. Additionally, the bearing is designed with oil drain channels and large spacing between pads and axial bearing covers to avoid oil accumulation in the bearing except for the lubrication gap. The two lower pads are each equipped with additional 'trailing edge grooves' (TEG) for active cooling of the trailing edge. A safety mechanism mounted in the pivot area to keep the pads in place hinders the oil feed through the pivot. Hence, oil is fed in via the pads axial edges (Figures 2 and 3). Both supply grooves, LEG and TEG are supplied separately via several internal galleries between bearing ring and housing (Figure 3).

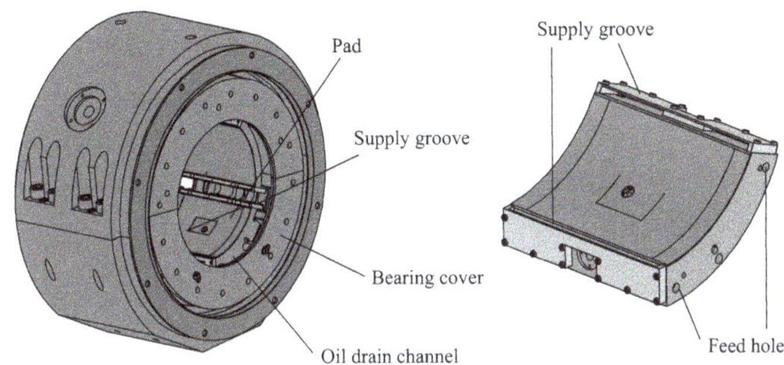

Figure 2. 3D-Model of the test bearing and a single pad.

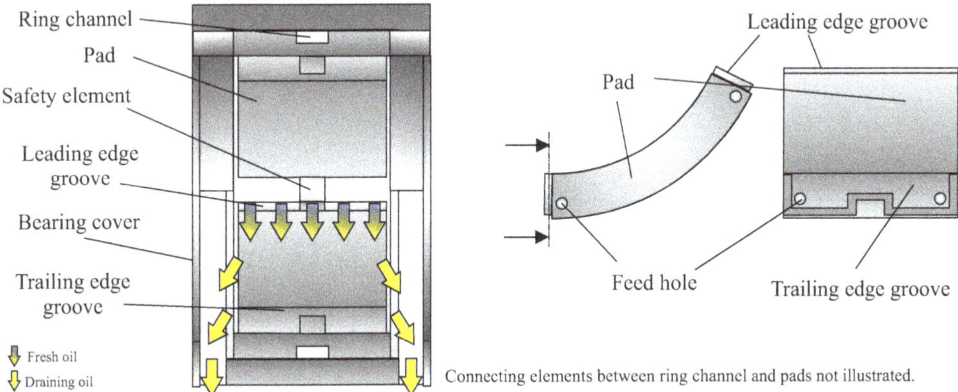

Figure 3. Principle drawing of the test bearing and a pad with supply grooves.

3. Experimental Results and Discussion

The experimental results in hydrodynamic operation are presented below. Tests were carried out under purely stationary operating conditions. The main oil supply for the bearing was 7.1 L/s at a feed temperature of 50 °C at all operating points. The results are compared in each case without and with feeding of the additional groove at the end of the two loaded pads. The following total cooling volume flows are examined:

- 0.5 L/s (0.25 L/s per TEG)
- 1.0 L/s (0.50 L/s per TEG)
- 1.5 L/s (0.75 L/s per TEG)
- 2.0 L/s (1.00 L/s per TEG)

The temperature of the cooling oil supplied is also 50 °C. It is assumed that the volume flow is divided equally between both grooves at the pad ends.

The comparison to a bearing without cooling is aimed to investigate the direct effect of an additional groove at the end of thermally highly loaded pads. This is given if the additional grooves are not supplied.

The axial concave profiling has been determined by Kukla [30] for a specific bearing load of 2.75 MPa. However, the experimental investigations show that a residual profiling remains under the thermal conditions that arise at this bearing load. This axially concave residual profiling decreases with increasing load. Therefore, for the following consideration of the experimental data of the test bearing, an operating point at a speed of 3000 rpm and a specific bearing load of 3.00 MPa is selected.

The graphs of the oil film thickness measured in the center of the bearing at a specific bearing load of 3.00 MPa and a speed of 3000 rpm are shown in Figure 4 without and with additional cooling.

A comparison of the three curves shows that the experimentally recorded oil film thicknesses differ only slightly. The gaps of the two loaded pads 3 and 4 are completely filled, while those of the unloaded pads 1, 2 and 5 do not show complete filling—recognisable from the non-steady course of the measured values. An influence of the additional cooling on the radial pad deformation in the area of the pad ends (pad 3: approx. 172°; pad 4: approx. 238°) cannot be determined from the measurement data. No influence on the axial deformations could be detected during the tests either.

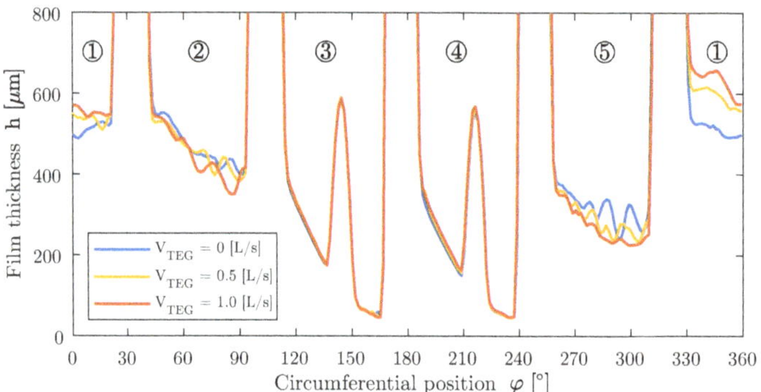

Figure 4. Experimentally determined oil film thickness (µm) over the circumference in the bearing center (pads 1–5) with and without feeding of the TEG. Operating point: n = 3000 rpm and $\bar{p}$ = 3.00 MPa.

The effect of the additional trailing edge cooling becomes obvious (Figure 5), if the curves of the temperatures in the center of the bearing for the loaded pads 3 and 4 are considered: While only the temperature level at the end of the pad is influenced for pad 3, the entire temperature level is reduced for pad 4.

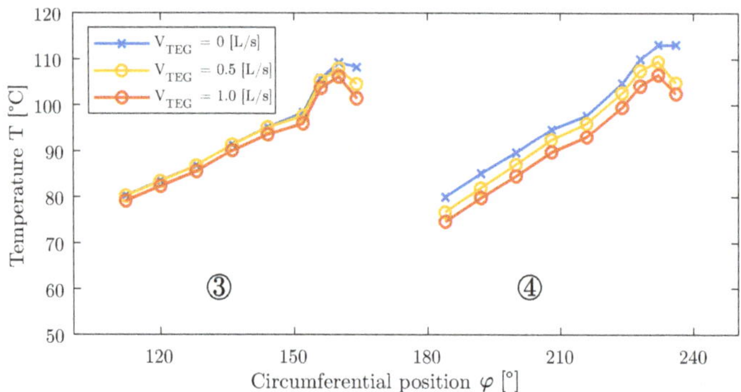

Figure 5. Measuring point temperatures (°C) of pads 3 and 4 in the axial center of the bearing (z = 0 mm) with and without feeding of the TEG. Operating point: n = 3000 rpm and $\bar{p}$ = 3.00 MPa.

The influence on the thermal behavior of pad 4 is due to a cooling oil carry-over. The effect is caused by the additional feeding at the trailing edge of pad 3. In addition to the hot oil leaving the end of pad 3, a portion of the supplied cooling oil is carried over to the beginning of pad 4. This provides a larger amount of fresh oil to the pad. The increased volume flow and the reduced gap inlet temperature result in a lower overall temperature level. Nevertheless, the qualitative temperature curves of pad 3 and 4 are similar for the two cases considered with additional cooling. The effect of trailing edge cooling can only be seen in isolation for pad 3 due to the mixing of the effects described for pad 4. The improved heat dissipation at the end of the pad can reduce the maximum temperature and the temperatures at the trailing edge.

In general, the additional supply to the low-pressure TEG provides the bearing with a larger quantity of lubricant, even if this is concentrated on the respective rear edges of the loaded pads (see Figure 6). A larger volume flow inevitably leads to an improvement in certain properties, with first and foremost the maximum temperature that is achieved. In order to be able to estimate the influence of a higher total volume flow rate on the thermal properties, the experimentally determined maximum bearing temperatures (left) and the corresponding power loss (right) are plotted versus the total oil flow rate in Figure 7. For this purpose, the same total volume flow is supplied to the bearing via the leading edge grooves as in the tests at a nominal oil flow (7.1 L/s) with additional cooling via the TEG at pads 3 and 4.

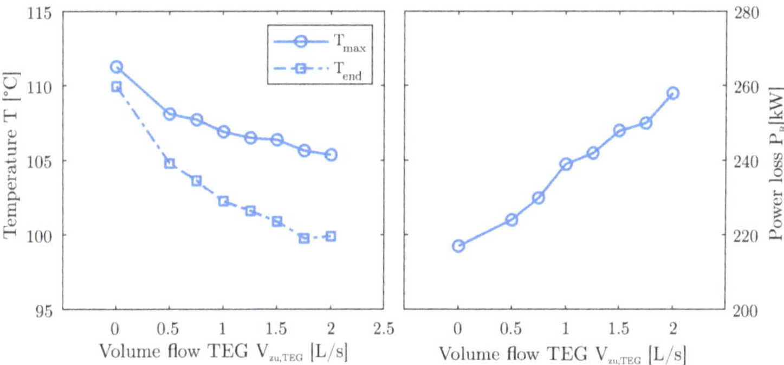

Figure 6. Experimentally determined maximum and final temperatures (°C) of pad 3 as well as the power loss depending on the supplied volume flow of the two cooling grooves. Operating point: n = 3000 rpm and $\bar{p}$ = 3.00 MPa.

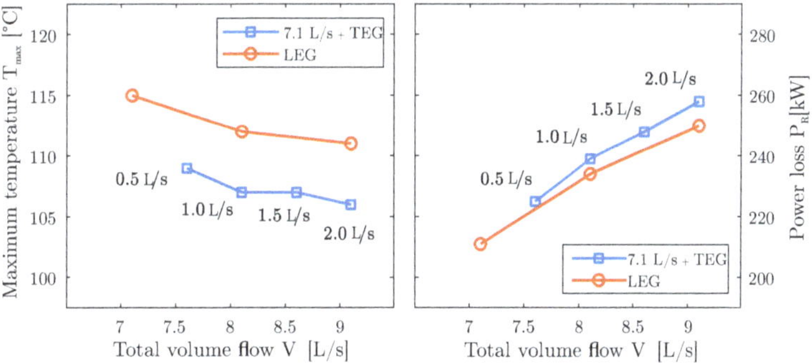

Figure 7. Experimentally determined maximum measuring point temperature and friction power as a function of the total volume flow supplied. Operating point: n = 3000 rpm and $\bar{p}$ = 3 MPa.

As expected, a higher oil flow rate has a positive effect on the maximum temperature of the bearing. The measured maximum pad temperatures are on average 5 K above those with additional cooling of the trailing edges of the pads. It can be shown that the improvement of the thermal condition is due to the selective feeding at the trailing edges of the highly loaded pads and not to the increased oil flow rate. When the nominal flow rate is increased, all pads benefit from the increased flow rate. The thermal properties of individual pads improve specifically, mainly those of the two lower thermally highly loaded pads, if the additional cooling grooves are fed. The measured power losses of the two variants show qualitatively corresponding curves, whereby that of the bearing with trailing edge cooling is slightly higher. However, this is plausible due to the additional shear stresses occurring in the two TEG at the pads 3 and 4.

4. Numerical Procedure

With the model, we introduce an approach to simulate the influence of the additional TEG on the thermal pad characteristics. In the further course, the model is used for the theoretical investigations and the simulation results are compared with the test data. The modeling can be subdivided as follows:

- CFD calculation of a single pad with additional supply groove at the end of the pad (OPENFOAM)
- Thermohydrodynamic journal bearing calculation considering thermomechanical pad deformations (COMBROS-CALCULIX)

CFD analyses are used to determine the dissipated heat flows at the end of a pad. The open source software package OPENFOAM is used. The coupling of structure (pad) and flow space is done with the CHT (conjugated heat transfer) method, in which equal temperatures and heat flows are assumed at phase boundaries. Thermomechanical pad deformations are not calculated within the CFD, but are taken into account by specifying the flow space based on the measured data.

For the determination of the steady-state operating characteristics, a coupling of the journal bearing calculation tool COMBROS with the open source finite element (FE) tool CALCULIX based on the work of Kukla [30] is modified and used.

4.1. CFD Calculation with OpenFOAM

The chtMultiRegionSimpleFoam solver of the simulation software package OPEN-FOAM-4.x (Open Source Field Operation and Manipulation) is used as CFD solver running within the Windows porting blueCFD-Core 2016-2. The solver calculates a stationary, turbulent, incompressible flow with the finite volume method. Flow, pressure and temperature fields, the heat exchange between fluid and solid and the temperature distribution in the solid are determined. Only single-phase flow is considered. Cavitation areas are not considered. Turbulence influences are modeled with the k-ω-SST model.

A consideration of the dissipative term is indispensable for the correct determination of the heat entry into the pad or the heat generation in the gap flow. Therefore the energy equation of the solver (Equation (1)) is extended by the dissipative term taking into account the shear forces according to Equation (2) to determine the temperature.

$$c \cdot \rho \cdot \left(u\frac{\partial T}{\partial x} + v\frac{\partial T}{\partial y} + w\frac{\partial T}{\partial z} \right) = \frac{\partial}{\partial x}\left(\lambda\frac{\partial T}{\partial x} \right) + \frac{\partial}{\partial y}\left(\lambda\frac{\partial T}{\partial y} \right) + \frac{\partial}{\partial z}\left(\lambda\frac{\partial T}{\partial z} \right) + \Phi \tag{1}$$

$$\Phi = \eta\left[\left(\frac{\partial u}{\partial y}\right)^2 + \left(\frac{\partial w}{\partial y}\right)^2 \right] \tag{2}$$

Included in the equation are the specific heat capacity c, density ρ, velocity components uvw, thermal conductivity λ, viscosity η and temperature gradients $\partial T/\partial x_i$. For the temperature-dependent dynamic viscosity, the relation given by FALZ [33,34] is implemented as a function of a reference temperature T_0, a reference viscosity η_0 and the FALZ exponent I, which is a specific property of the respective ISO-VG (International Organization for Standardization Viscosity Grade) class (Equation (3)).

$$\eta(T) = \eta_0 \cdot \left(\frac{T}{T_0}\right)^{-I} \tag{3}$$

Figure 8 shows two exemplary models that are built with a reduced number of nodes for the sake of clarity. The left part of the figure shows a model without an additional groove. The right part of the figure shows an example of a pad model with an additional groove at the end of the pad, the flow space is not shown. The models are designed assuming symmetry to the center of the bearing.

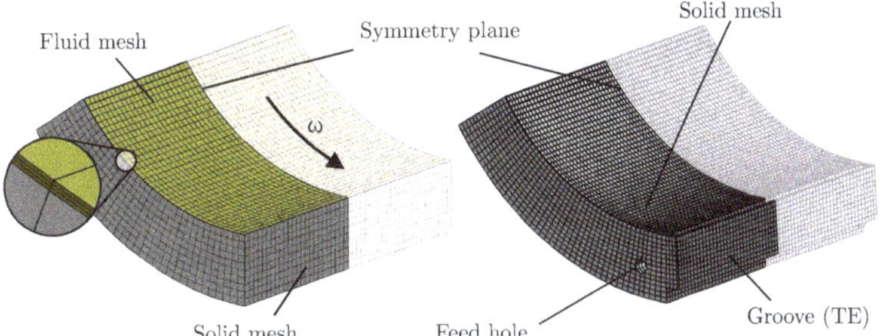

Figure 8. Illustration of the CFD model with reduced number of nodes. Left: Tilting-pad (grey) and fluid mesh (yellow) without additional supply groove. Right: Tilting-pad (grey) with TEG (fluid mesh not shown).

Due to computing times and data capacities, the simulation only considers a single pad without the surrounding spaces. Furthermore, the axial symmetry of the bearing is utilized when creating the model. However, this means that boundary conditions such as the shaft temperature must be specified. The boundary conditions are chosen based on the results of the experimental investigations. In summary, the following boundary conditions are used for the CFD:

- A single pad is considered, without the surrounding space.
- Symmetry to the middle plane of the pad in axial direction.
- Preset of the shaft temperature based on the experimentally collected data.
- Specification of the heat transfer coefficients (literature-based) at pad surfaces that do not belong to the running surface or supply groove.
- Specification of the ambient temperature at pad surfaces that do not count as part of the running surface or supply groove, based on the experimentally recorded data.
- Presetting of the circumferential speed of the shaft based on the shaft speed.
- No-slip condition on the bearing surface.
- No-slip condition on the shaft surface.
- Presetting of the pressure boundary conditions.
- Presetting of the volume flow of the TEG.
- Specification of the supply temperature.

4.2. Thermohydrodynamic Journal Bearing Calculation with COMBROS

For the determination of the steady-state bearing operating characteristics, the state-of-the-art journal bearing calculation program COMBROS version 1.3.0 (Developed at Institute of Tribology and Energy Conversion Machinery, Clausthal University of Technology, Germany) is used. The calculation is based on the simultaneous, iterative solution of the REYNOLDS and energy equations for the pressure and temperature distribution within the bearing. A comprehensive description of the calculation and function scope can be found in [5,7]. The main features of the program can be characterized as follows:

- Solution of the extended and generalized REYNOLDS differential equation.
- Simultaneous solution of the 3D energy equation for lubricating film, shaft and pad or bearing.
- Consideration of the occurrence of Taylor vortices and turbulence.
- Consideration of cavitation regions (Elrod algorithm, two-phase model).

- Realistic oil feed model.
- Approximate consideration of thermomechanical tread deformations.
- Interface for coupling of structural mechanics software.
- Consideration of 2D profiling of the running surface.
- Consideration of multi-lobe and tilting-pad bearings.
- Possibility to specify different heat transfer coefficients for different pad surfaces.

Different oil supply systems can be modeled, including direct lubrication via a LEG. The bearing is modeled without additional supply grooves at the end of the pads, as these cannot currently be represented in COMBROS. The TEG are modeled via the heat transfer on the corresponding surfaces of the pads.

The experimental results indicate that the three unloaded pads 1, 2 and 5 show a lack of lubrication at the relevant operating points. No hydrodynamic pressure build-up is generated. According to Hagemann [7], a lack of lubrication leads to instabilities in the numerical calculation of the equilibrium and thus to a negative influence on the convergence behavior. Due to this fact, the tilting mobility of the three pads is blocked in the calculation of the operating properties. There is also no calculation of the thermomechanical deformations of these pads.

4.3. Thermomechanical Deformation Calculation with CALCULIX

For the consideration of thermomechanical pad deformations, the structural mechanics tool CALCULIXversion 2.17 (Dhondt and Wittig, Germany) is used. The tool is coupled with the journal bearing calculation tool COMBROS. Functionality and scope can be found in the program documentation of Dhondt [35]. The basic coupling of the two programs is based on the procedure described by Kukla [30] with the aid of the development environment MATLAB version 2016a (The MathWorks, Inc., Natick, MA, USA).

Based on COMBROS input data, a geometric model of a tilting-pad is built up. The consideration of hydrostatic jacking grooves and a running surface profiling is done by a corresponding offset field, which is imposed on the nodes of the running surface. Required fields such as temperature or deformation data are transferred to the corresponding discretization within MATLAB. The following boundary conditions are used for the FE model:

- Blocking of the displacements of the nodes in the contact zone of pad and bearing ring.
- Specification of heat transfer coefficients and side flow temperatures at the pad surfaces.
- Pressure and temperature distribution on the running surface are taken from COMBROS.

4.4. Modeling the Influence of Active Cooling via TEG

The heat flow supplied to or dissipated from $\dot{Q}$ a surface A is defined according to Equation (4). In addition to the heat transfer coefficient α, this depends notably on the difference between the wall temperature T_W and the temperature of the surrounding fluid T_F.

$$\dot{Q} = \alpha \times A \times (T_W - T_F) \tag{4}$$

The heat transfer at pad surfaces is described by the heat transfer coefficients at the respective surfaces and the ambient temperature. As the ambient temperature the mean side flow temperature is assumed for all surfaces except the running surface [36].

In order to model the influence of the active trailing edge cooling within the bearing calculation, the heat transfer coefficient is iteratively adjusted in the FE-coupled calculation until the dissipated heat flows match those of the CFD simulations. Figure 9 shows the flow chart for the adjustment of the heat transfer coefficient. Starting from an initial value of the heat transfer coefficient at the trailing edge, a journal bearing calculation is performed. After running through the calculation loop, the temperature field at the trailing edge of pad 3 as well as the average side flow temperature are read out. The temperature field is additionally averaged. Subsequently, the heat flow dissipated over the surface is determined according to Equation (4). If the dissipated heat flow corresponds to that of the CFD simulation within a tolerance range, the iteration loop is exited. Otherwise, the heat transfer coefficient is modified and another calculation run is performed.

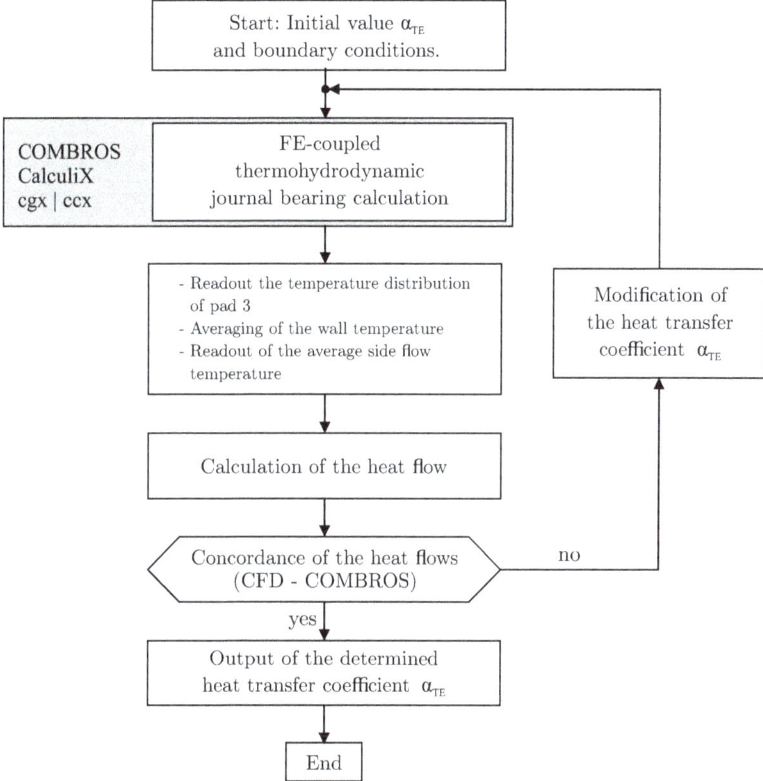

Figure 9. Flow chart for modifying the heat transfer coefficient at the trailing edge of pad 3.

5. Simulation Results and Discussion

Figure 10 shows the results of the iterative modification of the heat transfer for the specific bearing loads of 2.50 MPa up to 4.00 MPa. The dissipated heat fluxes at the trailing edge of pad 3 resulting from the bearing calculation with COMBROS are plotted against the heat transfer coefficient used. In the diagrams, the heat flows of the CFD simulation are displayed for the various supplied volume flows as reference degrees. In addition, the amount of heat dissipated for a weak heat transfer at a heat transfer coefficient of $250\,\text{W}/(\text{m}^2\,\text{K})$ is drawn in each case.

For the considered load cases, all heat flows of the CFD calculations can be mapped with corresponding heat transfer coefficients within the FE-coupled thermohydrodynamic bearing calculation. The heat transfer coefficients are in a range of $450\,\text{W}/(\text{m}^2\,\text{K})$ up to $1800\,\text{W}/(\text{m}^2\,\text{K})$.

The evaluation of the modeling is done by comparing the simulation results with the experimentally recorded data. The iteratively determined heat transfer coefficients for the different volume flows are used to map the TEG at the pad ends. As an exemplary case, the curves of the oil film thicknesses and measuring point temperatures are considered for a supplied volume flow of 2.0 L/s. The calculated and measured film thicknesses in the center of the bearing are shown in Figure 11. In the rear area of the loaded pads 3 and 4 behind the hydrostatic jacking grooves, there is a good agreement between the calculated and measured film thicknesses. At the beginning of the gap of the two pads, the curves of the measurement and the calculation differ. Since the tilting mobility of the unloaded pads 1, 2 and 5 is blocked and the pads are not taken into account in the deformation calculation, it is not possible to assess the simulated film thickness on the basis of the measured values.

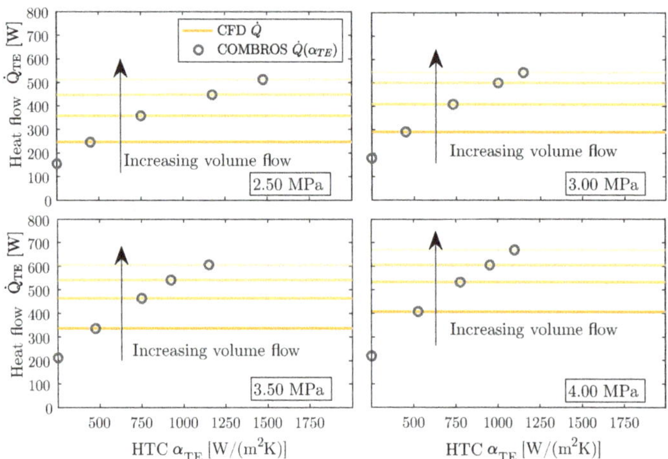

Figure 10. Calculated heat flow (W) at the trailing edge of pad 3 as a function of the heat transfer coefficient (W/(m² K) at different specific bearing loads for the volume flows 0.5 L/s, 1.0 L/s, 1.5 L/s, and 2.0 L/s. Speed: n = 3000 rpm.

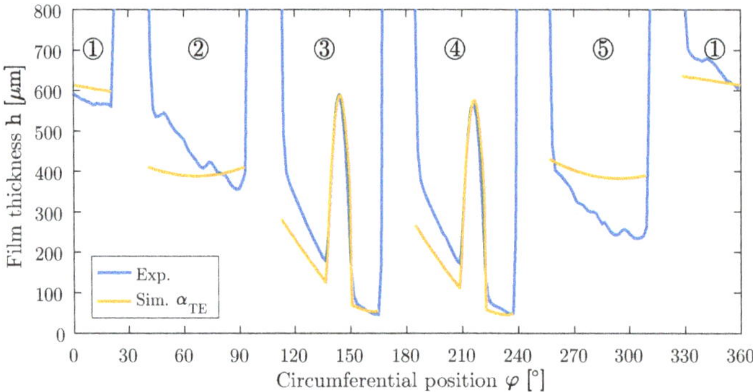

Figure 11. Measured and calculated film thickness (μm) over the circumference in the center of the bearing (pads 1–5) with an additional cooling oil volume flow of 2.0 L/s. Operating point: n = 3000 rpm and $\bar{p}$ = 3.00 MPa.

Figure 12 shows the calculated measuring point temperatures in comparison with the experimental values. Qualitatively, the curves agree well. The calculated temperatures of pad 3 match the experimentally measured values. The calculated temperature at the trailing edge of pad 3 is almost identical to the actual measured temperature. This can also be observed for the maximum measuring point temperature. On the other hand, the temperature curves determined for pad 4 differ more clearly from the measured values. This discrepancy is due to the already described cooling oil carry-over from the TEG of pad 3 to pad 4. This reduces the gap entry temperatures at the beginning of the pad. This effect cannot currently be modeled in COMBROS, which is the reason for the differences in the curves.

Figure 13 shows the calculated and measured temperature curves of the two loaded pads for different cooling oil volume flows at the trailing edge using the determined heat transfer coefficients in COMBROS.

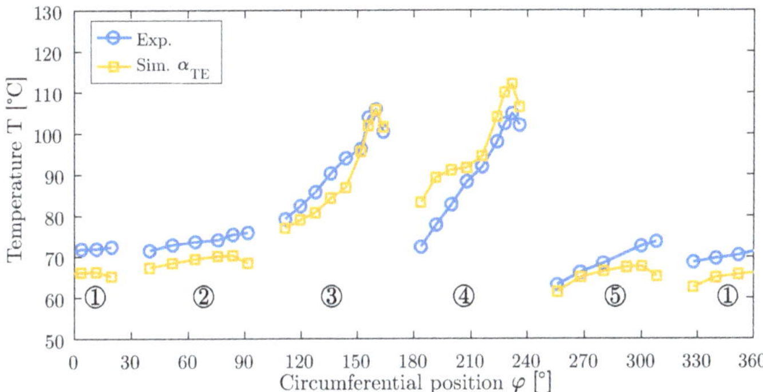

Figure 12. Measured and calculated measuring point temperatures (°C) over the circumference in the center of the bearing (pads 1–5) with an additional cooling oil volume flow of 2.0 L/s. Operating point: n = 3000 rpm and $\bar{p}$ = 3.00 MPa.

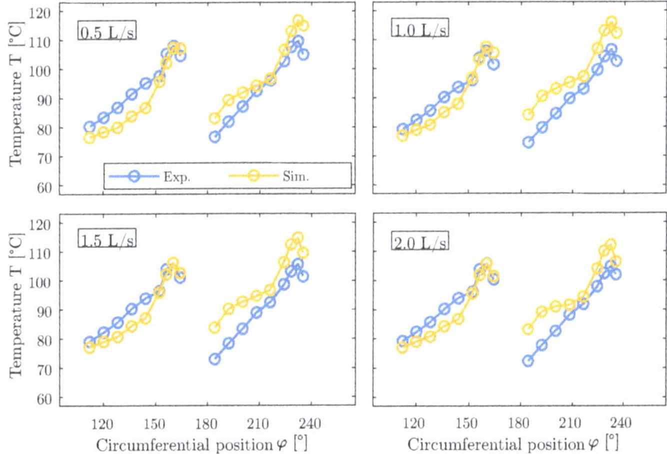

Figure 13. Measured and calculated measuring point temperatures (°C) of pad 3 and 4 over the circumference in the center of the bearing for the different cooling oil volume flow rates. Operating point: n = 3000 rpm and $\bar{p}$ = 3.00 MPa.

The pad temperatures at the trailing edge deviate noticeably from the measured values at 0.5 L/s and 1.0 L/s. With increasing volume flow, the results agree very well with the measured temperatures. However, the end temperatures are consistently predicted slightly too high. Although the cooling oil carry-over cannot be modeled, the comparison between calculation and measurement results shows that the temperature course at the trailing edge can be qualitatively reproduced. The temperature curve in this area indicates that a similarly large amount of heat is dissipated via the trailing edge of pad 4.

6. Conclusions

The test results show that at high thermal loads, the maximum measuring point temperature of the two loaded pads can be considerably reduced with the help of the additional TEG. A small amount of oil is sufficient (see Figure 5) to influence the temperature of the first loaded pad 3. As the fresh oil flow of the cooling increases, the effect on pad 4 increases. This is mainly due to the extra cooling oil being carried over from the groove at the trailing edge of pad 3 to pad 4. The entire temperature level of pad 4 is thereby lowered. Nevertheless, the improved heat transfer at the trailing edge of pad 4 can be seen by the temperature drop towards the end of the pad.

The measured data indicate a kind of saturation of the heat dissipation with increasing cooling oil volume flow (cf. Figure 6). The benefit of trailing edge cooling must therefore always be considered in relation to the power loss of the bearing. In general, it is possible to influence or specifically improve the thermal behavior of a pad. For example, the maximum temperature can be reduced by up to 3 K at the same load. However, the reduction of the pad temperature at the trailing edge is more notable. This temperature can be reduced by up to 8 K by feeding the TEG. An influence of the additional grooves on the deformation of the pads was not observed.

The use of the procedure to consider the cooling influence provides heat transfer coefficients with which the heat flows determined with the CFD can be adequately transferred to the COMBROS model. The procedure can be used to map the influence of the TEG on the thermal behavior of a pad. However, the procedure is only valid for pad 3. The measurement data show an influence of the additional grooves on the respective downstream pads, which increases with increasing volume flow. As things stand, it is not possible to model the described cooling oil carry-over, so that the temperature curves of pad 4 do not match the experimental data.

In summary, due to the agreement of the simulation and measurement results, the bearing modeling and the representation of the influence of the TEG can be considered valid within the model limits.

Author Contributions: Conceptualization, N.B. and M.S.; methodology, N.B. and B.B.; theoretical investigation, N.B.; experimental investigation, N.B. and M.S.; writing—original draft preparation, N.B. and M.S.; writing—review and editing, N.B., M.S. and B.B.; supervision, B.B.; project administration, N.B.; funding acquisition, N.B. All authors have read and agreed to the published version of the manuscript.

Funding: This research was funded by the German Federal Ministry of Economic Affairs and Energy. The financial support was assigned by the Industrial Research Association (AiF e. V.) in the project "Verbesserte Effizienz großer, schnelllaufender Radialkippsegmentgleitlager für verlustleistungskritische Anwendungen".

Acknowledgments: We acknowledge support by the Open Access Publication Funds of the Ruhr-Universität Bochum. The authors would like to thank D. Schüler for his valuable advice and GTW Alpen, especially C. Weißbacher, for the support of the experimental investigation and project administration.

Conflicts of Interest: The authors declare no conflict of interest.

Abbreviations

A	Area
B	Bearing length
c	Specific heat capacity
C_R	Radial clearance
D	Nominal diameter
h	Film thickness
I	FLAZ-exponent
m_p	Pad preload
n	Rotor speed

$\bar{p}$	Specific bearing load
P_R	Power loss
$\dot{Q}$	Heat flow
r	Radius
t	Thickness
T	Temperature
u, v, w	Flow velocities
V	Volume flow
x, y, z	Cartesian coordinates
α	Heat transfer coefficient
η	Dynamic viscosity
λ	Thermal conductivity
ρ	Density
φ	Angular coordinate
Φ	Dissipation
Ψ	Relative bearing clearance
ω	Rotation speed
Ω	Angular pad length
CFD	Computational fluid dynamics
FE	Finite elements
HTC	Heat transfer coefficient
LEG	Leading edge groove
TEG	Trailing edge groove
TPJB	Tilting-pad journal bearing

References

1. Mittwollen, N. *Betriebsverhalten von Radialgleitlagern Bei Hohen Umfangsgeschwindigkeiten und Hohen Thermischen Belastungen—Theoretische Untersuchungen*; VDI-Verlag: Dusseldorf, Germany, 1990; Volume 187.
2. Gerdes, R.; Fuchs, A. *Kippsegmentlager-Nachgiebigkeit: Nachgiebigkeit in der Segmentabstützung von Radial-Kippsegmentlagern und Deren Einfluss Auf die Lagerkennwerte und das Schwingungsverhalten Schnelllaufender Rotor-Lager-Systeme*; FVA-Heft: Frankfurt, Germany, 1997; Volume 511.
3. Fuchs, A. Schnelllaufende Radialgleitlagerungen Im Instationären Betrieb. Ph.D. Thesis, TU Braunschweig, Braunschweig, Germany, 2002.
4. Waltermann, H. *Optimierte Thermo-Elasto-Hydrodynamische Berechnungsverfahren für Gleitlager*; Reihe Maschinenbau, Shaker: Aachen, Germany, 1992.
5. Hagemann, T. *Verbesserte Radialgleitlagerberechnung: Stationär und Instationär hoch Belastete Radialgleitlager für Schnelllaufende Rotoren Bei Berücksichtigung der Lagerdeformationen*; FVA-Heft: Frankfurt, Germany, 2011; Volume 996.
6. Hagemann, T. Ölzuführungseinfluss bei schnell laufenden, hoch belasteten Radialgleitlagern unter Berücksichtigung des Lagerdeformationsverhaltens. In *Fortschrittsberichte des Instituts für Tribologie und Energiewandlungsmaschinen*; Shaker: Aachen, Germany, 2012; Volume 17.
7. Hagemann, T. *Dokumentation Radialgleitlagerberechnungsprogramm COMBROS R: Version 1.3.0*; ITR, TU Clausthal: Clausthal, Germany, 2017.
8. Fuchs, A.; Glienicke, J.; Schlums, H. *Programmdokumentation zu ALP3T, Version 4.3*; FVV Forschungsvorhaben Nr. 662 Robuste Lagerungen: Frankfurt am Main, Germany, 2005.
9. Ettles, C.M.; Anderson, H.G. Three-Dimensional Thermoelastic Solutions of Thrust Bearings Using Code Marmac1. *J. Tribol.* **1991**, *113*, 405. [CrossRef]
10. Ettles, C.M. The Analysis of Pivoted Pad Journal Bearing Assemblies Considering Thermoelastic Deformation and Heat Transfer Effects. *Tribol. Trans.* **1992**, *35*, 156–162. [CrossRef]
11. Heinrichson, N.; Santos, I.F.; Fuerst, A. The Influence of Injection Pockets on the Performance of Tilting-Pad Thrust Bearings—Part I: Theory. *J. Tribol.* **2007**, *129*, 895. [CrossRef]
12. Heinrichson, N.; Santos, I.F.; Fuerst, A. The Influence of Injection Pockets on the Performance of Tilting-Pad Thrust Bearings—Part II: Comparison Between Theory and Experiment. *J. Tribol.* **2007**, *129*, 904. [CrossRef]
13. Wodtke, M.; Fillon, M.; Schubert, A.; Wasilczuk, M. Study of the Influence of Heat Convection Coefficient on Predicted Performance of a Large Tilting-Pad Thrust Bearing. *J. Tribol.* **2013**, *135*, 021702. [CrossRef]
14. Papadopoulos, C.I.; Kaiktsis, L.; Fillon, M. Computational Fluid Dynamics Thermohydrodynamic Analysis of Three-Dimensional Sector-Pad Thrust Bearings With Rectangular Dimples. *J. Tribol.* **2014**, *136*, 011702. [CrossRef]

15. Taniguchi, S.; Makino, T.; Takeshita, K.; Ichimura, T. A Thermohydrodynamic Analysis of Large Tilting-Pad Journal Bearing in Laminar and Turbulent Flow Regimes With Mixing. *J. Tribol.* **1990**, *112*, 542. [CrossRef]
16. Hagemann, T.; Kukla, S.; Schwarze, H. Measurement and Prediction of the Static Operating Conditions of a Large Turbine Tilting-Pad Bearing Under High Circumferential Speeds and Heavy Loads. In Proceedings of the ASME Turbo Expo 2013: Turbine Technical Conference and Exposition, San Antonio, TX, USA, 3–7 June 2013. [CrossRef]
17. Kukla, S.; Hagemann, T.; Schwarze, H. Measurement and Prediction of the Dynamic Characteristics of a Large Turbine Tilting-Pad Bearing Under High Circumferential Speeds. In Proceedings of the ASME Turbo Expo 2013: Turbine Technical Conference and Exposition, San Antonio, TX, USA, 3–7 June 2013; p. V07BT30A020. [CrossRef]
18. Kukla, S.; Buchhorn, N.; Bender, B. Design of an axially concave pad profile for a large turbine tilting-pad bearing. *Proc. Inst. Mech. Eng. Part J J. Eng. Tribol.* **2017**, *231*, 479–488. [CrossRef]
19. Sano, T.; Magoshi, R.; Shinohara, T.; Yoshimine, C.; Nishioka, T.; Tochitani, N.; Sumi, Y. Confirmation of performance and reliability of direct lubricated tilting two pads bearing. *Proc. Inst. Mech. Eng. Part J J. Eng. Tribol.* **2015**, *229*, 1011–1021. [CrossRef]
20. Hagemann, T.; Zeh, C.; Schwarze, H. Heat convection coefficients of a tilting-pad journal bearing with directed lubrication. *Tribol. Int.* **2019**, *136*, 114–126. [CrossRef]
21. Arihara, H.; Kameyama, Y.; Baba, Y.; San Andrés, L. A Thermoelastohydrodynamic Analysis for the Static Performance of High-Speed—Heavy Load Tilting-Pad Journal Bearing Operating in the Turbulent Flow Regime and Comparisons to Test Data. *J. Eng. Gas Turbines Power* **2019**, *141*, 021023. [CrossRef]
22. Hagemann, T.; Schwarze, H. Theoretical and Experimental Analyses of Directly Lubricated Tilting-Pad Journal Bearings with Leading Edge Groove. *J. Eng. Gas Turbines Power* **2019**, *141*, 051010. [CrossRef]
23. Chen, G.; Wang, Q.; Cao, Y.; Tso, C. Development of an Isothermal Journal Bearing Employing Heat-Pipe Cooling Technology. *Tribol. Trans.* **1999**, *42*, 401–406. [CrossRef]
24. Chen, G.; Wang, Q.; Cao, Y. A tribological experimental investigation of a heat-pipe cooled isothermal journal bearing. *Tribol. Trans.* **2001**, *44*, 35–40. [CrossRef]
25. Nicholas, J.C. Tilting Pad Journal Bearings with Spray-Bar Blockers and By-Pass Cooling for High Speed, High Load Applications. In Proceedings of the 32nd Turbomachinery Symposium, College Station, TX, USA, 8–11 September 2003; pp. 9–11.
26. Martsinkovsky, V.; Yurko, V.; Tarelnik, V.; Filonenko, Y. Designing Radial Sliding Bearing Equipped with Hydrostatically Suspended Pads. *Procedia Eng.* **2012**, *39*, 157–167. [CrossRef]
27. Mermertas, Ü.; Brumbi, F.; Winkler, A.; Böckel, F.; Hagemann, T. Application of an enhanced 900 mm Tilting Pad Bearing in Large Steam Turbines. In Proceedings of the 14th EDF/Pprime Workshop: "Influence of Design and Materials on Journal and Thrust Bearing Performance", Futuroscope, France, 8–9 September 2015; Volume 2015.
28. Mermertas, Ü.; Hagemann, T.; Brichart, C. Optimization of a 900 mm Tilting-Pad Journal Bearing in Large Steam Turbines by Advanced Modeling and Validation. *J. Eng. Gas Turbines Power* **2018**, *141*, 021033. [CrossRef]
29. Hermes, G.J. Verbesserung der Betriebssicherheit hydrodynamischer Kippsegment-Radialgleitlager. *Tribol. Schmier.* **2016**, *63*, 51–59.
30. Kukla, S. Erhöhung Der Tragfähigkeit Großer Radialkippsegmentlager Durch Axiale Profilierung Der Segmentlauffläche. Ph.D. Thesis, Ruhr-Universität Bochum, Bochum, Geramny, 2018.
31. Kukla, S.; Buchhorn, N.; Bender, B. Increase of Operational Safety of Tilting-Pad Journal Bearings by Extraction of Hot Oil at the Trailing Edge. In Proceedings of the ASME 2015 Power Conference Collocated with the ASME 2015 9th International Conference on Energy Sustainability, the ASME 2015 13th International Conference on Fuel Cell Science, Engineering and Technology, and the ASME 2015 Nuclear Forum, San Diego, CA, USA, 28 June–2 July 2015; p. V001T13A007. [CrossRef]
32. Buchhorn, N.; Kukla, S.; Bender, B. Increased Load Carrying Capacity of Large Tilting-Pad Journal Bearings by Injection of Cold Oil. In Proceedings of the ASME Turbo Expo: Turbine Technical Conference and Exposition, Seoul, Korea, 13–17 June 2016; The American Society of Mechanical Engineers: New York City, NY, USA, 2016; p. V07BT31A026. [CrossRef]
33. Falz, E. *Grundzüge der Schmiertechnik*; Springer: Berlin/Heidelberg, Germany, 1931.
34. Han, D.C. Statische und Dynamische Eigenschaften von Gleitlagern Bei Hohen Umfangsgeschwindigkeiten und Bei Verkantung. Ph.D. Thesis, Universität Karlsruhe, Karlsruhe, Germany, 1979.
35. Dhondt, G. CalculiX CrunchiX USER'S MANUAL: Version 2.10. 2016. Available online: http://www.dhondt.de/ccx_2.10.pdf (accessed on 1 March 2021).
36. Hagemann, T.; Pfeiffer, P.; Si, X.; Zeh, C.; Schwarze, H. *Einfluss der Ölzuführung Auf Die Hydraulischen, Energetischen und Mechanischen Vorgänge in Schnell Laufenden und Hoch Belasteten Radialkippsegmentlagern*; FVA-Heft: Frankfurt, Germany, 2016; Volume 1184.

 *lubricants*

Article

Improvement of Tilting-Pad Journal Bearing Operating Characteristics by Application of Eddy Grooves

Eckhard Schüler * and Olaf Berner

Research & Development, Miba Industrial Bearings Germany GmbH, 37079 Göttingen, Germany; olaf.berner@miba.com
* Correspondence: eckhard.schueler@miba.com

Abstract: In high speed, high load fluid-film bearings, the laminar-turbulent flow transition can lead to a considerable reduction of the maximum bearing temperatures, due to a homogenization of the fluid-film temperature in radial direction. Since this phenomenon only occurs significantly in large bearings or at very high sliding speeds, means to achieve the effect at lower speeds have been investigated in the past. This paper shows an experimental investigation of this effect and how it can be used for smaller bearings by optimized eddy grooves, machined into the bearing surface. The investigations were carried out on a Miba journal bearing test rig with Ø120 mm shaft diameter at speeds between 50 m/s–110 m/s and at specific bearing loads up to 4.0 MPa. To investigate the potential of this technology, additional temperature probes were installed at the crucial position directly in the sliding surface of an up-to-date tilting pad journal bearing. The results show that the achieved surface temperature reduction with the optimized eddy grooves is significant and represents a considerable enhancement of bearing load capacity. This increase in performance opens new options for the design of bearings and related turbomachinery applications.

Keywords: sliding bearings; tilting pad journal bearings; flow transition; laminar flow; turbulent flow; non-laminar regime; temperature reduction; load capacity increase

Citation: Schüler, E.; Berner, O. Improvement of Tilting-Pad Journal Bearing Operating Characteristics by Application of Eddy Grooves. *Lubricants* **2021**, *9*, 18. https://doi.org/10.3390/lubricants9020018

Received: 22 December 2020
Accepted: 7 February 2021
Published: 10 February 2021

Publisher's Note: MDPI stays neutral with regard to jurisdictional claims in published maps and institutional affiliations.

1. Introduction

In high-performance turbomachinery, the maximum bearing temperatures can be a crucial parameter. If these temperatures approach the critical values for the bearing metal and the lubricating oil, the operational safety and long-term reliability of the whole system may be at risk. The continuous demand for increased bearing loads, often in combination with rising rotor speeds, leads to an increase of the maximum bearing temperatures which can be the limiting factor for this development trend. This situation creates the motivation to improve fluid-film bearings by reducing their maximum temperatures and thereby push the limits for turbomachinery development.

In contrast to other means for achieving lower bearing temperatures, the phenomenon of the laminar-turbulent flow transition in high load regions of the fluid-film directly reduces the maximum oil film temperatures and thereby also the maximum bearing metal temperatures considerably, as for example measured in [1,2].

Typically, the laminar-turbulent flow transition occurs if a critical speed is exceeded, under otherwise constant operating conditions. The most important effect of this transition is the occurrence of radial flow velocities, approximated by an eddy, which means an increase of the apparent heat conductivity of the lubricating oil. Figure 1 shows an exemplary temperature distribution radially across the oil film at the position of the highest bearing temperature for two speed cases, one with laminar flow and the other one, after the flow transition, with Taylor vortex and turbulent flow, predicted for a sliding speed difference of only 1 m/s. The eddy conductivity at Taylor vortex and turbulent flow leads to a relatively even temperature distribution, in contrast to the high temperature gradients at laminar flow. As depicted, non-laminar flow can cause a reduction of maximum oil film and bearing metal temperatures.

In practice, a significant drop in the maximum bearing temperature can be observed when the fluid-film flow at the bearing hot-spot becomes non-laminar [1,2]. In this context, the term 'hot-spot' refers to the position of the highest bearing metal temperature. To take this temperature reducing effect into account in the design phase, the bearing calculation tool must predict and consider the local flow regime reliably for the given operational parameters of a bearing. State-of-the-art thermo-elasto hydrodynamic lubrication (TEHL) tools, like the ones described in [3,4], have implemented turbulence models for the local flow regime prediction, for example based on [5–8].

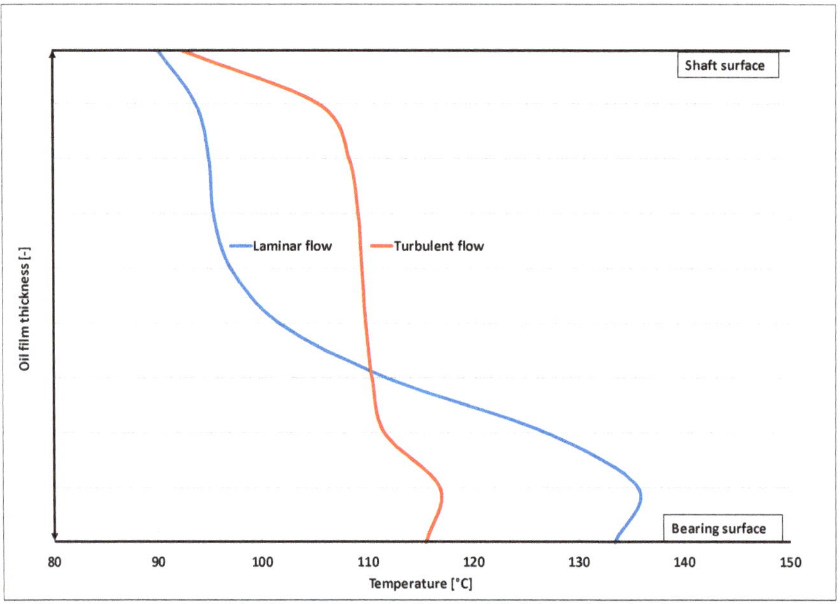

Figure 1. Exemplarily predicted laminar and turbulent flow profiles across the oil film at the position of the highest bearing temperature.

Due to its correlation with the fluid-film thickness, the local state of the lubricant flow is, for a given siding speed, fundamentally depending on the bearing diameter and, due to its effect on the local film thickness and local viscosity, on the bearing load. Figure 2 depicts these relations. As expected, a decrease of the bearing size and a rise of the bearing load, lead to an increase of the flow transition speed.

According to these predictions and based on the experience that a significant temperature drop is practically observed only in high-speed turbomachinery with large rotor diameters >≈450 mm, it can be concluded that the natural flow transition speeds of smaller bearings operating at high loads are above most operating conditions.

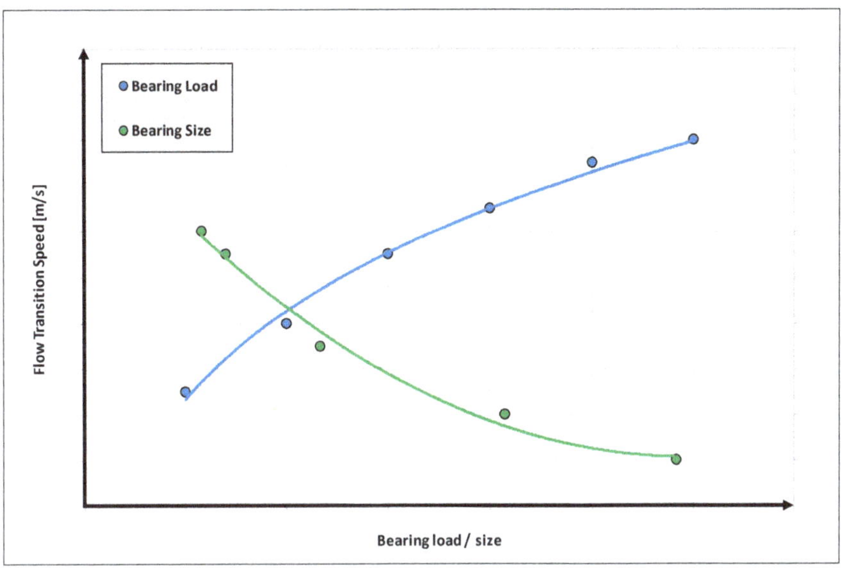

Figure 2. Predicted qualitative influence of the bearing load and the bearing size on the laminar-turbulent transition speed.

To use the temperature reducing effect of non-laminar flow nevertheless, an approach to reduce the flow transition speed is necessary. In [9] this idea was realized by milling a high number of specific grooves into the surface of a Ø500 mm bearing to disturb the laminar flow and thereby cause the desired increase of radial flow components at reduced sliding speeds, respectively reduced local Reynolds-numbers.

The experiment was successful. The maximum temperature measurements of the 4-tilting pad test bearing were considerably reduced, though not evenly for both loaded pads and by trend less effective at increased loads. However, based on this investigation, the flow regime in the fluid-film of a bearing can be influenced and the desired reduction of flow transition speed can be achieved.

However, for a broader use of this concept, several questions were left unanswered. The most important aspect is the scope of application, regarding the bearing size, sliding speed and specific bearing load. Other open points are the definition of an optimum groove geometry in terms of performance and manufacturing complexity and the specification of an effective arrangement of single grooves to a complete pattern.

To provide a contribution to these open questions, an experimental test project was decided and carried out on a Miba Industrial Bearings test rig.

2. Test Equipment

2.1. Test Rig

For the experimental investigations, the Ø120 mm high-performance journal bearing test rig depicted in Figure 3, was used. The electric motor drives the shaft (1) via a spur gear and a curved-tooth coupling (2). The shaft is supported by two fixed-profile support bearings (3, 4) located in the test rig housing (5). The test bearing (6) is held by the bearing carrier (7) and is located centrally between the support bearings. The bearing carrier is connected to the test rig housing by chains (8). The chains serve to align and axially fix the test bearing. The load is applied by a metal bellow (9) which is pressurized with compressed air. The resulting load is transferred to the bearing carrier via the traverse (10) and the connections rods (11) and is measured by a load cell (12). The vibration exciters (13) for generating dynamic loads were not required for the project described.

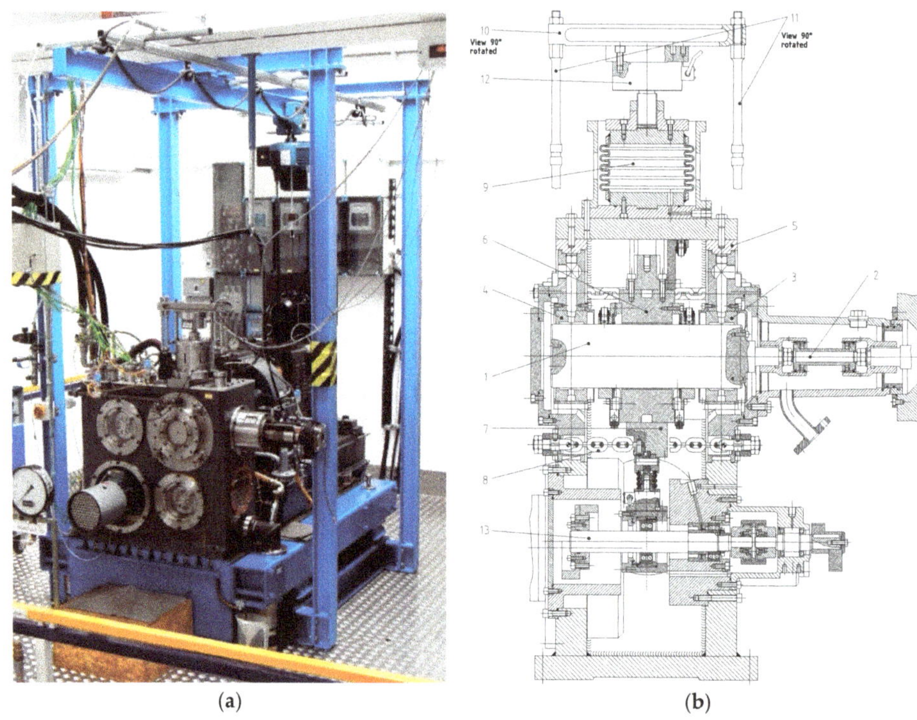

(a) (b)

Figure 3. (**a**) Journal bearing test rig front view photo; (**b**) test rig core sectional drawing.

The main data of the rig is shown in Table 1.

Table 1. Test rig main data.

Drive power [kW]	315
Shaft Ø [mm]	120
Maximum speed [rpm]	>20,000
Maximum stationary load [N]	63,000
Maximum lube oil flow rate [L/min]	130
Maximum dynamic load [N]	9000
Stationary measuring systems	Temperatures, rate of oil flow, bearing load
Dynamic measuring systems	Bearing load, shaft displacement

The maximum speed of 20,000 rpm corresponds to a sliding speed of 125 m/s and the maximum load of 63 kN results in a maximum specific bearing load of 7 MPa for an axial bearing length of 75 mm.

2.2. Test Bearing

As much of today's high-performance turbomachinery is equipped with tilting pad bearings, an up-to-date 5-tilting pad bearing was selected for the experimental investigations. Table 2 includes the bearing properties.

Table 2. Test bearing main data.

Number of tilting pads [-]	5
Nominal bore Ø [mm]	120
Axial pad length [mm]	75
Effective angular pad length [°]	52
Total Pad thickness [mm]	17.5
White metal thickness [mm]	2.0
Pivot position [-]	50% (center)
Diametrical bearing clearance, nominal [mm]	0.270
Preload, nominal [-]	0.31
Type of lubrication [-]	Directed, open end-plates
Number of temperature probes in loaded pads [-]	14 (each)
Materials [-]	Steel/Tegostar

The bearing was installed with load between pads (LBP) orientation. For an accurate measurement of the maximum pad temperatures, the loaded pads were equipped with a dense arrangement of thermocouples near the trailing edge and close to the axial midplane as shown in Figure 4, and thus in the area where the highest temperatures and temperature gradients occur.

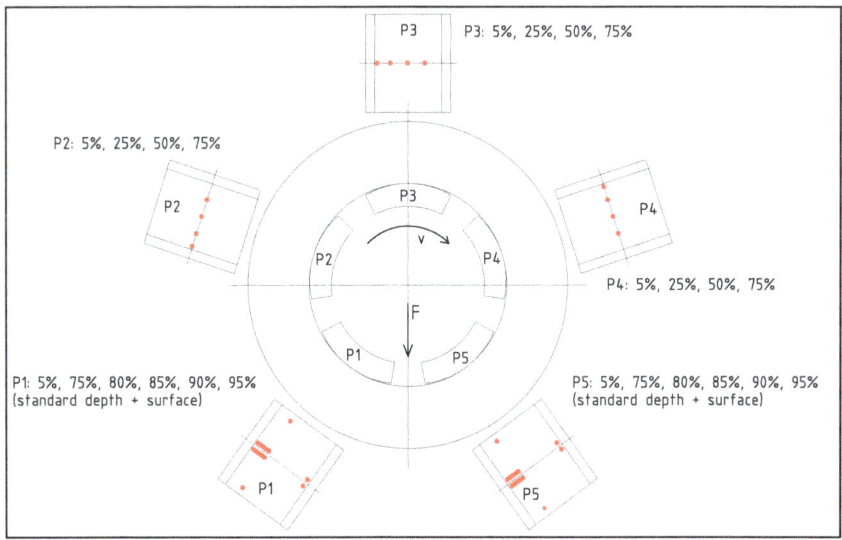

Figure 4. Bearing orientation and thermocouple locations.

At each defined circumferential position, two thermocouples were placed, one soldered into the bearing metal, with approximately 0.3 mm radial distance to the sliding surface and a second 1 mm below the white metal, corresponding to approximately 4 mm below the sliding surface for this bearing size, the same distance as the typical 75% sensor position of customer bearings. Thus, one temperature sensor in each pad of the test bearing is placed very similar to the standard sensor of customer bearings.

The sensors located 1 mm below the white metal were used for a comparison with reference data from older investigations, which were carried out without surface temperature sensors and as a backup solution if problems with the more sensitive surface temperature measurement would have occurred. In addition, information is obtained on how the temperature situation at the bearing surface affects the 75% standard position or other measuring points at the same depth, since in the field, measurements directly at the bearing surface are not practicable, mainly for safety reasons.

For the bearing alignment control, additional temperature probes were installed near the axial edges of the loaded pads.

Figure 5 shows the assembled test bearing.

Figure 5. Assembled test bearing with plain pad surfaces.

2.3. Eddy Grooves

After the completion of the test program with plain pads, the loaded pads were re-machined and equipped with eddy grooves. According to the details described in [10], eddy grooves are physical grooves in a specific area of a pad with a particular shape, arranged in a particular pattern, as shown in Figure 6a for the test bearing. The bearing assembly with eddy grooves is shown in Figure 6b.

The grooves are designed to disturb the laminar oil flow regime, which is typically present in this area of a bearing, causing high bearing temperatures, especially at higher loads. In circumferential direction, the adjacent grooves maintain the flow disturbance to keep-up the non-laminar status. In axial direction, the grooves are interrupted twice and end with a distance to the edges to prevent a drop of hydrodynamic pressure or load carrying capacity, respectively.

It should be mentioned that, from the authors' point of view, the described eddy grooves and their influence on the physics of hydrodynamic lubrication are not part of the topic of surface texturing as introduced in [11] and further investigated, for example, in [12]. The distinction is made by the shape, the position and the significantly larger geometric dimensions of the grooves, e.g., the depth is considerably higher than the minimum oil film thickness at higher loads. Accordingly, the mode of action is not aimed at increasing the hydrodynamic pressure or the reduction of friction losses but rather to influence the flow regime at typical turbomachinery steady-state points of operation at full fluid friction.

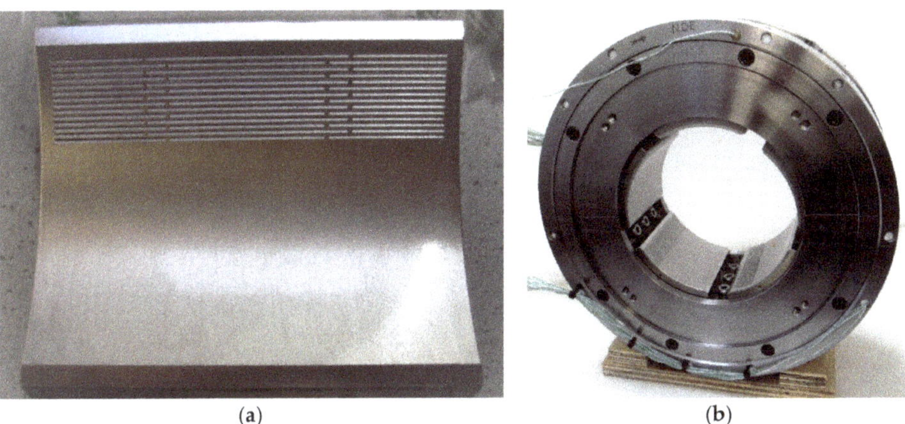

(a)   (b)

Figure 6. (**a**) Pad with eddy groove arrangement and (**b**) assembled test bearing.

2.4. Test Program

The tests program was defined according to Table 3.

The rate of oil flow was individually set for each sliding speed. To investigate the influence of the lube oil flow rate, medium, low, and high rates were defined, based on the oil temperature difference between outlet and inlet. For the medium flow rate, an oil temperature increase of 20 K–22 K was defined. For the high flow rate, 15 K–17 K, and for the low flow rate, 25 K–28 K, oil temperature increases were defined.

The actual number of test points was approximately 200 for the test bearing with plain pads as well as for the eddy groove bearing, as not all parameter combinations could be tested, mainly due to the limitations of the oil supply system.

Table 3. Test program.

Sliding speed range [m/s]	50–110; step size: 10
Specific load range [MPa]	0–4.0; step size: 0.5
Oil type [-]	VG 32, mineral oil
Oil inlet temperatures [°C]	45; 55

3. Results and Discussion

3.1. Plain Bearing

To verify the prediction of the natural flow transition speed, an initial test was carried out in which the shaft speed was incrementally increased, with otherwise constant parameters. The result of the temperature measurements is shown in Figure 7. The predicted and measured transition speeds are in good agreement, while the predicted temperature drop is significantly higher than the experimentally observed value. Despite this difference, the test confirms that the flow transition occurs, even at this moderate load of 1.5 MPa, at speeds considerably above normal operating conditions.

The further focus of the plain bearing investigation was on the creation of a reference data base for the comparison with the eddy groove bearing.

In addition, the comparison between the soldered-in surface probe measurements and the probe measurements 1 mm below the white metal was carried out also for the plain bearing, although it is more important for the eddy groove bearing, because its results cannot be predicted with current calculation tools.

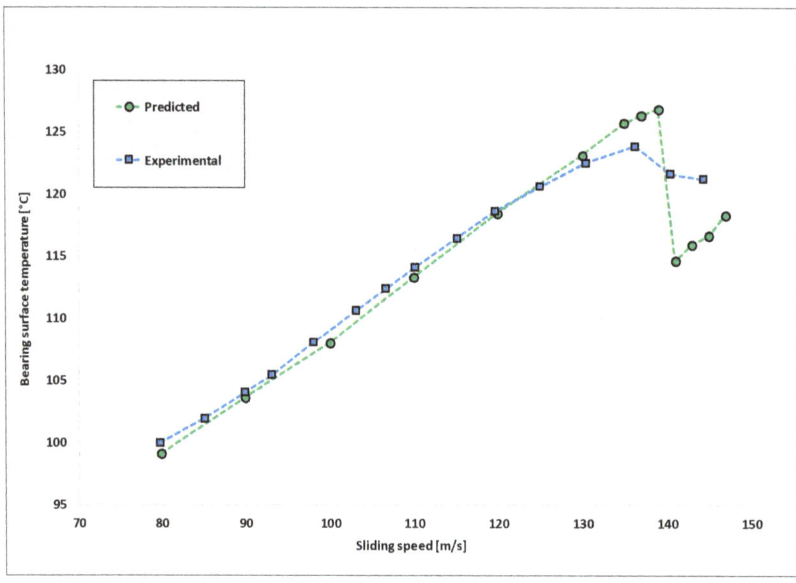

Figure 7. Experimental and predicted temperature development at stepwise rotor speed-up for the 5-tilting pad test bearing at T_{in} = 56 °C, Q = 77 L/min and 1.5 MPa specific load.

Figure 8 shows a typical measured temperature profile over all five pads, with separate data for the surface and standard depth probes in the loaded pads: while the maximum pad temperatures at the standard depth are 104 °C and 112 °C, respectively, the corresponding maximum surface temperatures are 113 °C and 126 °C.

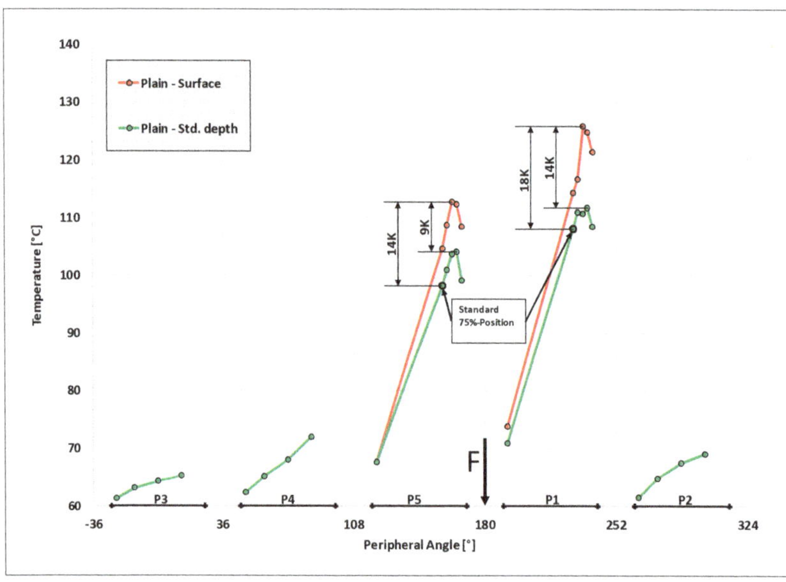

Figure 8. Plain surface bearing: measured temperatures from surface probes (red graph) and standard depth probes (green graph) at v = 90 m/s, 3.0 MPa specific load, Tin = 45 °C and medium oil flow rate.

A comparison of these values with the temperatures measured at the standard 75%-position, 1 mm below the white metal, underlines the importance of taking the surface temperatures into account for the evaluation of the operational safety of the bearing: while the maximum temperatures at the standard depth are only 5 K/4 K above the 75%-position measurement, the difference to the surface is 14 K/18 K for this point of operation.

Under comparable conditions, Hagemann et al. [13] also reported significant differences between the maximum surface temperature and the 75%-position below the white metal.

Figure 9 summarizes the differences between surface and standard depth temperatures, individually for both the max temperature and the 75%-position temperature as a function of the specific bearing load by averaging all measurements taken at the same bearing load at rotor speeds between 50 m/s–110 m/s.

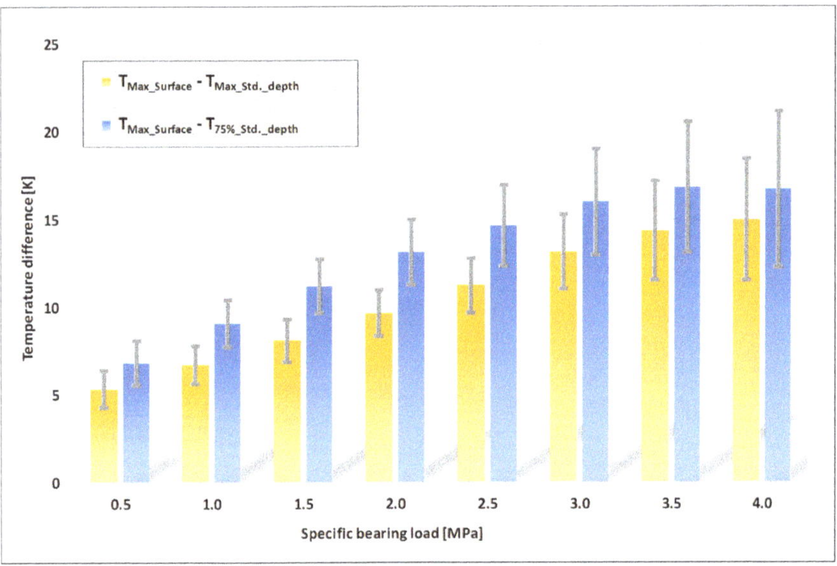

Figure 9. Plain surface bearing: Load-dependent mean difference between the maximum surface temperature and the maximum standard depth temperature (yellow bars) and between the maximum surface temperature and the standard 75%-position temperature (blue bars) including the standard deviations (grey bars).

As expected, the temperature difference increases with rising bearing load. Due to the speed dependent oil flow rates according to the defined oil temperature rise, the influence of the rotor speed is insignificant.

At the standard sensor depth, 1 mm below the white metal, the difference between the maximum temperature and the 75%-position temperature is relatively moderate and constant, so that in practice a temperature sensor at the 75%-position can meet the requirements for measuring the relevant bearing temperature for condition monitoring.

3.2. Eddy Groove Bearing

While an impact of eddy grooves on the bearing's power loss was not measurable, the attempt to reduce the maximum temperatures of the test bearing by provoking the transition of the otherwise laminar flow in the area upstream the lowest film thickness was successful. Due to the non-laminar flow regime induced by the bearing-specific eddy grooves, the maximum surface temperatures dropped considerably.

The effect was clearly visible at all tested speeds and increased with the bearing load and sliding speed. Figure 10 shows a measured temperature comparison over all five pads for a high speed, high load case: the eddy groove bearing develops 16 K/17 K lower maximum surface temperatures than the plain bearing. The temperature reduction at the standard depth is 13 K/10 K, which is lower than at the surface, but still significant.

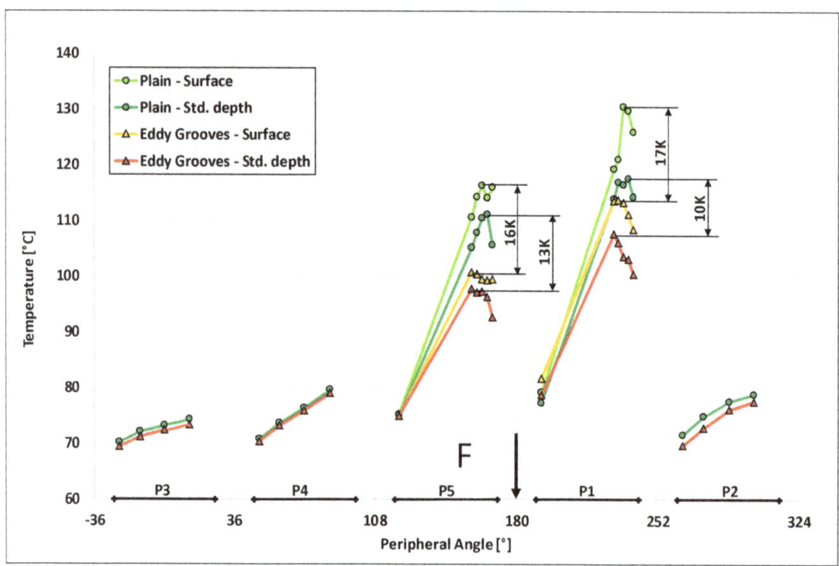

Figure 10. Temperature distributions at the surface and at standard sensor depth for the eddy groove and the plain surface bearing at v = 100 m/s, 3.0 MPa specific bearing load, T_{in} = 55 °C and medium oil flow rate.

By averaging all measurements taken at the same load at rotor speeds between 50 m/s–110 m/s, Figure 11 summarizes the reduction of maximum surface temperatures and the standard 75%-position temperatures due to eddy grooves compared to the plain bearing: the eddy groove cooling effect at the bearing surface is strong, especially at high bearing loads, while the temperature reduction at the 75%-position is only moderate. The reduced impact on the 75%-position standard thermocouple measurement result was expected because the eddy groove effect takes place in the oil film near the bearing surface and naturally becomes smaller with increasing distance. However, the surface temperature reduction has the clear potential to be used to increase the safety margin or power density of a bearing.

This potential is illustrated in Figure 12: for the given operational parameters, the bearing with plain pads develops a maximum surface temperature of 110 °C, indicated by point 1 in the chart. Regarding accelerated oil aging and bearing issues based on oil carbonization, this temperature is not far from critical values.

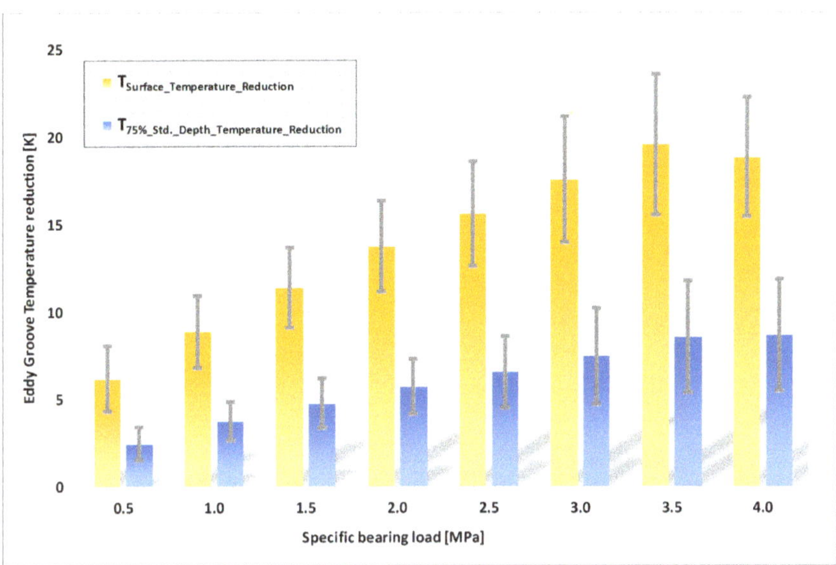

Figure 11. Temperature reduction by eddy grooves compared to the plain surface: Load-dependent mean differences for the maximum surface temperatures (yellow bars) and for the 75%-position standard depth temperatures (blue bars), including the standard deviations (grey bars).

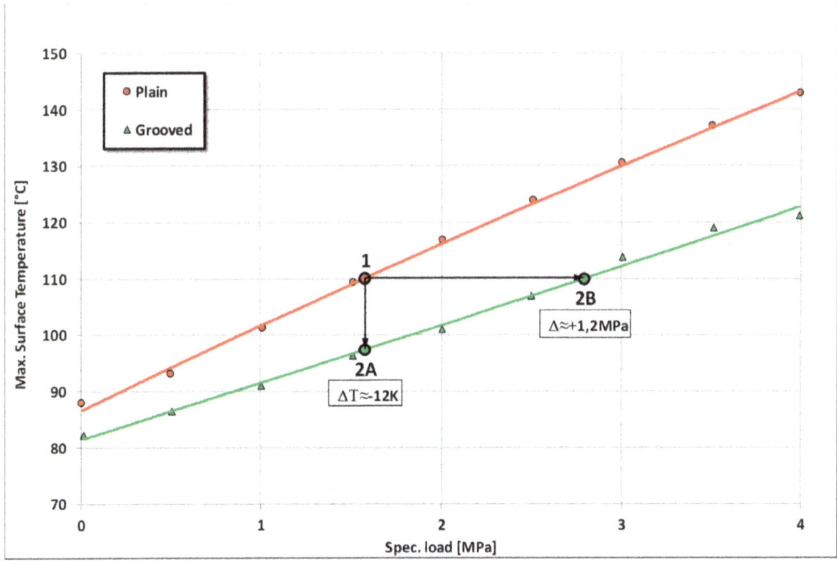

Figure 12. Eddy groove application example at v = 100 m/s, T_{in} = 55 °C and medium oil flow rate.

With the eddy groove modification of this bearing, opportunities arise to either use the benefit solely for a temperature reduction by 12 K and thereby increase the operational safety of the bearing, indicated by point 2A, or to use the full benefit to increase the bearing load at a constant maximum temperature, shown as point 2B. A new point of operation between 2A and 2B would result, compared to point 1, in a reduced bearing temperature in combination with an increased bearing load.

Regarding the service life of eddy groove bearings, the investigation showed no indication that a reduction must be assumed. The bearing evaluation after the comprehensive test program with high loads and temperatures revealed no negative effects on the grooves, such as geometric changes caused by creeping. This may also be a consequence of the low creeping values of the used bearing material, but the reduced level of maximum bearing temperatures generally reduces creeping tendencies. In practical use, the usual mild wear, e.g., due to particle abrasion, is also tolerable, since the depth of the grooves is considerably above the permissible wear values.

Figure 13 summarizes the increase of bearing load capacity investigated in this project: with identical maximum surface temperatures, the bearing can carry a considerably higher load if it is equipped with eddy grooves.

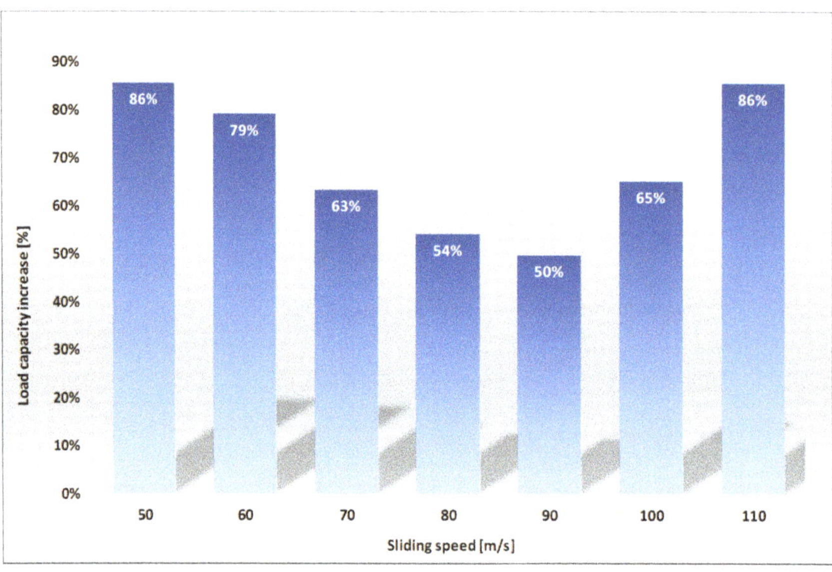

Figure 13. Summarized temperature-related relative increase of load capacity by eddy grooves, compared to the plain surface. The figures represent the mean increase of bearing load capacity at the given sliding speeds for identical maximum surface temperatures.

4. Conclusions

For the improvement of operational safety and performance of fluid-film bearings, Miba developed and tested the patent pending eddy groove technology:

In highly loaded, high-speed fluid-film bearings, a laminar flow regime is typically present in the area upstream the lowest film thickness. Because of the low thermal conductivity of lubricating oil, this leads to high radial temperature gradients near the bearing surface, resulting in high bearing temperatures, while the gradients and thus the temperatures radially towards the shaft are considerably lower.

When exceeding a threshold speed, the transition of the laminar flow to Taylor vortex and turbulent flow creates additional radial fluid flow components which reduce the high temperature gradients. This results in a considerable temperature reduction near and at the bearing surface. The transition of the laminar flow thus significantly reduces the maximum bearing temperatures, which can have a very positive effect on the operational safety and load capacity of the bearing. However, the flow change is practically only observed in very large bearings, so that this natural effect is typically not achievable for most applications.

With Miba eddy grooves, the onset of non-laminar flow with significant radial flow components can be shifted to lower sliding speeds, so that the reduction of maximum temperatures, respectively the increase of the bearing load capacity can be achieved in the technically relevant sliding speed range of smaller bearings. The effectiveness of this technology was proved experimentally on a Miba journal bearing test rig on a Ø120 mm, 5-tilting pad bearing: in the entire test range at speeds between 50 m/s–110 m/s, and loads between 0 MPa–4 MPa, a significant reduction of maximum bearing temperatures due to eddy grooves was achieved, especially at higher bearing loads, under otherwise identical operating conditions. There was no noticeable influence on the power loss.

Temperature measurements directly at the bearing surface show the full potential of this technology to reduce the decisive maximum oil film and bearing metal temperatures, which is especially useful for high performance applications. Thereby, eddy groove bearings push the limits of safe operation considerably and open new technical possibilities.

The reduced bearing temperatures, induced by eddy grooves, can be used to:

1. Increase the safety margin of the bearing and slow down the lubricating oil aging
2. Operate bearings at higher loads without an increase of maximum temperatures
3. Decrease the bearing diameter or the axial length to reduce the power loss and the necessary rate of oil supply without an increase of maximum bearing temperatures

Regarding the bearing size, eddy grooves have been successfully tested at bore diameters of 120 mm and 500 mm [9]. Above this range, the effectiveness of eddy grooves is as good as certain, if the laminar-turbulent transition hasn't already occurred naturally. Below this range, it is assumed that eddy grooves also reduce the maximum temperatures of smaller bearings, while an accurate lower size limit is not yet determined. Miba eddy grooves can also be used for other bearing designs, like fixed-profile journal bearings. Further investigations of the eddy groove technology are planned.

Author Contributions: O.B. prepared the test rig and the test bearings. O.B. and E.S. conducted the experiments. E.S. analyzed the test data, wrote the draft and revised the manuscript. All authors have read and agreed to the published version of the manuscript.

Funding: This research received no external funding.

Data Availability Statement: Data only available on request due to protection of IP.

Acknowledgments: The authors would like to thank T. Hagemann and D. Schüler for their helpful advice and support of the manuscript revision.

Conflicts of Interest: The authors declare no conflict of interest.

References

1. Hopf, G.; Schüler, D. Investigations on Large Turbine Bearings Working Under Transitional Conditions between Laminar and Turbulent Flow. *J. Tribol.* **1989**, *111*, 628–634. [CrossRef]
2. Taniguchi, S.; Makino, T.; Takeshita, K.; Ichimura, T. A Thermohydrodynamic Analysis of Large Tilting-Pad Journal Bearing in Laminar and Turbulent Flow Regimes with Mixing. *J. Tribol.* **1990**, *112*, 542–548. [CrossRef]
3. He, M. Thermoelastohydrodynamic Analysis of Fluid Film Journal Bearings. Ph.D. Thesis, University of Virginia, Charlottesville, VA, USA, 2003.
4. Hagemann, T. Ölzuführungseinfluss bei schnell laufenden und hoch belasteten Radialgleitlagern unter Berücksichtigung des Lagerdeformationsverhaltens. Ph.D. Thesis, TU Clausthal, Clausthal-Zellerfeld, Germany, 2011.
5. Ng, C.-W.; Pan, C.H.T. A Linearized Turbulent Lubrication Theory. *J. Basic Eng.* **1965**, *87*, 675–682. [CrossRef]
6. Elrod, H.G.; Ng, C.W. A Theory for Turbulent Fluid Films and Its Application to Bearings. *J. Lubr. Technol.* **1967**, *89*, 346–362. [CrossRef]
7. Hirs, G.G. A Bulk-Flow Theory for Turbulence in Lubricant Films. *J. Lubr. Technol.* **1973**, *95*, 137–145. [CrossRef]
8. Constantinescu, V.N. Basic Relationships in Turbulent Lubrication and Their Extension to Include Thermal Effects. *J. Lubr. Technol.* **1973**, *95*, 147–154. [CrossRef]
9. Hopf, G. Experimentelle Untersuchungen an großen Radialgleitlagern für Turbomaschinen. Ph.D. Thesis, Ruhr-Universität Bochum, Bochum, Germany, 1989.
10. Schüler, E.; Berner, O. Hydrodynamic Sliding Bearing. Patent WO2021004803A1, 14 January 2021.
11. Hamilton, D.B.; Walowit, J.A.; Allen, C.M. A Theory of Lubrication by Microirregularities. *J. Basic Eng.* **1966**, *88*, 177–185. [CrossRef]

12. Henry, Y.; Bouyer, J.; Fillon, M. Experimental analysis of the hydrodynamic effect during start-up of fixed geometry thrust bearings. *Tribol. Int.* **2018**, *120*, 299–308. [CrossRef]
13. Hagemann, T.; Zemella, P.; Pfau, B.; Schwarze, H. Experimental and theoretical investigations on transition of lubrication conditions for a five-pad tilting-pad journal bearing with eccentric pivot up to highest surface speeds. *Tribol. Int.* **2020**, *142*, 106008. [CrossRef]

Article

CFD Analysis of Journal Bearing with a Heterogeneous Rough/Smooth Surface

Mohammad Tauviqirrahman [1,*], J. Jamari [1], Arjuno Aryo Wicaksono [1], M. Muchammad [1,2], S. Susilowati [3], Yustina Ngatilah [3] and Caecilia Pujiastuti [3]

[1] Laboratory for Engineering Design and Tribology, Department of Mechanical Engineering, Diponegoro University, Jl. Soedharto SH, Tembalang, Semarang 50275, Indonesia; j.jamari@gmail.com (J.J.); arjunowicaksono@gmail.com (A.A.W.); m_mad5373@yahoo.com (M.M.)

[2] Laboratory for Surface Technology and Tribology, Faculty of Engineering Technology, University of Twente, Drienerlolaan 5, Postbus 217, 7500 AE Enschede, The Netherlands

[3] Faculty of Engineering, University of Pembangunan Nasional "Veteran" East Java, Jl. Raya Rungkut Madya Gunung Anyar, Surabaya 60294, Indonesia; zuzisukasno@gmail.com (S.S.); yustinangatilah@gmail.com (Y.N.); caeciliapujiastuti@gmail.com (C.P.)

* Correspondence: mohammad.tauviqirrahman@ft.undip.ac.id

Abstract: In the present study, a computational investigation into acoustic and tribological performances in journal bearings is presented. A heterogeneous pattern, in which a rough surface is engineered in certain regions and is absent in others, is employed to the bearing surface. The roughness is assumed to follow the sand-grain roughness model, while the bearing noise is solved based on broadband noise source theory. Three types of heterogeneous rough/smooth journal bearings exhibiting different placement and number of the rough zone are evaluated at different combinations of eccentricity ratio using the CFD method. Numerical results show that the heterogeneous rough/smooth bearings can supply lower noise and larger load-carrying capacity in comparison with conventional bearings. Moreover, the effect on the friction force is also discussed.

Keywords: acoustic; computational fluid dynamics (CFD); lubrication; roughness

Citation: Tauviqirrahman, M.; Jamari, J.; Wicaksono, A.A.; Muchammad, M.; Susilowati, S.; Ngatilah, Y.; Pujiastuti, C. CFD Analysis of Journal Bearing with a Heterogeneous Rough/Smooth Surface. *Lubricants* **2021**, *9*, 88. https://doi.org/10.3390/lubricants9090088

Received: 30 July 2021
Accepted: 31 August 2021
Published: 7 September 2021

Publisher's Note: MDPI stays neutral with regard to jurisdictional claims in published maps and institutional affiliations.

1. Introduction

Journal bearing is one of the most critical friction pairs in machine elements, in which the applied force is fully supported by the pressure of the lubricating film. The main function of the bearing is to keep the shaft always rotating about its axis, smoothing the rotary motion, reducing friction between the two surfaces, and dampening vibrations due to the rotating motion of the shaft and motor [1]. Within recent decades, a large quantity of research focusing on surface modification by texturing has been and continues to be performed. This is mainly because surface texturing has become a feasible way to improve journal bearing performance. Tala-Ighil, et al. [2] presented a detailed study relating to the effect of promoting a surface texture in the form of a cylindrical dimple by varying the location of the texture arrangement. The results of their study indicated that the application of texture on the entire bearing surface produces a detrimental effect, while on the other hand, the application of partial surface texture can improve the performance of journal bearings. Brizmer and Kligerman [3] found a potential benefit of micro-texture with laser surface texturing (LST) on the inner surface of bearings on the load-carrying capacity of journal bearings. Their finding was also confirmed by Ji et al. [4]. Later, Meng and his group [5–7] studied more deeply the effect of compound groove texture in various forms on tribological and acoustic performance through the computational fluid dynamics (CFD) method. Their main results stated that the optimal dimple compound can reduce noise levels and increase load-carrying capacity and frictional forces. This result was also experimentally verified by the same author [8]. Wang et al. [9] revealed that the provision of a texture in the form of a convex–concave spherical texture configuration on the bearing surface was able to significantly reduce the friction coefficient. An interesting finding was reported by Manser et al. [10] who found that by combining the effects of micro-texturing and micro-polar non-Newtonian lubricant, an enhancement load support but low friction of journal bearing was achieved. In recent lubrication, Saleh et al. [11] revealed that to get a high load-support, a convex texture with a perpendicular direction of curvature was recommended.

Furthermore, considerably more investigations into bearing performances related to surface roughness are also available. Based on the Reynolds equation, Javorova [12] demonstrated the significance of the surface roughness inclusion for the bearing performance analysis. Using Cristensen's stochastic roughness theory, Hsu et al. [13] explored the effect of two types of surface roughness directions, namely longitudinal and transversal, under a magnetic field, on the operational performance of bearings. They showed that by promoting longitudinal roughness, the load-carrying capacity can be enhanced. On the other hand, the opposite effect was observed when transverse roughness is employed. For the bearing with slip/no-slip pattern, Kalavathi et al. [14] derived the generalized Reynolds equation by considering roughness nature by employing Christensen's stochastic theory. They identified that the influence of roughness plays a notable role on the load-carrying capacity. It was confirmed that the load-carrying capacity increases with surface roughness. Later, the effects of surface roughness on the transient behavior of hydrodynamic journal bearings during startup were explored by Cui et al. [15]. They found that the longitudinal surface configuration has a quite significant effect on reducing the hydrodynamic force. Al-Samieh [16] explored the effect of surface roughness in sinusoidal waviness terms for Newtonian and non-Newtonian lubricants. It was observed that as the amplitude of the waviness increases, more fluctuations of pressure distribution occur. Later, Tauviqirrahman et al. [17] revealed that hydrodynamic pressure and load-carrying capacity decrease with surface roughness. In their case, it was assumed that the roughness was applied to the whole bushing surface. Recently, Gu et al. [18] reported that the surface roughness should be taken into account in the optimization of the surface texture. In general, as well as the surface texture, the surface roughness has a significant role in altering the tribological performance of the journal bearing.

In the present work, an investigation into the enhancement of the performance of journal bearings via an engineered rough surface, with emphasis on improving tribological indices and enhancing the acoustical performance (i.e., low noise), is studied utilizing the computational fluid dynamics (CFD) approach. By designing an engineered heterogeneous bearing surface, on which the roughness is applied to a certain area and is absent in others, the lubrication performance can be improved. The so-called heterogeneous rough/smooth pattern introduced here is inspired by the construction of heterogeneous slip/no-slip surface. As reported by numerous researchers, for example [14,19–24], the heterogeneous slip/no-slip configuration was proven to increase the load-carrying capacity and reduce the friction force significantly. It is believed that a configuration of rough/smooth regions will lead to enhanced journal bearing properties. In this analysis, to capture the cavitation phenomena in a more proper way that may occur in the bearing, the multi-phase cavitation approach is adopted as discussed by many workers [25–28]. Moreover, based on the survey literature, remarkable progress in the research of roughened journal bearings is mainly concerned with tribological characteristics. Few studies have been devoted to investigating the effect of surface roughness on bearing noise. Therefore, in this study, in addition to the tribological performance, the acoustic characteristic of bearing is of particular interest.

2. Theory

2.1. Governing Equations

In the present work, the flow behavior induced by the surface motion is solved by calculating the Navier–Stokes instead of the Reynolds theory for incompressible flow. In this study, the Reynolds-averaged Navier–Stokes (RANS) equation coupled with the mass conservation equation is used. To simplify the computational processing, the isothermal lubricant conditions are assumed.

The RANS equation (momentum equation) is:

$$\frac{\partial}{\partial x_i}(\rho u_i u_j) = -\frac{\partial p}{\partial x_i} + \frac{\partial}{\partial x_j}\left[\mu \frac{\partial u_i}{\partial x_j} - \overline{\rho u_i' u_j'}\right] \tag{1}$$

The mass conservation equation is:

$$\frac{\partial}{\partial x_i}(\rho u_i) = 0 \tag{2}$$

where ρ is to the lubricant density; $u_i(u_j)$ is the average velocity of the lubricant along the coordinates $x_i(x_j)$, that is the coordinate X, Y, or Z; p is the hydrodynamic pressure; μ is the viscosity; u'_i and u'_j are the fluctuation velocities; $\overline{\rho u_i' u_j'}$ is the Reynolds stress. In the present study, the standard turbulent kinetic energy k and turbulent dissipation rate ε_d models [29] are employed to solve the Reynolds stress.

Once the hydrodynamic pressure is calculated through Equations (1) and (2), in terms of the tribological performances, the load-carrying capacity of the bearing can be calculated by integrating the hydrodynamic pressure acting on the bearing surface, while the friction force exerted by the lubricant on the surface is obtained by integrating the shear stress over the surface area.

In the acoustic analysis studied here, the bearing noise is of particular interest. The noise tends to exist during the bearing operation due to the turbulence in the lubricant. In this work, a computational approach to solving the noise produced in the lubricant utilizes the broadband noise source model [29]. Here, the acoustic power level P_A is expressed as follows [29]:

$$P_A = a\rho \left(\frac{u_i^3}{l}\right)\frac{u_i^5}{a_o^5} \tag{3}$$

where u_i and l are turbulence velocity and length scales, respectively and a_o is the speed of the sound which is set to 1480 m/s. In Equation (3), a is a model constant. Furthermore, Equation (3) can be reduced in terms of k and ε_d as follows:

$$P_A = a_\varepsilon \rho \varepsilon_d \left(\frac{\sqrt{2k}}{a_0}\right)^5 \tag{4}$$

Here, the rescaled constant a_ε is set to 0.1 [29].

2.2. Cavitation Modeling

In this work, the mixture model of cavitation is employed as provided by CFD software. The mixture model represents vapor–liquid two-phase flow by considering that the liquid phase becomes vapor phase when the lubricant film pressure falls below the saturation pressure. Based on this approach, the growth of gas bubbles which often accompanies the cavitation process is also calculated. In this study, the multi-phase cavitation model of Zwart–Gelber–Belamri is used [29,30]. In cavitation, the liquid–vapor mass transfer (evaporation and condensation) is governed by the vapor transport equation:

$$\frac{\partial}{\partial t}(\alpha_v \rho_v) + \nabla.(\alpha_v \rho_v \mathbf{v}) = R_g - R_c \tag{5}$$

where α_v represents vapor volume fraction and ρ_v refers to vapor density. R_g and R_c account for the mass transfer between the liquid and vapor phases in cavitation. For the

Zwart–Gelber–Belamri model, assuming that all the bubbles have the same size in a system, the final form of the cavitation is as follows [29,30]:

$$p \le p_{sat}, \quad R_g = F_{evap}\frac{3\alpha_{\mathrm{nuc}}(1 - \alpha_v)\rho_v}{R_B}\sqrt{\frac{2}{3}\frac{p_{sat} - p}{\rho}} \tag{6}$$

$$p \ge p_{sat}, \quad R_c = F_{cond}\frac{3\alpha_v\rho_v}{R_B}\sqrt{\frac{2}{3}\frac{p - p_{sat}}{\rho}} \tag{7}$$

where F_{evap} = evaporation coefficient = 50, F_{cond} = condensation coefficient = 0.01, R_B = bubble radius = 10^{-6} m, α_{nuc} = nucleation site volume fraction = 5×10^{-4}, ρ = liquid density and p_{sat} = saturation pressure.

2.3. Roughness Modeling

In the present study, the sand-grain model as shown in Figure 1a is adopted to characterize the roughness profile of the rough surface of the heterogeneous rough/smooth bearing. Here, a close-packed monolayer of spheres with diameter K_s is used to cover the surface uniformly. For modeling the surface roughness, the modified law-of-the-wall for mean velocity is employed. This equation can be expressed as follows [29]:

$$\frac{u_p u^*}{\tau_w/\rho} = \frac{1}{\kappa}\ln\left(E\frac{\rho u^* y_p}{\mu}\right) - \Delta B \tag{8}$$

where $u^* = C_\mu^{1/4}k^{1/2}$ and $\Delta B = (1/\kappa)\ln f_r$. For sand-grain roughness, ΔB is affected by the physical roughness height K_s, while the height is assumed constant per surface [29].

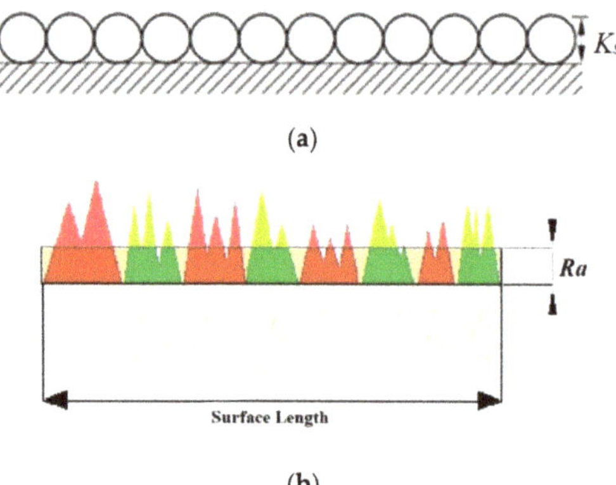

Figure 1. (**a**) Uniform sand-grain roughness model; (**b**) Roughness profile.

It should be noted that the roughness height K_s is the equivalent sand-grain roughness height and is not equal to the geometric roughness height of the surface. Therefore, it is necessary to use the conversion factor to convert the geometric roughness height of the surface into an equivalent sand-grain roughness. In this work, the R_a as shown in Figure 1b, is chosen as a parameter to represent the roughness height K_s (Figure 1a). The R_a represents the arithmetic average of the roughness profile, and in reality it is measured by the profilometer.

For all computations here, the R_a value will be an input to specify the roughness level of the heterogeneous rough/smooth bearing. According to the experiment performed by Adams et al. [31], the correlation between K_s and R_a can be defined as follows:

$$K_s = 5.863R_a \tag{9}$$

In FLUENT, to model the roughness effect, two roughness parameters must be specified, that is, the roughness constant C_s and the roughness height K_s. Here, because the k-ε_d turbulence model is used and the uniform sand-grain is assumed, the default roughness constant ($C_s = 0.5$) is employed as suggested by ANSYS FLUENT [29].

3. Simulation Method

3.1. Model

The basic geometry of the journal bearing used here adopts the geometry as presented by Meng, et al. [6]. In this study, the concept of the heterogeneous rough/smooth bearing is introduced in which the rough condition is applied on certain areas while the others are smooth. From a numerical perspective, the heterogeneous roughness distribution is made by applying the surface boundary condition on the chosen area by inputting the sand-grain roughness value K_s to model the roughness. Here, the film thickness of the lubricant will follow the surface profile as the input in the CFD program.

Three patterns of the heterogeneous rough/smooth journal bearing with various rough-smooth locations, namely 1 L, 2 L, and 3 L patterns, as shown in Figure 2 are studied and then compared with conventional (smooth) journal bearing (denoted as S pattern in this case). The journal bearing geometry and the characteristics of the lubricating fluid can be seen in Table 1.

Table 1. Parameters of the model.

Parameter	Symbol	Value	Unit
Bearing radius	R	50	mm
Width-diameter ratio	B/D	0.8	[[–]
Radial clearance	c	0.152	mm
Eccentricity ratio	ε	0; 0.1; 0.2; 0.3; 0.4; 0.5; 0.6; 0.7; 0.8	[[–]
Attitude angle	Φ	54	Deg
Fluid density	ρ	998.2	kg/m^3
Fluid viscosity	μ	0.001005	Pa.s
Rotational speed	n	2000	rpm
Saturation pressure	p_{sat}	2340	Pa
Vapor density	ρ_v	0.5542	kg/m^3
Vapor viscosity	μ_v	1.34×10^{-5}	Pa.s
Roughness level	R_a	25	μm

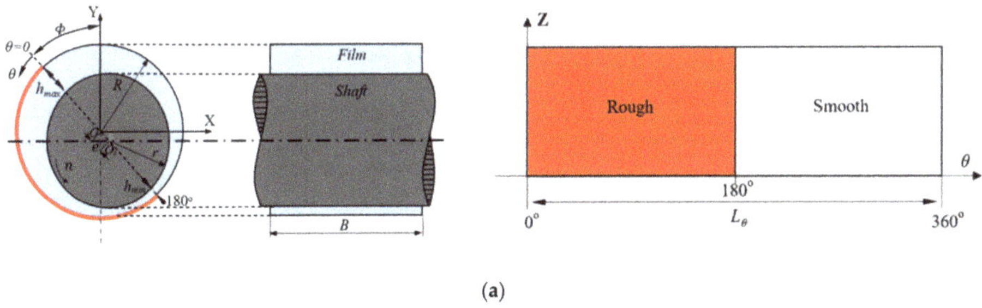

(a)

Figure 2. *Cont.*

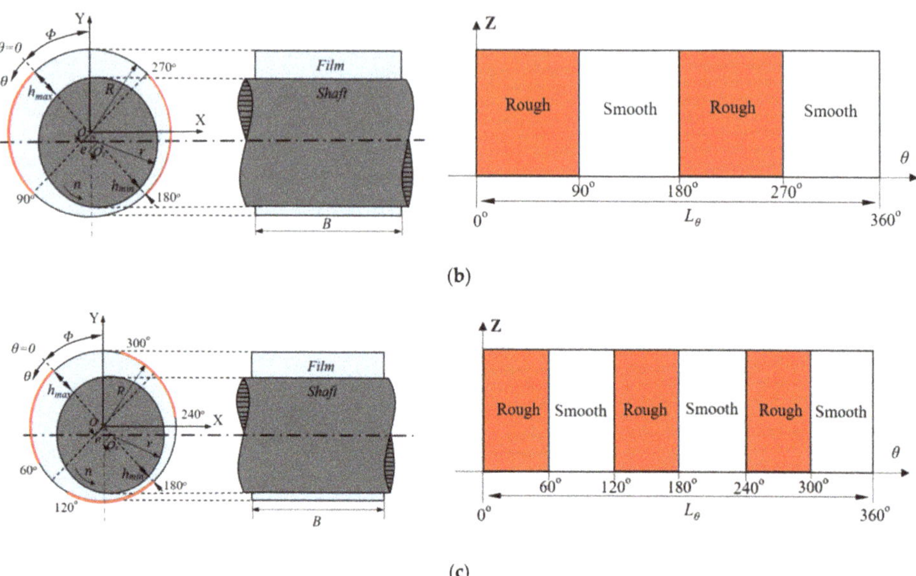

Figure 2. Three types of heterogeneous rough/smooth bearing with different artificial roughness zones, (**a**) one-rough zone (1 L); (**b**) two-rough zones (2 L); (**c**) three-rough zones (3 L).

To ensure that the flow regime in the studied journal bearing is turbulent, the calculation is performed by comparing the critical Reynolds number $(Re_c = (\rho \times 2\pi r \times n \times c \times (1-\varepsilon))/60)$ and the real Reynolds number $(Re_r = (\rho \times 2\pi r \times n \times c)/(60 \times \mu))$. For all values of eccentricity ratio considered here, the calculated real Reynolds number, Re_r is always much larger than the critical one, Re_c. For example, for the case of $\varepsilon = 0.8$, the Re_r is 1575 which is much larger than the Re_c of 0.32. From the physical framework, it means that turbulence may occur in the fluid film and thus, from the numerical framework, such turbulence phenomena must be modeled during the lubrication analysis to achieve accurate results.

3.2. Meshing

In this study, the mesh used consists of a uniform hexahedral grid. The face-meshing, edge-sizing and sweep-method features are used to form a mesh configuration. This results in the corresponding grid distribution being employed in the radial, circumferential and axial direction: $12 \times 400 \times 60$. For all journal bearing configurations studied here, the mesh distribution is based on independent mesh results. As a note, the division of fluid layers taken after sensitivity analysis revealed that several values (12, 14, 16) of layers division change the chosen main parameter (load-carrying capacity in this case) by less than 2% in the CFD model. To conclude, the 12-layer division of fluid domain is employed for all simulations because it provides a reasonable computational time with a feasible level of independent mesh. In detail, the mesh configuration and the criteria of the mesh formed are shown in Figure 3 and Table 2 below. From Table 2, it can be observed that for all cases here, during the grid generation the average skewness is much lower than 0.25. This indicates that based on the skewness mesh metrics spectrum [29], the following mesh distributions are categorized as excellent and thus the discretization error due to the mesh generation can be prevented.

3.3. Assumption and Boundary Condition

In this study, the journal moves with the shaft rotational speed n relative to the stationary bearing surface. Simulations are carried out using pressure-inlet and pressure-outlet boundary conditions. The values of the pressure at the inlet and outlet are taken as the ambient pressure, i.e., zero pressure. For moving wall boundary conditions, the surface is set to a rotating speed of 2000 rpm. In this research, the no-slip boundary condition is applied to the entire surface. In detail, Table 3 shows the boundary conditions used for the entire simulation case, whereas the setup of boundary conditions for the computational domain is depicted in Figure 4.

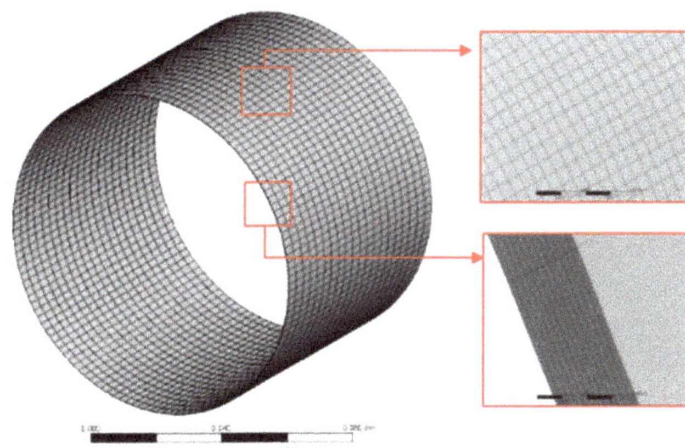

Figure 3. Mesh of the computational domain.

Table 2. Specification of the domain meshing.

Mesh Criteria	Value
Edge sizing 1	400 division
Edge sizing 2	60 division
Face Meshing	12-layers of division
Method	Sweep
Element number	288,000
For case $\varepsilon = 0$	
Maximum skewness	9.137×10^{-2}
Minimum skewness	7.194×10^{-3}
Average skewness	5.703×10^{-2}
For case $\varepsilon > 0$	
Maximum skewness	0.155
Minimum skewness	5.019×10^{-3}
Average skewness	5.604×10^{-2}

Table 3. Boundary condition.

Boundary Condition	Setup
Inlet	Pressure inlet (0 Pa)
Outlet	Pressure outlet (0 Pa)
Stationary wall	No-slip
Moving wall	No slip, $n = 2000$ rpm

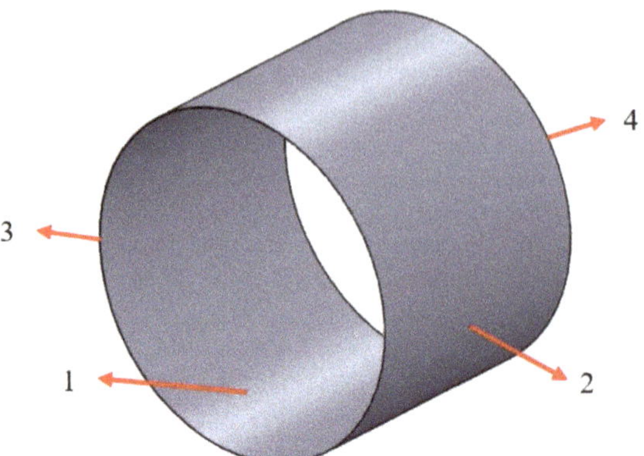

Figure 4. Boundary condition of the computational domain: 1—moving wall, 2—stationary wall, 3–inlet, 4—outlet.

3.4. Solution Setup

In the present study, the governing equations for the fluid domain are discretized by the finite volume method using ANSYS FLUENT. To obtain an accurate pressure, the SIMPLE scheme is employed for the velocity–pressure coupling. For the momentum and volume fraction equations, a first-order upwind discretization scheme is employed. For spatial discretization of the turbulent kinetic energy and turbulent dissipation rate, the second-order upwind discretization scheme is chosen.

4. Results and Discussion

4.1. Validation

To confirm that the developed CFD model and its solution setup are valid with appropriate accuracy, in this section, a comparison study between the present study and the reference is conducted in terms of the Sommerfeld number S in which $S = (r/c)^2(2\mu nrB/W)$. Here, the result is compared with the numerical and experimental data of Gao et al. [32] under the same input conditions and computed operational parameters (i.e., $\varepsilon = 0.55$, $\varepsilon = 0.685$, and $\varepsilon = 0.95$, $D = 80$ mm, $B = 80$ mm, $c = 0.08$ mm, $\Phi = 63.95°$, $n = 500–4000$ rpm, $\mu = 0.001$ Pa.s, $\rho = 998.2$ kg/m^3), as reflected in Figure 5. It can be found that the obtained values from the CFD code developed here are very close to the published ones both from the numerical and experimental results. Their deviations are less than 4% as indicated in Figure 5b, suggesting validation of the developed CFD code.

4.2. At Varied Eccentricity Ratio

In application, the bearing performance is notably affected by the eccentricity ratio, representing the magnitude of the loading during operation. Thus, in this work, the prediction of the acoustic and tribological performance is made for different eccentricity ratios ε, i.e., 0, 0.1, 0.2, 0.3, 0.4, 0.5, 0.6, 0.7, and 0.8. The range of eccentricity ratio chosen here may accommodate the range of bearing loading from very low to heavy loadings. All computational results presented here are evaluated at a rotational speed of 2000 rpm with a surface roughness level R_a of 25 μm. As a note, the surface with a value R_a of 25 μm is categorized as "rough" surface (R_a = 12.5–100 μm) [33]. As demonstrated by Tauviqirrahman et al. [17], the surface class of "rough" has the strongest effect on the tribological performance [17] in comparison to other classes such as precision (R_a = 0.1–0.2 μm), fine (R_a = 0.4–0.8 μm), and medium (R_a = 1.6–6.3 μm).

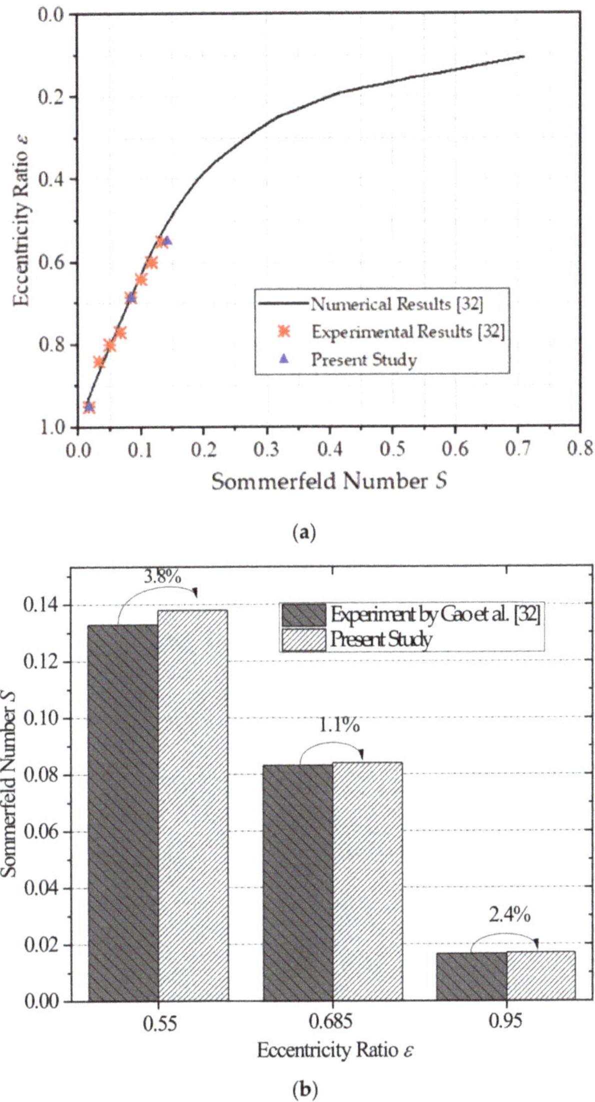

Figure 5. (**a**) Comparison between the result of the present study and the literature [32], and (**b**) histogram of deviation between the present result and the experiment of Gao et al. [32].

To show the effect of an engineered rough pattern on bearing characteristics, a plot of eccentricity ratio versus Sommerfeld number for both a conventional bearing ($R_a = 0$ and denoted as S pattern) and the bearing with the heterogeneous rough/smooth area (i.e., 1 L, 2 L, 3 L patterns) is reflected in Figure 6. From Figure 6, several characteristics can be seen. First, for all cases, an increase in the eccentricity ratio will decrease the Sommerfeld number S. The decrease in the Sommerfeld number occurs significantly when the eccentricity ratio ε is greater than 0.2. Secondly, when $\varepsilon = 0$ to 0.6, the heterogeneous rough/smooth bearing pattern with two-rough zones (2 L) gives the lowest Sommerfeld Number value when compared with the conventional smooth bearing (S) pattern and other heterogeneous rough/smooth bearing patterns. However, for eccentricity ratios of

0.7 and 0.8, heterogeneous rough/smooth bearing with one-rough zone (1 L) gives the lowest Sommerfeld Number. In the other words, although not superior to all eccentricity ratios studied here, the 2 L pattern gives the best performance in reducing the Sommerfeld number, which means that the enhanced load-carrying capacity can be achieved. From the results depicted in Figure 6, it can also be observed that the heterogeneous rough/smooth bearing, irrespective of the rough patterns, can generate the load-carrying capacity for all values of eccentricity ratio including for the concentric position. As is known, for the conventional bearing, no load-carrying capacity is produced when the concentric condition is applied due to the absence of the hydrodynamic pressure. This finding is interesting, and hence, the heterogeneous rough/smooth bearing can be compared to the heterogeneous slip/no-slip pattern. Based on the reference [19,22], even though there is no wedge effect in the case of concentric journal bearing, the heterogeneous slip/no-slip pattern could produce a relatively high load-carrying capacity. It indicates that the behavior of a "rough" surface can be correlated to the wettability of the surface (in particular the surface with hydrophobic coating) inducing the slip boundary. This is understandable because as discussed by Patankar [34] and Jung and Bhushan [35], the wettability of a surface is a function of its roughness.

As observed in Figure 6, in terms of the Sommerfeld number, the 2 L pattern is superior for the case of low to medium loading, while the 1 L pattern is more appropriate to the case of higher loading. To further explore the positive effect of the application of an engineered rough surface of 1 L and 2 L patterns, the comparison of the performance between the conventional (smooth) bearing and the heterogeneous rough/smooth one is presented. Here, in the following computation, the performance ratio is introduced and defines the ratio of the hydrodynamic parameters (i.e., load-carrying capacity, friction force, and acoustic power level) predicted for the heterogeneous rough/smooth surface against that of a classical (smooth) surface. Figures 7–9 summarize the ratio of the hydrodynamic performance parameters with a heterogeneous rough/smooth surface to that without a rough zone.

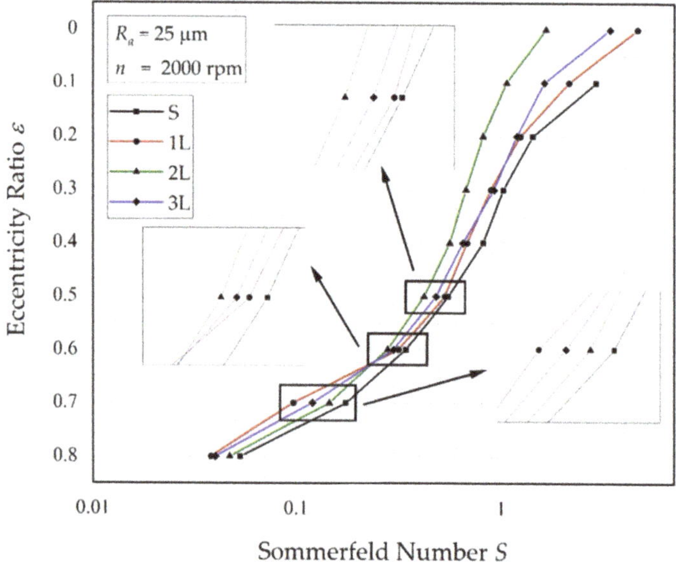

Figure 6. Eccentricity ratio vs. Sommerfeld number. Note: R_a = 25 μm.

Figure 7 shows the performance ratio of the load-carrying capacity for 1 L and 2 L patterns varying by eccentricity ratios. As a note, in Figure 7 the 3 L pattern is excluded because as depicted in Figure 6, the 3 L has a severe behavior in terms of Sommerfeld number compared to the other patterns. From Figure 7, the following features can be drawn. First, the values of the performance ratio for the heterogeneous rough/smooth bearings (1 L and 2 L) have a value above one which indicates that the load-carrying capacity of the two models is better when compared to the conventional (smooth) model for all eccentricity ratios considered here. It can also be drawn from Figure 7 that the benefit of heterogeneous rough/smooth patterns decreases with increased eccentricity ratio. Once again, this behavior seems to be similar to the behavior of the journal bearing with a heterogeneous slip/no-slip pattern. As discussed by several researchers focusing on the application of the heterogeneous slip/no-slip bearing, for example, Fortier and Salant [19], and Cui et al. [23], the eccentricity ratio reduces the positive effect of the applied engineered slip surface. In other words, the larger effect of the heterogeneous rough/smooth pattern can be achieved if the wedge effect is reduced. Second, the ratio of the load-carrying capacity of the 2 L model to the smooth model is higher at an eccentricity ratio from 0.1 to 0.6. For the eccentricity ratios of 0.7 and 0.8, the 1 L model has a higher performance ratio. This result is consistent with the previous finding as shown in Figure 6 in terms of Sommerfeld Number. This is understandable because the value of the load-carrying capacity of a journal bearing is inversely proportional to the value of the Sommerfeld Number. Third, it should be noted that as reflected in Figure 7, for the case of concentric position ($\varepsilon = 0$), the conventional (smooth) bearing cannot support the load, and as a consequence the value of the performance ratio of the load-carrying capacity for two models becomes infinite. From Figure 7, it is confirmed that the heterogeneous rough/smooth pattern in the concentric position of the journal bearing can support a load of 6.2 N for 1 L, and 17.53 N for 2 L. Once again, this indicates that the application of rough/smooth patterns can be promising in low loading operational condition.

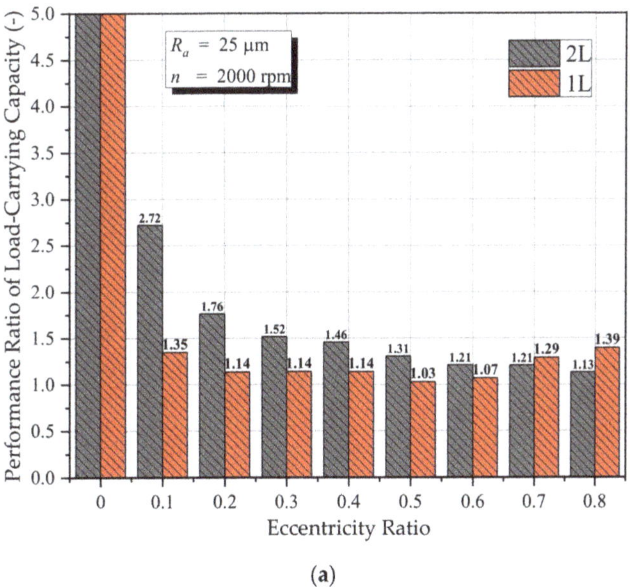

Figure 7. *Cont.*

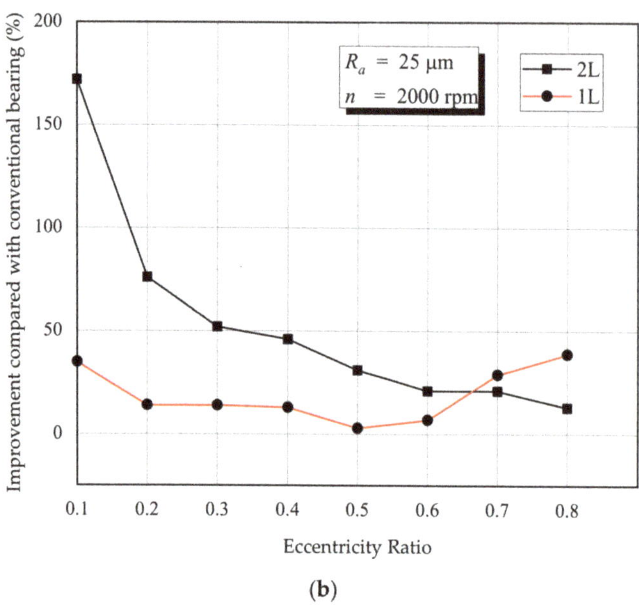

(**b**)

Figure 7. Effect of the arrangement of the rough zone on the load-carrying capacity under several eccentricity ratios, (**a**) lubrication performance ratio of load-carrying capacity, (**b**) improvement of the load-carrying capacity (compared with conventional bearing).

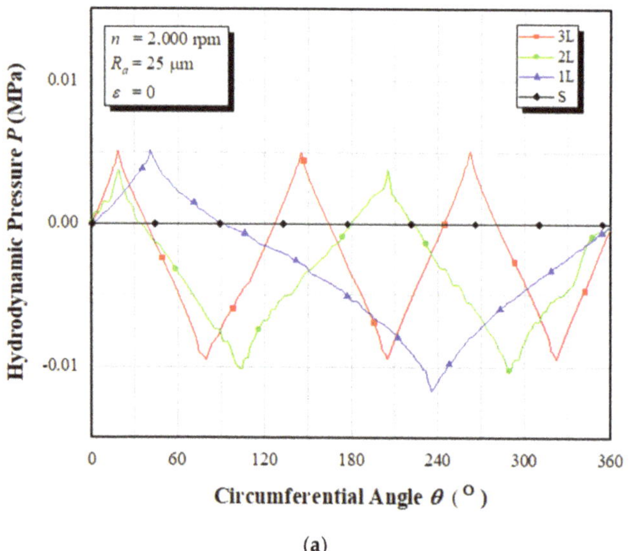

(**a**)

Figure 8. *Cont.*

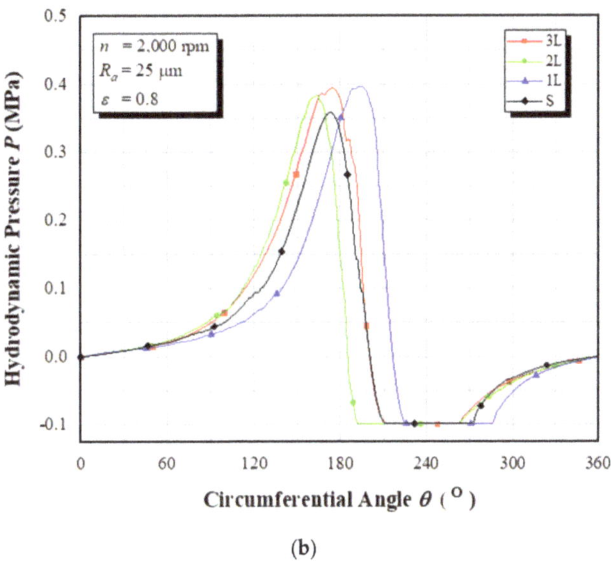

(b)

Figure 8. Hydrodynamic pressure distributions of the heterogeneous rough/smooth bearings for (**a**) $\varepsilon = 0$, and (**b**) $\varepsilon = 0.8$. The results are evaluated at the mid-plane of the bearing.

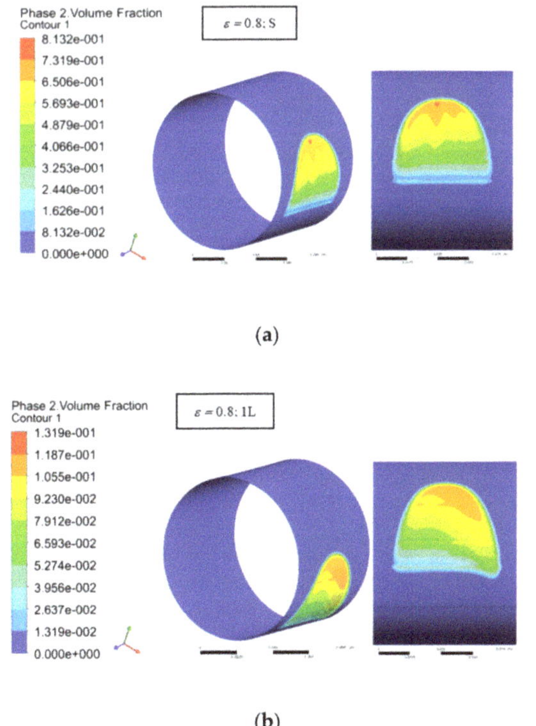

(a)

(b)

Figure 9. *Cont.*

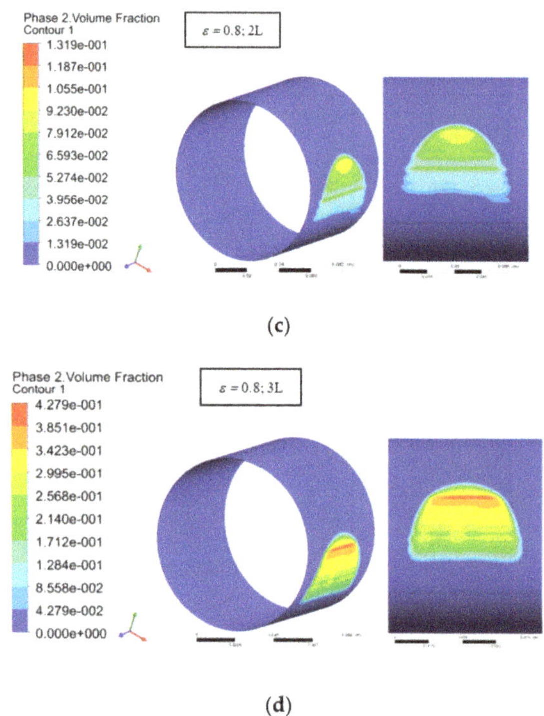

(c)

(d)

Figure 9. Comparison of the contour of vapor volume fraction between (**a**) the conventional (smooth) bearing, (**b**) the heterogeneous slip/no-slip bearing with 1L pattern, (**c**) the heterogeneous slip/no-slip bearing with 2 L pattern, (**d**) the heterogeneous slip/no-slip bearing with 3 L pattern.

Fourth, the differences in the placement and number of the rough zone for the design of the heterogeneous rough/smooth patterns have an impact on the load-carrying capacity of a journal bearing. For $\varepsilon = 0.1$, the performance ratio of the 2 L model is 101.8% higher than that of the 1 L model. At the medium loading position, say $\varepsilon = 0.4$, the 2 L model has a higher performance ratio of 54.8% when compared to the 1 L model. At the heavy loading position, i.e., $\varepsilon = 0.8$, the 1 L model has an increase in bearing performance which is 23.1% higher than the 2 L model. This result strengthens the findings highlighted in Figure 6, that the 2 L model is superior for the eccentricity ratios from 0 to 0.6.

The question "why do the different patterns of heterogeneous rough/smooth bearings bring out the different conclusions of load-carrying capacity" arises. The main contribution of the load-carrying capacity generations may give us a further understanding of this behavior. Figure 8 depicts the hydrodynamic pressure distribution for either conventional bearings or heterogeneous rough/smooth bearings for the concentric and non-concentric situations. Here, a high eccentricity ratio ($\varepsilon = 0.8$) is employed to explore the correlation of the wedge effect and the roughness effect. Through observation of Figure 8, one can observe that while the conventional bearing does not generate the lubrication performance in concentric condition, at the same condition the heterogeneous rough/smooth bearing can support the load irrespective of the rough zone. The capability of the heterogeneous rough/smooth bearing to build up the pressure is also found for the operation with a high eccentricity ratio. It is interesting to note that for the case of $\varepsilon = 0.8$, the profile of hydrodynamic pressure for the conventional bearing is similar to the heterogeneous rough/smooth bearings in the value of the peak pressure. The difference lies in the value of the maximum pressure and the width of the cavitation zone. However, such a difference is relatively small, as can also be observed in Figure 9, reflecting the comparison of the contour of vapor volume fraction between the conventional (smooth) bearing and the heterogeneous rough/smooth bearings. As mentioned earlier, the multi-phase cavitation model adopted here allows for phase change in a cavitation process. When the lubricant enters the divergent zone, the film pressure might fall below the saturation vapor pressure p_{sat}, and the lubricant would rupture. In the present study, the saturation pressure P_{sat} used is 2340 Pa (as shown in Table 1), and the pressure at the inlet and outlet boundaries are taken as the ambient pressure, i.e., zero pressure. As a consequence, for each value of local pressures in the computational domain, FLUENT will reduce them with the environmental pressure p_{atm} of 1 atm ($\approx$101,325 Pa). Therefore, when the cavitation occurs, the local pressure will be set to the saturation pressure (2340 Pa). By FLUENT, these values are converted to the negative value, i.e., $-98{,}985$ Pa (≈ -0.1 MPa), as depicted in Figure 8, to show that the cavitation exists. Based on Figures 8b and 9, when the eccentricity ratio is 0.8, the width of the cavitation zone does not change very much. It ranges from 50–70° depending on the bearing pattern. For example, in the case of 2 L pattern, the cavitation occurs at the circumferential angle θ of around 190°–236°.

Based on Figure 8, it is observed that for the case of a heterogeneous rough/smooth bearing with the 2 L pattern, the predicted maximum pressure is 11% higher compared with the conventional bearing. It indicates that for the same eccentricity ratio, the bearing with an engineered rough pattern can sustain a higher load than a conventional bearing. Again, this result strengthens the hypothesis proposed that the concept of heterogeneous rough/smooth bearings is very similar to the heterogeneous slip/no-slip pattern concerning how the heterogeneous rough/smooth bearing produces the load-carrying capacity. Fortier and Salant [19] found that the bearing with slip yields a lower Sommerfeld number (or higher load-carrying capacity) than the corresponding conventional bearing with the same eccentricity ratio. It is also confirmed from Figure 8 that, unlike the case with high ε, the pressure profiles are strongly affected by the placement and the number of the rough zones for the case with low ε. For example, for the 3 L pattern, three peak pressures can be observed following the number of rough zones. However, when the ε is increased to be very high ($\varepsilon = 0.8$ in this case), the 3 L pattern produces only one peak pressure for high. This means that the effect of the heterogeneous rough/smooth surface decreases with increased wedge effect. Again, this characteristic can be compared to the one with a heterogeneous slip/no-slip surface.

In terms of friction force, the effects of the application of 1 L and 2 L heterogeneous rough/smooth bearings are reflected in Figure 10. Specific features can be observed based on Figure 10. First, unlike the load-carrying capacity, the engineered rough zone leads to greater friction force compared to the conventional bearing. Compared to conventional journal bearings, the friction force of the heterogeneous rough/smooth bearing is 11–32% higher depending on the eccentricity ratio and the rough pattern (i.e., location and number). In other words, the application of the rough zone leads to a negative effect on friction. Second, concerning the eccentricity ratio effect, the performance ratio of frictional force for the 2 L and 1 L models does not change very much as the eccentricity ratio increases. The numerical results indicate that the use of the 2 L pattern consistently generates around 12% lower friction force compared with the 1 L bearing irrespective of the eccentricity ratio.

Concerning the question of "which is the best pattern to be used in journal bearing", it seems that, based on the tribological point of view, the 2 L pattern gives the most benefits of performance. Although greater friction is observed at the 2 L configuration, in comparison with the conventional bearing, the benefit in enhancing the load-carrying capacity is very significant. As a note, as is reflected in Figure 7b, a 180% improvement in load-carrying capacity can be achieved by the heterogenous rough/smooth bearing with the 2 L pattern. However, as is known in reality, the condition of a lubricated bearing with high load-carrying capacity and low friction is the main indicator of a "good" bearing including the heterogeneous rough/smooth bearing. Therefore, for future work, the issue of how to reduce the friction force by heterogeneous rough/smooth bearing will be explored more extensively.

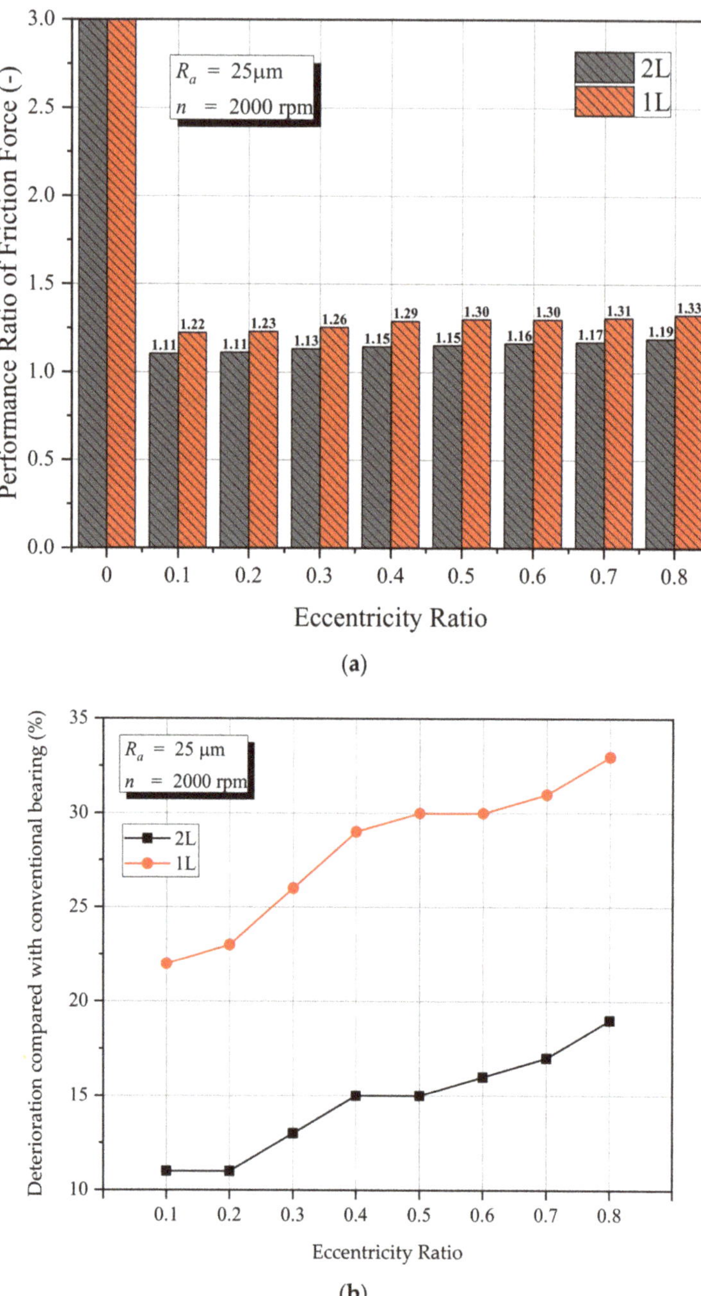

(a)

(b)

Figure 10. Effect of the arrangement of the rough zone on friction force under several eccentricity ratios, (**a**) lubrication performance ratio of friction force, (**b**) deterioration of the friction force (compared with conventional bearing).

Concerning the acoustic performance, it is interesting to explore the characteristics of the main indicator of the bearing noise, namely, the average acoustic power level. For the researchers, the reduction in the average acoustic power level is considered as standard for enhancing the acoustic performance of contacting pairs. Is the average acoustic power level affected by the engineered rough surface? Is it significant or not? To answer these questions, the histogram of the performance ratio of the average acoustic power level versus the eccentricity ratio is presented in Figure 11a, while in Figure 11b, the enhancement of the acoustic power level by the heterogeneous rough/smooth patterns is shown. It should be noted that for an eccentricity ratio of 0 (i.e., concentric position), the conventional (smooth) bearings do not have a level of acoustic strength inducing the infinite value of the performance ratio for the two models. Based on Figure 11, it can be observed that the heterogeneous rough/smooth bearings bring out the improvement of the acoustic performance in terms of average acoustic power level irrespective of the rough zone placement. Compared with the conventional bearing, a 4–12% lower noise level can be achieved. Further, the numerical results show that the 2 L pattern produces a 5% lower average acoustic power level compared to the 1 L pattern for all eccentricity ratios. This indicates that reasonably engineering a rough zone position is essential in reducing the bearing noise. Additionally, these results also strengthen the previous result concerning the positive effect of the 2 L pattern application in enhancing load-carrying capacity.

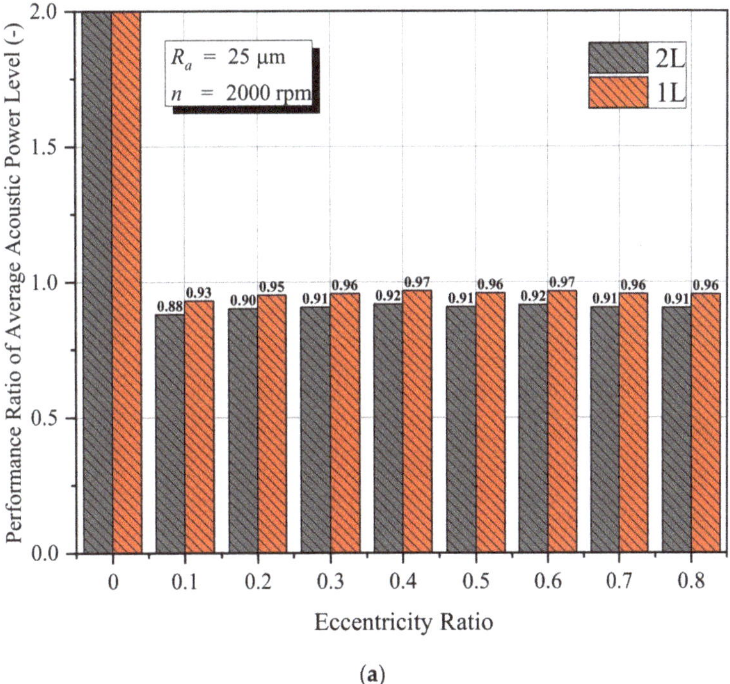

Figure 11. *Cont.*

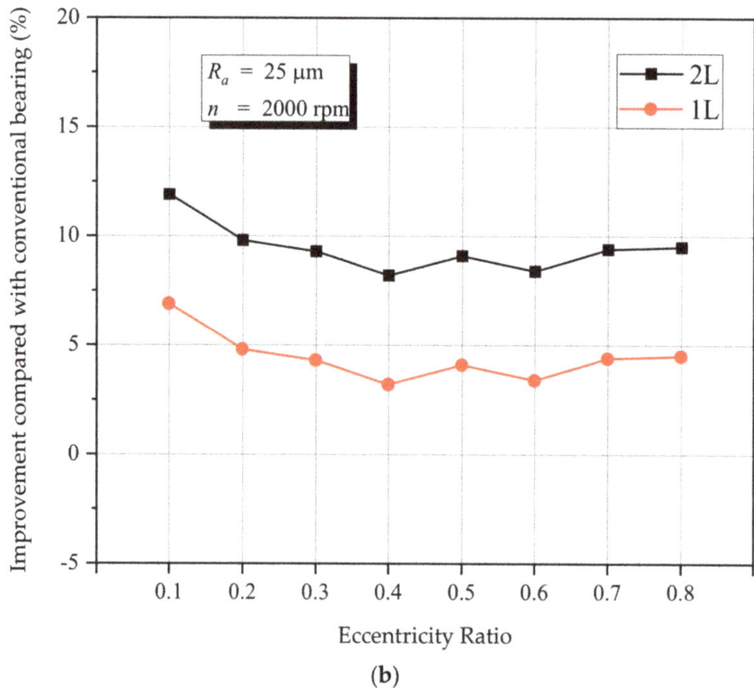

(b)

Figure 11. Effect of the arrangement of the rough zone on average acoustic power level under several eccentricity ratios, (**a**) lubrication performance ratio of average acoustic power level, (**b**) improvement of the average acoustic power level (compared with conventional bearing).

5. Conclusions

The current work presents the benefit of a well-chosen heterogeneous rough/smooth surface on the acoustic and tribological performance of the journal bearing. Three alternatives of placement of a rough-smooth zone to create the heterogeneous rough/smooth bearing are studied in terms of average acoustic power level, load-carrying capacity, and friction force. In general, the results show that a journal bearing with a heterogeneous rough/smooth pattern has an advantage over a conventional bearing irrespective of the rough pattern in terms of load-carrying capacity and average acoustic power level. The effect of heterogeneous rough/smooth patterns is more dominant for low and moderate eccentricity ratios. Specifically, more enhanced load-carrying capacity and reduced average acoustic power level can be obtained through the two-rough zones (2 L) pattern. However, the numerical results also confirmed the drawback of the application of heterogeneous rough/smooth patterns. These patterns lead to up to 30% higher friction force in comparison to the conventional bearing. Therefore, for future work, the exploration of the optimal parameters of heterogeneous rough/smooth patterns focusing on how to reduce such friction should be performed.

Author Contributions: Conceptualization, M.T. and M.M.; Data curation, A.A.W.; Funding acquisition, J.J.; Project administration, S.S. and Y.N.; Software, M.M.; Supervision, J.J.; Validation, A.A.W. and C.P.; Visualization, M.T. and A.A.W.; Writing—review and editing, M.T. All authors have read and agreed to the published version of the manuscript.

Funding: This research is fully funded by RPI Grant, No. 233-29/UN7.6.1/PP/2021.

Data Availability Statement: The data presented in this study are available on request from the corresponding author.

Acknowledgments: The authors fully acknowledged Institute for Research and Community Services (LPPM), Diponegoro University for the approved fund which makes this important research viable and effective.

Conflicts of Interest: The authors declare no conflict of interest.

Nomenclature

a_o	Local speed of the sound
B	Bearing width
c	Radial clearance
C_s	Roughness constant
D	Bearing diameter
E	Empirical constant
f_r	Roughness function
F_{evap}	Evaporation coefficient
F_{cond}	Condensation coefficient
h_{min}	Minimum film thickness
h_{max}	Maximum film thickness
K_s	Roughness height
k	Turbulent kinetic energy
l	Length scale
L_θ	Circumferential length of the bearing
n	Rotational speed
p	Hydrodynamic pressure
P_A	Acoustic power level
p_{sat}	Saturation pressure
W	Load-carrying capacity
r	Shaft radius
R	Bearing radius
R_a	Arithmetic average of the roughness profile
R_B	Bubble radius
Re_c	Critical Reynolds number
Re_r	Real Reynolds number
R_g, R_c	Mass transfer between the liquid and vapor phase
u_p	Mean velocity of the fluid at the near-wall node P
u^*	dimensionless velocity
y_p	Distance from point P to the wall
α_{nuc}	Nucleation site volume fraction
α_v	Vapor volume fraction
ε	Eccentricity ratio
ε_d	Turbulent dissipation rate
κ	von Karman constant
θ	Circumferential angle
μ	Lubricant viscosity
μ_v	Vapor viscosity
ρ	Lubricant density
ρ_v	Vapor density
Φ	Attitude angle

References

1. Malcolm, E.; Leader, P.E. *Understanding Journal Bearings*; Applied Machinery Dynamics, Co.: Durango, CO, USA, 2001.
2. Tala-Ighil, N.; Fillon, M.; Maspeyrot, P. Effect of textured area on the performances of a hydrodynamic journal bearing. *Tribol. Int.* **2011**, *44*, 211–219. [CrossRef]
3. Brizmer, V.; Kligerman, Y. A laser surface textured journal bearing. *J. Tribol.* **2012**, *134*, 031702. [CrossRef]
4. Ji, J.; Fu, Y.; Bi, Q. Influence of geometric shapes on the hydrodynamic lubrication of a partially textured slider with micro-grooves. *J. Tribol.* **2014**, *136*, 041702. [CrossRef]
5. Meng, F.M.; Zhang, L.; Liu, Y.; Li, T.T. Effect of compound dimple on tribological performances of journal bearing. *Tribol. Int.* **2015**, *91*, 99–110. [CrossRef]

6.   Meng, F.M.; Wei, Z.; Minggang, D.; Gao, G. Study of acoustic performance of textured journal bearing. *Proc. Ins. Mech. Eng. Part J J. Eng. Tribol.* **2016**, *230*, 156–169. [CrossRef]

7.   Meng, F.M.; Zhang, W. Effects of compound groove texture on noise of journal bearing. *J. Tribol.* **2017**, *140*, 031703. [CrossRef]

8.   Meng, F.; Yu, H.; Gui, C.; Chen, L. Experimental study of compound texture effect on acoustic performance for lubricated textured surfaces. *Tribol. Int.* **2019**, *133*, 47–54. [CrossRef]

9.   Wang, J.; Zhang, J.; Lin, J.; Ma, L. Study on lubrication performance of journal bearing with multiple texture distributions. *Appl. Sci.* **2018**, *8*, 244. [CrossRef]

10. Manser, B.; Belaidi, I.; Khelladi, S.; Chikh, M.A.A.; Deligant, M.; Bakir, F. Computational investigation on the performance of hydrodynamic micro-textured journal bearing lubricated with micropolar fluid using mass-conserving numerical approach. *Proc. Ins. Mech. Eng. Part J J. Eng. Tribol.* **2020**, *234*, 1310–1331. [CrossRef]

11. Saleh, A.M.; Crosby, W.; El Fahham, I.M.; Elhadary, M. The effect of liner surface texture on journal bearing performance under thermo-hydrodynamic conditions. *Ind. Lub. Tribol.* **2020**, *72*, 405–414. [CrossRef]

12. Javorova, J. EHD lubrication of journal bearings with rough surfaces. In Proceedings of the International Conference "Mechanical Engineering in XXI Century", Nis, Serbia, 25–26 November 2010.

13. Hsu, T.C.; Chen, J.H.; Chiang, H.L.; Chou, T.L. Lubrication performance of short journal bearings considering the effects of surface roughness and magnetic field. *Tribol. Int.* **2013**, *61*, 169–175. [CrossRef]

14. Kalavathi, G.K.; Dinesh, P.A.; Gururajan, K. Influence of roughness on porous finite journal bearing with heterogeneous slip/no-slip surface. *Tribo. Int.* **2016**, *102*, 174–181. [CrossRef]

15. Cui, S.; Gu, L.; Fillon, M.; Wang, L.; Zhang, C. The effects of surface roughness on the transient characteristics of hydrodynamic cylindrical bearings during startup. *Tribol. Int.* **2018**, *128*, 421–428. [CrossRef]

16. Abd Al-Samieh, M.F. Surface roughness effects for Newtonian and non-Newtonian lubricants. *Tribol. Ind.* **2019**, *41*, 56–63. [CrossRef]

17. Tauviqirrahman, M.; Ichsan, B.C.; Jamari, M. Influence of roughness on the behavior of three-dimensional journal bearing based on fluid-structure interaction approach. *J. Mech. Sci. Technol.* **2019**, *33*, 4783–4790. [CrossRef]

18. Gu, C.; Meng, X.; Wang, S.; Ding, X. Study on the mutual influence of surface roughness and texture features of rough-textured surfaces on the tribological properties. *Proc. Ins. Mech. Eng. Part J. J. Eng. Tribol.* **2021**, *235*, 256–273. [CrossRef]

19. Fortier, A.E.; Salant, R.F. Numerical analysis of a journal bearing with a heterogeneous slip/no-slip surface. *J. Tribol.* **2005**, *127*, 820–825. [CrossRef]

20. Lin, Q.; Wei, Z.; Zhang, Y.; Wang, N. Effects of the slip surface on the tribological performances of high-speed hybrid journal bearings. *Proc. Ins. Mech. Eng. Part. J J. Eng. Tribol.* **2016**, *230*, 1149–1156. [CrossRef]

21. Bhattacharya, A.; Dutt, J.K.; Pandey, R.K. Influence of hydrodynamic journal bearings with multiple slip zones on rotordynamic behavior. *J. Tribol.* **2017**, *139*, 061701. [CrossRef]

22. Wu, Z.; Ding, X.; Zeng, L.; Chen, X.; Chen, K. Optimization of hydrodynamic lubrication performance based on a heterogeneous slip/no-slip surface. *Ind. Lubr. Tribol.* **2019**, *71*, 772–778. [CrossRef]

23. Cui, S.; Zhang, C.; Fillon, M.; Gu, L. Optimization performance of plain journal bearings with partial wall slip. *Tribol. Int.* **2020**, *145*, 106–137. [CrossRef]

24. Tauviqirrahman, M.; Afif, M.F.; Paryanto, P.; Jamari, J.; Caesarendra, W. Investigation of the tribological performance of heterogeneous slip/no-slip journal bearing considering thermo-hydrodynamic effects. *Fluids* **2021**, *6*, 48. [CrossRef]

25. Dhande, D.Y.; Pande, D.W. Multiphase flow analysis of hydrodynamic journal bearing using CFD coupled fluid structure interaction considering cavitation. *J. King Saud Univ. Eng. Sci.* **2018**, *30*, 345–354. [CrossRef]

26. Morris, N.J.; Shahmohamadi, H.; Rahmani, R.; Rahnejat, H.; Garner, C.P. Combined experimental and multiphase computational fluid dynamics analysis of surface textured journal bearings in mixed regime of lubrication. *Lubr. Sci.* **2018**, *30*, 161–173. [CrossRef]

27. Sun, D.; Li, S.; Fei, C.; Ai, Y.; Liem, R.P. Investigation of the effect of cavitation and journal whirl on static and dynamic characteristics of journal bearing. *J. Mech. Sci. Technol.* **2019**, *33*, 77–86. [CrossRef]

28. Tauviqirrahman, M.; Jamari, J.; Wibowo, B.S.; Fauzan, H.M.; Muchammad, M. Multiphase computational fluid dynamics analysis of hydrodynamic journal bearing under the combined influence of texture and slip. *Lubricants* **2019**, *7*, 97. [CrossRef]

29. ANSYS. *ANSYS Fluent, Version 16.0: User Manual*; ANSYS, Inc.: Canonsburg, PA, USA, 2017.

30. Zwart, P.; Gerber, A.G.; Belamri, T. A two-phase flow model for predicting cavitation dynamic. In Proceedings of the Fifth International Conference on Multiphase Flow, Yokohama, Japan, 30 May 2004.

31. Adams, T.; Grant, C.; Watson, H. A Simple algorithm to relate measured surface roughness to equivalent sand-grain roughness. *Int. J. Mech. Eng. Mechatron.* **2012**, *1*, 66–71. [CrossRef]

32. Gao, G.; Yin, Z.; Jiang, D.; Zhang, X.; Wang, Y. Analysis on design parameters of water-lubricated journal bearings under hydrodynamic lubrication. *Proc. Inst. Mech. Eng. Part J J. Eng. Tribol.* **2016**, *230*, 1019–1029. [CrossRef]

33. Japanese Industrial Standard/Japanese Standards Association. *JIS B 0601:2013 Geometrical Product Specifications (GPS)—Surface Texture: Profile Method—Terms, Definitions and Surface Texture Parameters (Foreign Standard)*; Japanese Industrial Standard/Japanese Standards Association: Tokyo, Japan, January 2013.

34. Patankar, N.A. On the modeling of hydrophobic contact angles on rough surfaces. *Langmuir* **2003**, *19*, 1249–1253. [CrossRef]

35. Jung, Y.C.; Bhushan, B. Contact angle, adhesion and friction properties of micro and nanopatterned polymers for superhydrophobicity. *Nanotechnology* **2006**, *17*, 4970–4980. [CrossRef]

 *lubricants*

Article

Design and Analysis of an Aerostatic Pad Controlled by a Diaphragm Valve

Federico Colombo, Luigi Lentini *, Terenziano Raparelli, Andrea Trivella and Vladimir Viktorov

Politecnico di Torino, Corso Duca degli Abruzzi 24, 10129 Turin, Italy; federico.colombo@polito.it (F.C.); terenziano.raparelli@polito.it (T.R.); andrea.trivella@polito.it (A.T.); vladimir.viktorov@polito.it (V.V.)
* Correspondence: luigi.lentini@polito.it

Abstract: Because of their distinctive characteristics, aerostatic bearings are particularly suitable for high-precision applications. However, because of the compressibility of the lubricant, this kind of bearing is characterized by low relative stiffness and poor damping. Compensation methods represent a valuable solution to these limitations. This paper presents a design procedure for passively compensated bearings controlled by diaphragm valves. Given a desired air gap height at which the system should work, the procedure makes it possible to maximize the stiffness of the bearing around this value. The designed bearings exhibit a quasi-static infinite stiffness for load variation ranging from 20% to almost 50% of the maximum load capacity of the bearing. Moreover, the influence of different parameters on the performance of the compensated pad is evaluated through a sensitivity analysis.

Keywords: aerostatic bearings; compensation; infinite stiffness; diaphragm valve; design

Citation: Colombo, F.; Lentini, L.; Raparelli, T.; Trivella, A.; Viktorov, V. Design and Analysis of an Aerostatic Pad Controlled by a Diaphragm Valve. *Lubricants* **2021**, *9*, 47. https://doi.org/10.3390/lubricants9050047

Received: 6 April 2021
Accepted: 21 April 2021
Published: 27 April 2021

Publisher's Note: MDPI stays neutral with regard to jurisdictional claims in published maps and institutional affiliations.

1. Introduction

Because of their distinctive characteristics, aerostatic bearings are particularly suitable for high-precision applications. Their zero friction, cleanness and long life make aerostatic bearings successful in applications such as machine tools, measuring machines and power board testing [1]. In fact, this type of bearing significantly increases the measuring accuracy of this type of measuring system since it does not generate additional vibrations with respect to the more conventional ball bearings [2,3].

However, because of the compressibility of the lubricant, this kind of bearing is characterized by low relative stiffness and poor damping. For these reasons, many solutions have been proposed since the 1960s, which was a pivotal period in the history of gas lubrication [4].

The first attempts that were made to increase aerostatic bearing stiffness consisted of many sensitivity analyses concerning the type, location and size of the supply hole [5]. Additionally, this type of investigation was further enhanced with the aid of more accurate numerical models. Boffey et al. [6] and Colombo et al. [7] performed numerical and experimental studies to assess the effect of the supply pressure and orifice size and location on the performance of rectangular aerostatic pads. In these works, it was found that choosing a lower nominal air gap height led to slightly higher stiffness.

The same result was found by the adoption of porous or grooved surfaces. In fact, the higher performance that stems from the use of these kinds of feeding systems can be exploited in the presence of lower nominal air gap heights. Moreover, porous and grooved bearings present further relevant drawbacks. Groove geometries have to be suitably selected to avoid pneumatic hammer [8–10], whereas porous surfaces have some critical issues related to the choice of material, which should have both suitable porosity and impact toughness [11,12].

Subsequently, technological improvements in the field of manufacturing and electronics have made it possible to enhance aerostatic bearing performance by means of compensation methods. These methods consist of the use of additional devices and/or components that, when integrated with aerostatic pads, can enhance performance. Although there are many compensation strategies, e.g., modifying the feeding system of the pad or its geometry, the main goal is to compensate for air gap variations. Compensation methods can be active or passive. In passive compensation methods, bearings are integrated with devices and/or components that exploit only the energy associated with the supply pressure, e.g., pneumatic valves and compliant elements. Conversely, actively compensated bearings are integrated with elements such as sensors, controllers and actuators that require external sources of energy to function [13]. Looking at the current solutions, in actively compensated bearings, the presence of closed-loop systems makes it possible to obtain significant improvements in both static and dynamic performance. Because of their high dynamics, ease of integration and high power density, piezoelectric actuators are used in many active compensation systems. Al-Bender [14] and Aguirre [15] proposed an active compensation solution that makes it possible to exploit different types of conicity control. Here, a circular and centrally fed aerostatic pad is integrated with a (Proportional Integrative) PI controller, displacement capacitive sensors and three symmetrically and circumferentially placed piezo actuators. The integration of these elements with the pad makes it possible to suitably modify the conicity of the air gap in order to compensate for air gap variations. Similarly, Maamari et al. [16,17] proposed another type of conicity control by using a magnetic actuation and internal servomechanism. Adopting a geometrical compensation (also called a support control [14]) is another option that has been proposed by Colombo et al. [18] and Matsumoto et al. [19,20]. In this compensation strategy, a piezoelectric actuator is used in a closed-loop control to compensate for air gap variations by modifying the thickness of the pad. Moreover, in other solutions, piezoelectric actuators were also used to control the air flow exhausted/supplied from/to the bearing [21–23].

Despite their higher effectiveness, active compensation solutions are still too expensive to be integrated into most current industrial applications. By contrast, passive compensation solutions, notwithstanding their lower dynamics and effectiveness, represent an acceptable and cheap alternative. In fact, it was found that, if suitably selected, compliant elements and pneumatic valves can obtain higher and even quasi-static infinite stiffness over 20% of the bearing load range [24]. Regarding passive compensation solutions, Newgard and Kiang [25] proposed the use of elastic orifices to regulate the air flow supplied to the clearance in order to compensate for air gap variations. Another valuable solution proposed by different authors [24,26,27] consists of the use of a bearing with convergent and compliant gap geometries, effecting a passive conicity control. Chen and Lin [28] proposed an aerostatic pad with a disk spring compensator to increase the bearing damping. Ghodsiyeh et al. [29,30] and Colombo et al. [31,32] proposed a novel prototype of an aerostatic pad controlled by means of a diaphragm valve. It was found that this kind of passive compensation makes it possible to obtain quasi-static infinite stiffness and high damping in a low frequency range. Moreover, using pneumatic valves represents a low-cost solution characterized by ease of integration and simple construction. In [33], a prototype of a passively compensated pad was also proposed where a differential diaphragm valve was used.

This paper presents a procedure to suitably design aerostatic pads controlled by diaphragm valves depending on the desired air gap height at which the system has to work. The design procedure is described by considering a simple geometry of the pad, i.e., a circular and centrally fed aerostatic pad. Results demonstrate that this design procedure makes it possible to significantly increase the stiffness of the pad at the desired air gap height. Moreover, it shows how the performance of the compensated pad varies depending on the supply pressure, the ratio between the valve and pad supply hole diameters and the size of the pad.

2. The Compensated Pad

2.1. Description and Functioning

Figures 1–3 show the schemes of the diaphragm valve and the aerostatic pad considered in the proposed design procedure. As can be seen, the valve is composed of a main body, moving nozzle, upper body, micrometric screw, Belleville washer and a lower body. The nozzle is mounted coaxially with respect to the vertical hole drilled in the main body of the valve and it has an orifice of diameter d_p. The presence of the Belleville washer and the micrometric screw makes it possible to regulate its relative initial position (in the absence of supply pressure) (x_0) with respect to a diaphragm of diameter D_m and thickness s, which is clamped between the main and the lower body of the valve through an O-ring (see Figure 2a). As shown in Figure 2a,b, the initial position (in the absence of supply pressure) of the nozzle can be positive $(x_0 > 0)$ when it is not in contact with the diaphragm or negative $(x_0 < 0)$ when the nozzle imposes a preload on the diaphragm. When the pressure in the valve chamber p_1 increases, the diaphragm deflects, thus increasing the nozzle–diaphragm distance of $x = x_0 + x_v$. Figure 3 shows a scheme of the circular and centrally fed aerostatic pad that is considered for the proposed design procedure.

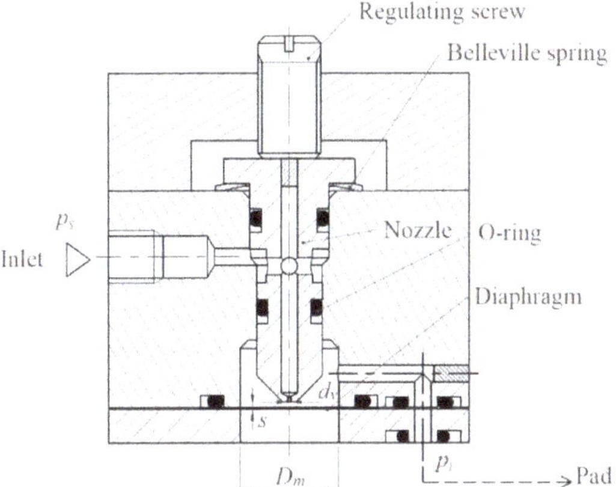

Figure 1. Scheme of the diaphragm valve.

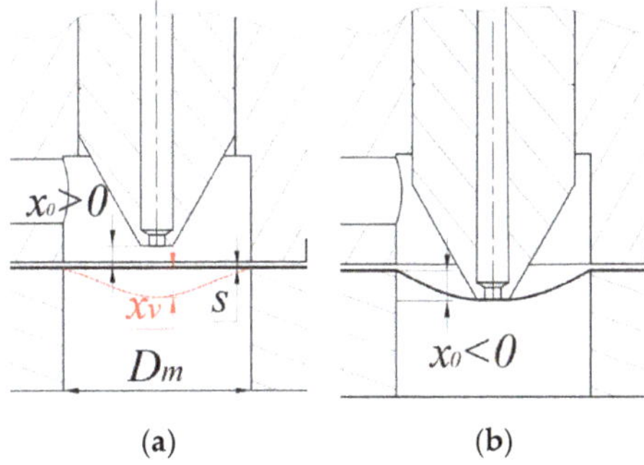

Figure 2. (**a**) Configuration for $x_0 > 0$. (**b**) Configuration for $x_0 < 0$.

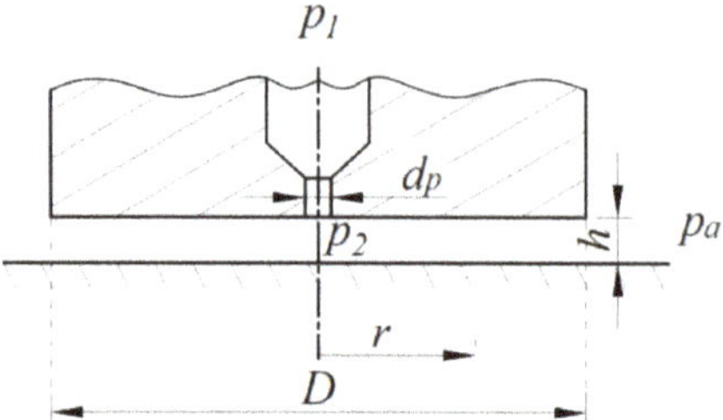

Figure 3. Scheme of the aerostatic pad.

The pad has an outer diameter D and a supply hole of diameter d_p. The supply pressure of the pad p_1 is provided through the outlet of the valve. Once the compressed air reaches the air pad, it passes through the supply hole and then is exhausted from the air gap. During its passage, the air pressure reduces, firstly, to p_2 due the pressure drop that occurs at the curtain area under the supply hole and, secondly, it gradually reduces up to ambient pressure p_a at the outer edge of the pad.

The compensated pad is modeled as a lumped pneumatic circuit composed of capacitances and resistances, as shown in Figure 4.

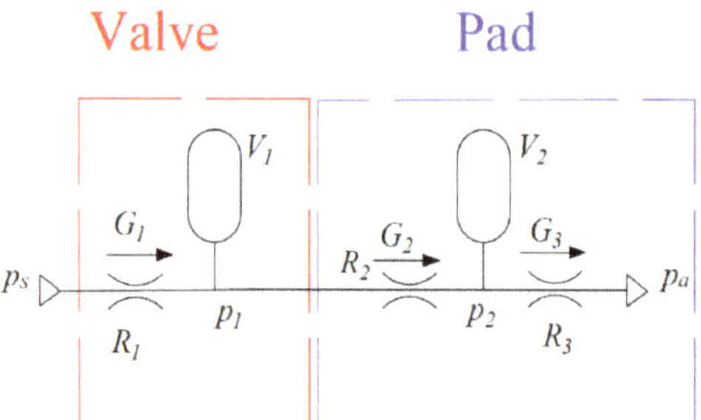

Figure 4. Lumped pneumatic circuit of the compensated pad.

where p_s is the supply pressure of the system, R_1, R_2 and R_3 are the lumped resistances related to the valve, the pad and the air gap. G_1, G_2 and G_3 are the correspondent air mass flow rates that are computed as:

$$G_1 = c_{d_1} \frac{p_s}{\sqrt{R_g T_s}} \pi d_v x \, \varphi\left(\frac{p_1}{p_s}\right)$$

$$\varphi\left(\frac{p_1}{p_s}\right) = \begin{cases} \left[\left(\frac{p_1}{p_s}\right)^{2/k} - \left(\frac{p_1}{p_s}\right)^{(k+1)/k}\right]^{1/2} & , \ \frac{p_1}{p_s} < \left(\frac{2}{k+1}\right)^{k/(k-1)} \\ \left[\left(\frac{2}{k+1}\right)^{2/(k-1)} - \left(\frac{2}{k+1}\right)^{(k+1)/(k-1)}\right]^{1/2} & , \ \frac{p_1}{p_s} \geq \left(\frac{2}{k+1}\right)^{k/(k-1)} \end{cases}$$

$$c_{d_1} = 1.05\left(1 - 0.3 e^{-0.005\,Re_1}\right)$$

$$Re_1 = \frac{G_1}{\pi d_v \mu}$$

\hfill (1)

$$G_2 = c_{d_2} \frac{p_1}{\sqrt{R_g T_1}} \pi d_p h \, \varphi\left(\frac{p_2}{p_1}\right)$$

$$\varphi\left(\frac{p_2}{p_1}\right) = \begin{cases} \left[\left(\frac{p_2}{p_1}\right)^{2/k} - \left(\frac{p_2}{p_1}\right)^{(k+1)/k}\right]^{1/2} & , \ \frac{p_2}{p_1} < \left(\frac{2}{k+1}\right)^{k/(k-1)} \\ \left[\left(\frac{2}{k+1}\right)^{2/(k-1)} - \left(\frac{2}{k+1}\right)^{(k+1)/(k-1)}\right]^{1/2} & , \ \frac{p_2}{p_1} \geq \left(\frac{2}{k+1}\right)^{k/(k-1)} \end{cases}$$

$$c_{d_2} = 1.05\left(1 - 0.3 e^{-0.005\,Re_2}\right)$$

$$Re_2 = \frac{G_2}{\pi d_p \mu}$$

\hfill (2)

$$G_3 = \frac{\pi h^3 \left(p_2^2 - p_a^2\right)}{12\mu R_g T_2 \cdot ln\left(\frac{D}{d_p}\right)} \tag{3}$$

where k, μ and R_g are the specific heat ratio, dynamic viscosity and gas constant of the air. T_s is the supply temperature of the air, c_{d_i} ($i = 1$, 2) are the discharge coefficients related to the supply holes of the nozzle and the pad (for further details, readers can refer to [34]). The expressions of the air mass flow rates G_j ($j = 1$, 2 , 3) are obtained by considering isentropic expansions through the nozzle and the supply hole and isothermal conditions in the air gap. It is worth noting that because of the large heat exchange between the metallic part of the system and the small volumes occupied by the fluid, all the temperatures are considered equal to the ambient temperature ($T_s = T_1 = T_2 = T_a$). The load capacity of the compensated pad can be obtained by integrating the expression of the air gap pressure distribution (for further details, see [35]):

$$p(r) = p_2 \sqrt{1 - \left(1 - \frac{p_a^2}{p_2^2}\right) \frac{ln\left(\frac{2r}{d_p}\right)}{ln\left(\frac{D}{d_p}\right)}} \tag{4}$$

$$F_P = p_2 \frac{\pi d_p^2}{4} \sqrt{\frac{\pi A}{8}} e^{\frac{2}{A}} \left[erf\left(\sqrt{\frac{2}{A}}\right) - erf\left(\sqrt{\frac{2}{A}} \cdot \frac{p_a}{p_2}\right)\right] \quad A = \frac{\left(1 - \frac{p_a^2}{p_2^2}\right)}{ln\left(\frac{D}{d_p}\right)} \tag{5}$$

The considered compensation method is based on the fact that, for each value of the external load applied upon a pad ($F^{ext} = F_p$), there exists a supply pressure p_1 that makes it possible to keep constant a desired air gap height h^*. In order to clarify the working principle of the compensation method, Figure 5a,b compare the operating principles of a conventional to that of the considered compensated pad. These figures compare how a conventional pad (Figure 5a) and a compensated pad (Figure 5b) work when the external load F increases. Here, continuous lines represent the air flow G_2 supplied by the supply holes of the pad for a given supply pressure p_1 and a variable downstream pressure p_2. Similarly, dotted lines represent the air flow G_3 exhausted from an air gap with a constant height. Given an initial condition (h_0, $G_0 = G_2 = G_3$, F_0), in a conventional pad, the air flow G and the air gap height h decrease as the applied load F increases (see Figure 5a). This is because the operating principle of the pad depends on the air flow supplied by its supply holes. The idea of the proposed compensation method is to exploit the presence of the diaphragm valve to compensate for these air gap height variations by increasing the supply pressure of the bearing. As can be seen from Figure 5b, if the diaphragm valve were able to suitably modify the supply pressure of the pad in accordance with external load variations, the compensated pad would work at a constant air gap height, thus providing quasi-static infinite stiffness.

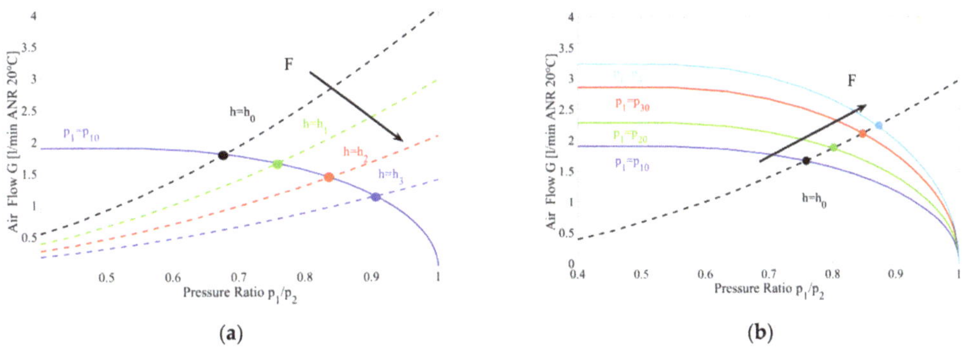

Figure 5. (**a**) Working conditions of a conventional pad. (**b**) Working conditions of a compensated pad.

The aim of this work is to show how to design the compensated pad (the valve and the pad) in order to obtain the highest stiffness in a neighborhood of a desired air gap height h^*.

2.2. Design Procedure

Once a desired air gap height h^* is decided, the most important parameters that have to be defined to maximize the stiffness of the pad are:

- the stiffness of the diaphragm (k_m);
- the initial position of the nozzle with respect to the diaphragm (x_0).

The first step in defining the values of these parameters is to compute the ideal values of the supply pressure of the pad $p_1{}^{id}$ that make it possible to obtain a constant air gap height h^*. These pressure values can be easily computed by applying the continuity equation at the air gap entrance:

$$\overbrace{c_{d_2}(Re,h^*)\frac{p_1{}^{id}}{\sqrt{R_g T_1}}\,\pi d_p h^*\,\varphi\left(\frac{p_2}{p_1{}^{id}}\right)}^{G_2} - \overbrace{\frac{\pi\,h^{*3}\left(p_2^2 - p_a^2\right)}{12\mu R_g T_s\cdot ln\left(\frac{d}{d_p}\right)}}^{G_3} = 0 \tag{6}$$

Based on the assumption that, in most cases, external loads applied upon the pad F are proportional with respect to the pressure downstream of its supply holes p_2, given the external loads and the desired air gap height h^*, Equation (6) can be used to compute the corresponding p_1^{id}. Being nonlinear, Equation (6) must be solved iteratively through the Regula Falsi Method. It is worth noting that discharge coefficients can be also considered as a function of the Reynolds number Re and the considered air gap height h^* using different formulations [34] (as shown in Equations (1) and (2)).

Once p_1^{id} has been computed for each value of the applied load F, it is necessary to limit the maximum value of p_1^{id} considering the maximum supply pressure that can be provided to the valve. In these instances, the maximum value was taken as equal to 0.7 MPa. This choice was made considering the supply pressures that are normally used in conventional gas bearing applications ([36], see page 228). Figure 6 shows the trend of the ideal and effective (considering the limit of 0.7 MPa) supply pressure p_1 as a function of the applied load (computed through Equation (5)). As can be seen, the range of compensation reduces as the maximum supply pressure is reduced.

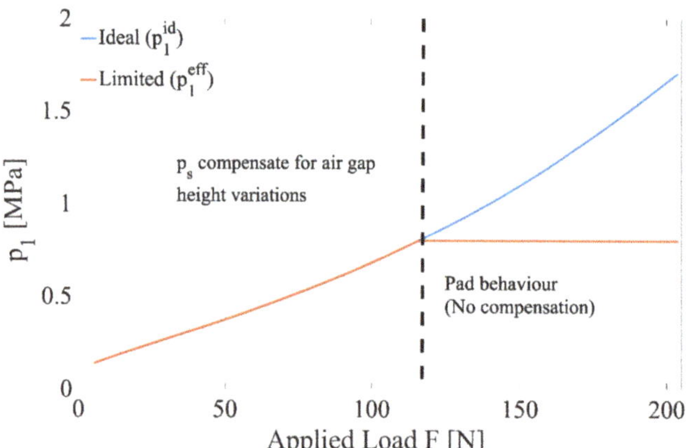

Figure 6. Trend of p_1^{id} and p_1^{eff} versus the applied load F^{ext}.

At this stage, given that

- the air mass flow rate $G = G_3$;
- the effective supply pressure that the valve must supply to the pad p_1^{eff},

it is possible to compute the correspondent valve opening law: the distance between the nozzle and the diaphragm x. Similarly to the previous step, this can be done by imposing the continuity equation at the valve chamber.

$$x^{id} = \frac{G_3 \sqrt{R_g T_s}}{\pi d_v p_s \ \varphi\left(\frac{p_1^{eff}}{p_s}\right)} \tag{7}$$

The only difference from the previous step is that the dimension of the nozzle diameter is unknown. In order to define a suitable value of this diameter d_v, x were computed considering different ratios of d_v/d_p: 0.5, 0. 25 and 0.1.

Figure 7 shows the trend of the ideal distance x^{id} versus the pressure in the valve chamber for different d_v/d_p ratios. Here, it is possible to see that adopting lower ratios of d_v/d_p results in more compliant diaphragms with a more nonlinear behavior. In view of this, a better choice would be to consider a ratio as large as possible to reduce the dimensions and the nonlinear behavior of the valve diaphragm. On the other hand, the smaller the diameter ratio, the more effective the compensating action of the valve. This can be argued considering the curtain areas downstream of the valve nozzle and the supply hole of the pad:

$$\pi d_p h < \pi d_v x \qquad \qquad \text{Pad behaviour (Case 1)}$$
$$\pi d_p h > \pi d_v x \qquad \qquad \text{Valve compensation (Case 2)}$$

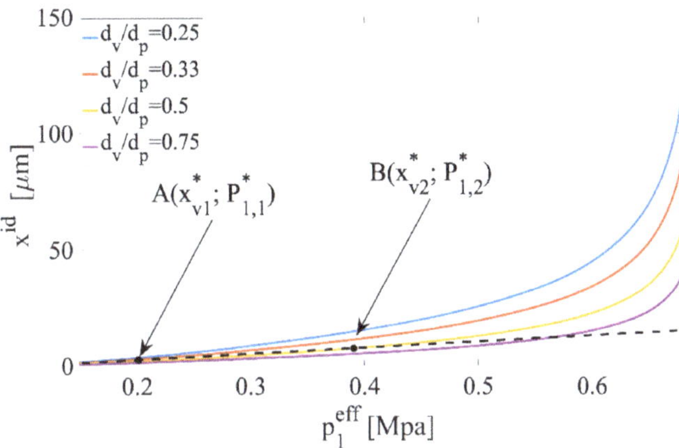

Figure 7. Trend of the ideal distance x^{id} versus the pressure in the valve chamber for different d_v/d_p ratios.

Depending on which of these curtain areas is the smaller one, the system behavior is governed by the pad (Case 1) or the compensating action of the valve (Case 2). After these considerations, the ratio d_v/d_p was taken as equal to 0.5 ($d_p = 1$ mm and $d_v = 0.5$ mm). At this point, on the basis of the curve obtained in Figure 7, it is necessary to approximate the ideal curve representing the valve opening law with a straight line. The better the straight line approximates the ideal curve, the closer the real behavior of the system will be to the ideal one. In view of this, the line was taken between the ideal curve at pressures equal to 0.2 MPa and $\left(\frac{2}{k+1}\right)^{k/(k-1)} p_s$.

Once these two points A ($x_{v,1}^*$; $p_{1,1}^*$) and B ($x_{v,2}^*$; $p_{1,2}^*$) are defined, by suitably manipulating the equation of the line, it is possible to obtain the values of the corresponding diaphragm stiffness k_m and its initial distance from the nozzle x_0:

$$x_0 = x_{v,1}^* - \left(\frac{x_{v,2}^* - x_{v,1}^*}{p_{1,1}^* - p_{1,2}^*} \right) \times p_{1,2}^* \tag{8a}$$

$$k_m = \frac{\pi D_m^2}{4 \left(\frac{x_{v,2}^* - x_{v,1}^*}{p_{1,2}^* - p_{1,1}^*} \right)} \tag{8b}$$

where D_m = 3 mm is the effective diameter of the diaphragm (the diameter corresponding to the active surface of the diaphragm) that, in order to reduce the valve dimensions, was taken as 1 mm larger than the outer diameter of the nozzle.

2.3. Computational Algorithm to Obtain the Bearing Features

When the parameters that characterize the behavior of the compensated pad were defined, namely k_m, x_0, d_v, d_p, p_s, h^*, the static performance was obtained through the lumped model depicted in Figure 4. Here, the equations related to the analytical solution of a circular and centrally fed pad were used to compute the load capacity (Equation (5)) and the air flow (Equation (3)) of the pad. The presence of the lumped volumes V_1 and V_2 was necessary to iteratively calculate the correspondent pressure p_1 and p_2. The structure of the algorithm to compute the static curves of the compensated pad is similar to that adopted in [31] and it has been implemented in the Matlab environment. The algorithm consists of two main parts. The first part is necessary to compute an initial static solution of the problem (h_0, F_{p_0}, G_0, P_{0_0}, P_{1_0}, P_{2_0}), where the load capacity F_{p_0} and the air mass flow rate G_0 of the pad are computed by considering the air gap height h_0 as an input parameter of the model. Once the convergence conditions on the load capacity and flow rate are simultaneously satisfied, the resulting physical parameters h_0, F_{p_0}, G_0, P_{0_0}, P_{1_0}, P_{2_0} are used as input variables in the second main part of the algorithm. In this second step, the air gap height values that guarantee the equilibrium of the pad are iteratively computed by simulating the application of a step force ΔF (in the time domain). Given the initial static condition (h_0, F_{p_0}, G_0, P_{0_0}, P_{1_0}, P_{2_0}) and the selected step force ΔF, the external load F^{ext} acting on the pad is computed as:

$$F^{ext}_{\ i} = F_{p_0} + i \cdot \Delta F \tag{9}$$

where, i is the number of iterative steps that have been already solved ($i = 0$, 1, 2, $\ldots, i_{max}$). In each iterative step (i), the equilibrium air gap height is computed by solving the equilibrium equation of the pad (Equation (10)) through Euler's explicit method:

$$F^{ext} = F_p - M\ddot{h} \tag{10}$$

where M is the moving mass of the pad. Once the equilibrium is reached ($M\ddot{h} \approx 0$ and the convergence conditions are satisfied), the obtained results are used as the new initial condition, and the new external load F^{ext} is obtained from the previous one by adding a further step force ΔF (see Equation (9)). As discussed in [31], it is necessary to use this kind of algorithm since the compensating action of the valve renders the load capacity F_p and the air mass flow rate G non-injective functions of the air gap height h (there is more than one load capacity and air mass flow rate for the same air gap height value).

3. Sensitivity Analysis

The lumped model and the proposed design procedure are used to perform a sensitivity analysis aimed at investigating the influence of the pad size (diameter D), the ratio between the pad and valve supply holes (d_v/d_p), the desired air gap height (h^*) and valve supply pressure (p_s). Figure 8 shows a scheme that summarizes what has been done in the present work. Firstly, the design procedure and the lumped model were applied to a reference case where the input parameters were chosen by considering the supply pressure (p_s = 0.7 MPa) and bearing size (D =40 mm) that are conventionally used in gas bearing applications and supply hole diameters that can favor the compensation action of the valve and the manufacturing of the valve nozzle (d_v = 0.5 mm and d_p = 1 mm). In fact, taking large supply hole diameters of the pad makes it possible to obtain a compensated system whose behavior is mainly governed by the valve nozzle and the air gap. Then, this analysis was extended to evaluate the effect of the influence of size of the pad, the ratio between the pad and valve supply holes, the desired air gap height and valve supply pressure.

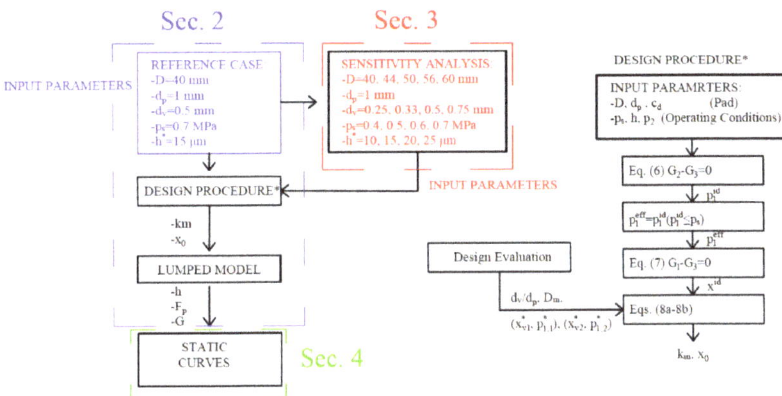

Figure 8. Flow charts summarizing the steps of the proposed design procedure and sensitivity analysis.

4. Results and Discussion

Figures 9–21 show the results of the performed sensitivity analysis.

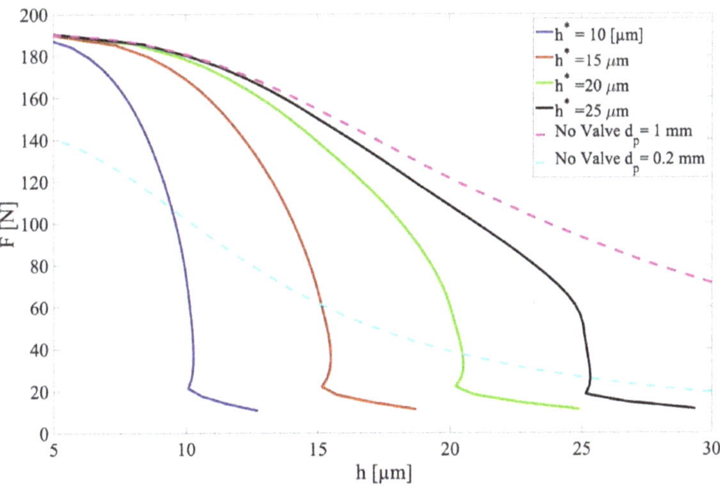

Figure 9. Comparison among the load capacities of the two benchmarks and compensated pad designed considering different desired air gap heights h^*, by considering the other parameters as constant: D = 40 mm, d_p = 1 mm, d_v = 0.5 mm and p_s = 0.7 MPa.

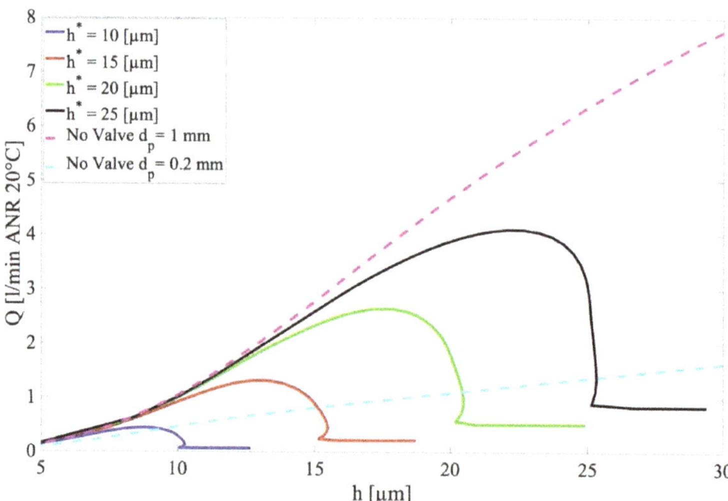

Figure 10. Comparison among the air consumptions of the two benchmarks and compensated pad designed considering different desired air gap heights h^*, by considering the other parameters as constant: $D = 40$ mm, $d_p = 1$ mm, $d_v = 0.5$ mm and $p_s = 0.7$ MPa.

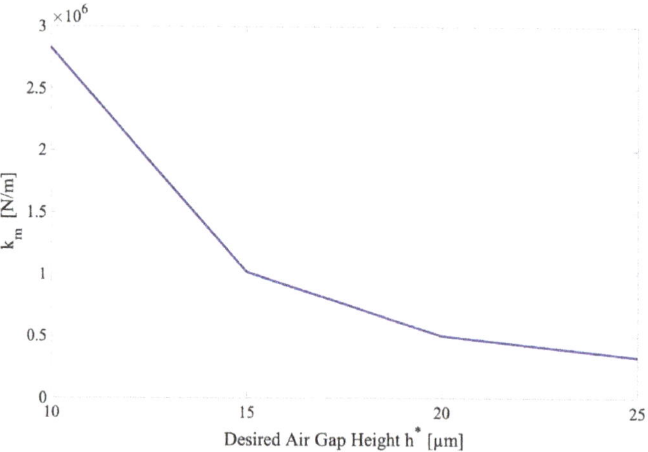

Figure 11. Trend of the optimal diaphragm stiffness expressed as a function of the desired air gap height h^*, by considering the other parameters as constant: $D = 40$ mm, $d_p = 1$ mm, $d_v = 0.5$ mm and $p_s = 0.7$ MPa.

4.1. Effect of the Selected Air Gap Height

Figures 9 and 10 show the load capacity and the air consumption of different compensated pads designed to work at different air gap heights $h^* = 10, 15, 20$ and 25 μm, whereas the other input parameters are chosen equal to those of the reference case indicated in Figure 8 ($p_s = 0.7$ MPa, $D = 40$ mm, $d_v = 0.5$ mm and $d_p = 1$ mm). The characteristic curves of the non-compensated pad and a "more conventional" pad ($d_p = 0.2$ mm) were used as benchmark curves to better figure out the efficiency of the proposed method. It is possible to see that the stiffness of the compensated pad is much higher than those of the benchmark curves over almost the entire analyzed load range and for all the desired air gap heights h^* (Figure 9). As regards the air consumption (Figure 10), for the compensated pad, it increases as the desired air gap height is increased too. However, by considering the common air gap heights used in industrial applications (around 10–15 μm), the air consumption is quite similar or lower than those of the second benchmark ($d_p = 0.2$ mm). Other relevant considerations can be made on the diaphragm stiffness that is required to obtain these compensated pads. As it is possible to see in Figure 11, this stiffness decreases in a nonlinear way as the desired air gap height is decreased.

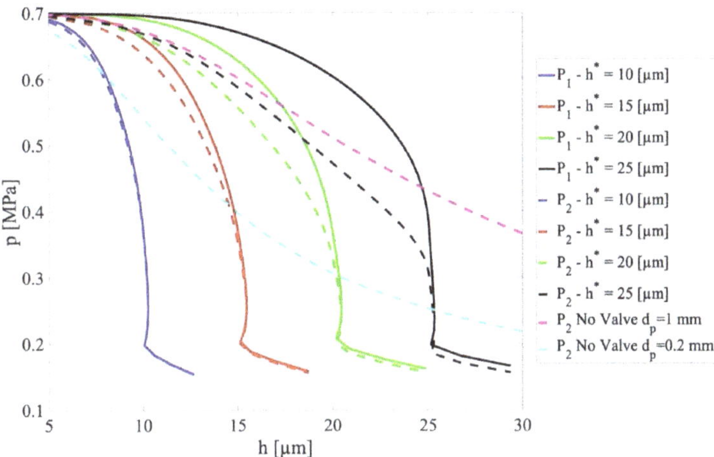

Figure 12. Comparison between the pressures upstream and downstream of the orifice of the pad considering different values of the desired air gap height h^*, by considering the other parameters as constant: $D = 40$ mm, $d_p = 1$ mm, $d_v = 0.5$ mm and $p_s = 0.7$ MPa.

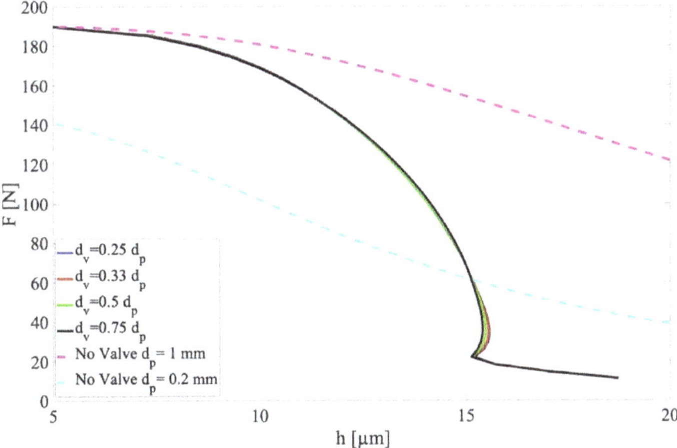

Figure 13. Comparison among the load capacities of the two benchmarks and compensated pad designed considering different ratios of the nozzle d_v diameters, by considering the other parameters as constant: $D = 40$ mm, $h^* = 15$ μm and $p_s = 0.7$ MPa.

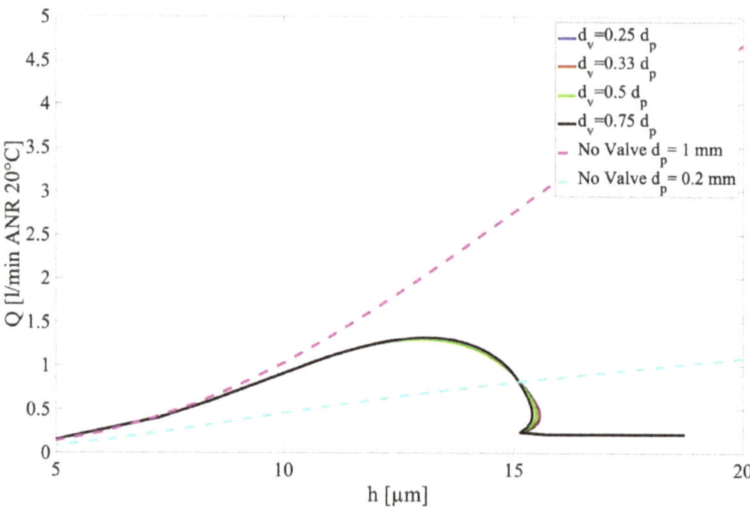

Figure 14. Comparison among the air consumptions of the two benchmarks and compensated pad designed considering different ratios of the nozzle d_v and orifice d_p diameters, by considering the other parameters as constant: $D = 40$ mm, $h^* = 15$ μm and $p_s = 0.7$ MPa.

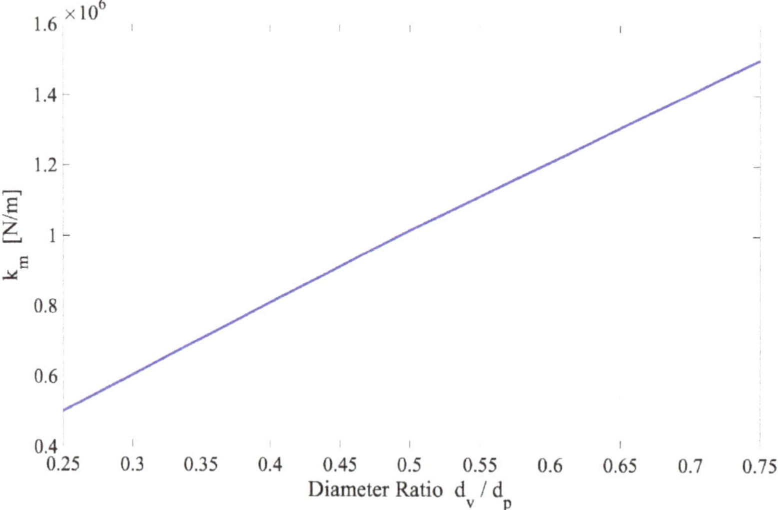

Figure 15. Trend of the optimal diaphragm stiffness expressed as a function of the diameter ratio d_v/d_p and orifice d_p diameters, by considering the other parameters as constant: $D = 40$ mm, $h^* = 15$ μm and $p_s = 0.7$ MPa.

Further insight into the functioning of the compensated pad can be obtained by considering the pressure drop occurring downstream of the supply hole of the pad. From Figure 12, it is clear that, thanks to the valve regulation, close to the desired air gap height, the pressure drop downstream of the supply hole of the pad is almost zero (the presence of this hole has no effect on the behavior of the pad). This effect becomes more relevant as the desired air gap height is decreased.

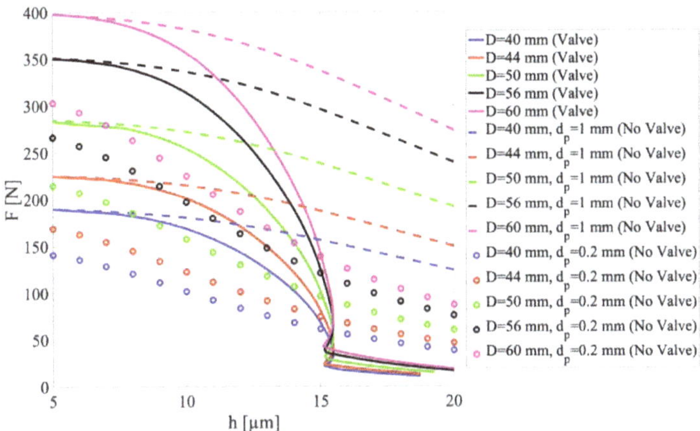

Figure 16. Comparison among the load capacities of the two benchmarks and compensated pad designed considering different outer diameters of the circular pad D, by considering the other parameters as constant: $d_p = 1$ mm, $d_v = 0.5$ mm, $h^* = 15$ μm and $p_s = 0.7$ MPa.

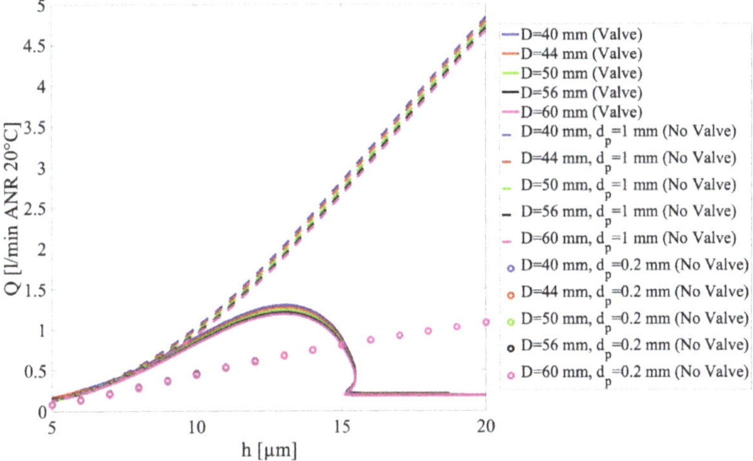

Figure 17. Comparison among air consumption of the two benchmarks and compensated pad designed considering different outer diameters of the circular pad D, by considering the other parameters as constant: $d_p = 1$ mm, $d_v = 0.5$ mm, $h^* = 15$ μm and $p_s = 0.7$ MPa.

4.2. Effect of the Diameter Ratio

Figures 13–15 report the results from the simulations aimed at investigating the effect of varying the supply hole of the valve d_v keeping constant the supply hole of the pad $d_p = 1$ mm. This because relatively large diameters for the supply hole of the pad are essential to obtain higher performance from this kind of compensation system. As it is possible to see from Figures 13 and 14, varying the diameter of the supply hole of the valve does not affect the performance of the compensation system. This can be easily understood by considering that, in this instance, the reduction in the conductance of the valve supply hole (due to the lower diameter) is compensated by an increase in the diaphragm stiffness (lower values of x).

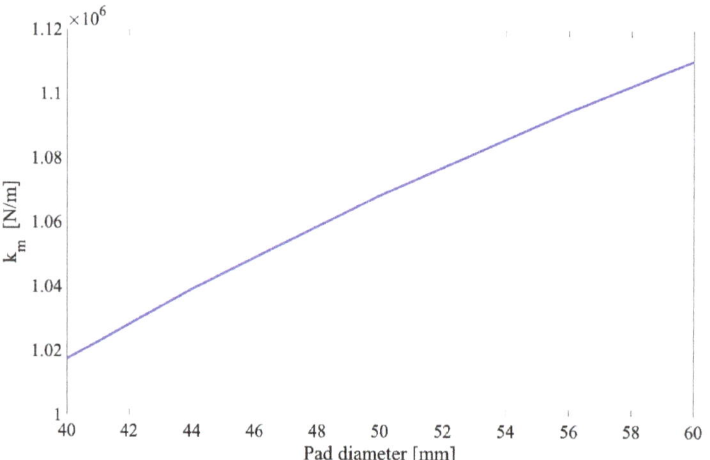

Figure 18. Trend of the optimal diaphragm stiffness expressed as a function of the outer diameter of the pad D, by considering the other parameters as constant: $d_p = 1$ mm, $d_v = 0.5$ mm, $h^* = 15$ μm and $p_s = 0.7$ MPa.

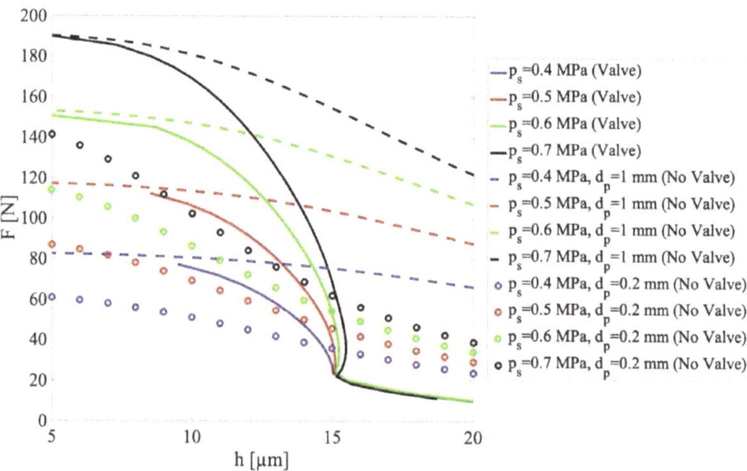

Figure 19. Comparison among the load capacities of the two benchmarks and compensated pad designed considering different values of the supply pressure p_s, by considering the other parameters as constant: $d_p = 1$ mm, $d_v = 0.5$ mm, $h^* = 15$ μm and $D = 40$ mm.

4.3. Effect of the Pad Size

The influence of increasing the pad size was investigated by considering the performance of different compensated pads with outer diameters that are larger than the reference case of 10, 25, 40 and 50%. As can be seen from Figure 16, the region of maximum stiffness (around the desired air gap height h^*) of the pad increases proportionally with respect to the size of the pad. Conversely, the air consumption is almost the same considering this increment in the pad size (Figure 17). Additionally, in this case, to obtain higher performance, the stiffness of the diaphragm has to suitably increase with the diameter of the pad. Hence, increasing the size of the pad integrated does not reduce the effectiveness of the method when this size variation is compensated by a suitable variation in the diaphragm stiffness (Figure 18).

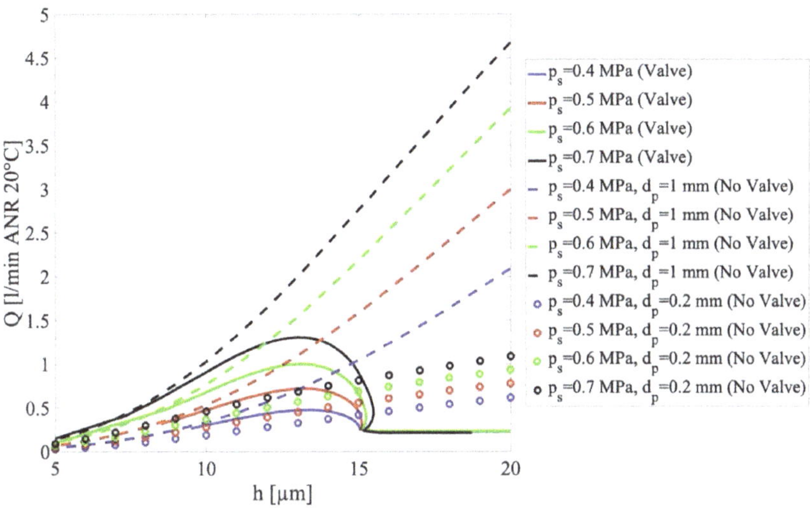

Figure 20. Comparison among the air consumptions of the two benchmarks and compensated pad designed considering different values of supply pressure p_s, by considering the other parameters as constant: $d_p = 1$ mm, $d_v = 0.5$ mm, $h^* = 15$ μm and $D = 40$ mm.

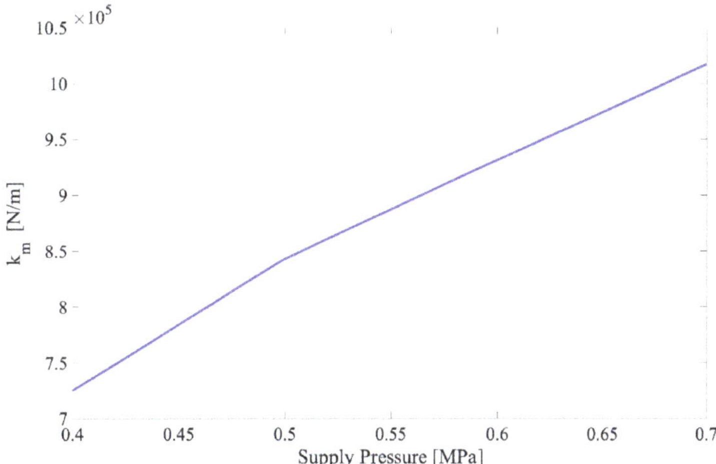

Figure 21. Trend of the optimal diaphragm stiffness expressed as a function of the supply pressure p_s, by considering the other parameters as constant: $d_p = 1$ mm, $d_v = 0.5$ mm, $h^* = 15$ μm and $D = 40$ mm.

4.4. Effect of the Supply Pressure

Figures 19 and 20 show the results of the simulations aimed at investigating the effect of reducing the supply pressure of the compensated pad (reference case, see Figure 8). As expected, reducing the supply pressure of the compensated pad reduces both the maximum load capacity and the region of maximum stiffness (Figure 19). As regards air consumption (Figure 20), it obviously reduces with the supply pressure and it is always of the same order of the second benchmark ($d_p = 0.2$ mm). As can be seen from Figure 21, the stiffness of the membrane reduces as the supply pressure is reduced.

5. Conclusions

This paper presents a design procedure for a circular and centrally fed aerostatic pad controlled by a diaphragm valve. As demonstrated by previous numerical and experimental works [29–31,33], this kind of compensation method makes it possible to significantly increase the static stiffness of air pads up to quasi-static infinite stiffness. Here, a design procedure is proposed to suitably select the geometry of this kind of passively compensated pad depending on the values of a desired air gap height at which the system has to work. The numerical results demonstrate that, thanks to this design procedure, it is always possible to obtain a quasi-infinite stiffness in a neighbor of the desired air gap height. Moreover, a sensitivity analysis was performed to study the effect of varying the desired air gap height, supply pressure, diameter of the valve supply hole and the pad size. It was found that the proposed design procedure is effective in the presence of all these variations. In particular, the most relevant results are that:

- reducing the value of the desired air gap height globally increases the stiffness of the system along with the compensation range;
- increasing the outer radius of the integrated pad or modifying the supply hole diameter of the valve does not reduce the effectiveness of the compensation method but it results in a different value of the required diaphragm stiffness;
- given the desired air gap height, the extent of the compensation range is almost proportional to the supply pressure.

Future works will focus on numerical simulation by considering bearings with multiple orifices and experimental tests to validate the preliminary results obtained herein.

Author Contributions: Conceptualization, L.L., T.R., V.V., F.C., A.T.; Data curation, L.L.; formal analysis, L.L.; funding acquisition, T.R.; Investigation, L.L., T.R., V.V., F.C., A.T.; methodology, L.L., T.R., V.V., F.C., A.T.; project administration, T.R., V.V.; software, L.L.; supervision, T.R., V.V., F.C., A.T. Validation, T.R., V.V., F.C., A.T.; writing—original draft, L.L.; writing—review and editing, L.L.;. All authors have read and agreed to the published version of the manuscript.

Funding: This research received no external funding.

Conflicts of Interest: Authors declare no conflict of interest.

Nomenclature

D Outer diameter of the pad (m)
D_m Diameter of the valve membrane (m)
F_p Load capacity of the pad (N)
F^{ext} Applied load (N)
G_1 Air flow rate through the valve nozzle (kg/s)
G_2 Air flow rate through the pad orifice(kg/s)
G_3 Air flow rate through the air gap (kg/s)
M Moving mass of the pad (kg)
Q Air flow rate (l/min ANR 20 °C)
R_g Air constant (J/(kg·K))
R_1 Pneumatic resistance of the valve nozzle ((s·Pa)/kg)
R_2 Pneumatic resistance of the pad orifice ((s·Pa)/kg)
R_3 Pneumatic resistance of the air gap ((s·Pa)/kg)
Re_1 Reynolds number of the valve nozzle (-)
Re_2 Reynolds number of the pad orifice (-)
T_a Ambient temperature (K)
T_s Supply temperature (K)
T_1 Valve chamber temperature (K)
T_2 Air gap temperature (K)

V_1 Volume of the valve chamber (m^3)
V_2 Volume at the air gap inlet (m^3)
c_{d_1} Valve nozzle discharge coefficient (-)
c_{d_2} Pad orifice discharge coefficient (-)
d_v Diameter of the valve nozzle (m)
d_p Diameter of the pad orifice (m)
k Ratio of the air specific heats (-)
k_m Diaphragm stiffness (N/m)
h Air gap height (m)
h^* Desired air gap height (m)
p_s Valve supply pressure (Pa)
p_a Ambient pressure (Pa)
p_1 Pressure at the valve chamber (Pa)
p_2 Pressure at the air gap inlet (Pa)
$p_{1,i}^*$ Supply pressure selected for the diaphragm design (Pa)
p_1^{id} Ideal supply pressure of the pad (Pa)
p_1^{eff} Effective supply pressure of the pad (Pa)
s Diaphragm thickness (m)
x Membrane-Nozzle distance (m)
x^{id} Ideal diaphragm deflection (m)
x_0 Initial membrane-nozzle distance (m)
x_v Membrane deflection due to the air pressure (m)
$x_{v,i}^*$ Deflection selected for the diaphragm design (m)
μ Dynamic viscosity (Ns/m^2)

References

1. Gao, Q.; Chen, W.; Lu, L.; Huo, D.; Cheng, K. Aerostatic Bearings Design and Analysis with the Application to Precision Engineering: State-of-the-Art and Future Perspectives. *Tribol. Int.* **2019**, *135*, 1–17. [CrossRef]
2. Zmarzły, P. Multi-Dimensional Mathematical Wear Models of Vibration Generated by Rolling Ball Bearings Made of AISI 52100 Bearing Steel. *Materials* **2000**, *13*, 5440. [CrossRef] [PubMed]
3. Ali, S.H. Method of Optimal Measurement Strategy for Ultra-High-Precision Machine in Roundness Nanometrology. *Int. J. Smart Sens. Intell. Syst.* **2015**, *8*, 896–920.
4. Lentini, L.; Moradi, M.; Colombo, F. A Historical Review of Gas Lubrication: From Reynolds to Active Compensations. *Tribol. Ind.* **2018**, *40*, 165–182. [CrossRef]
5. Kazimierski, Z.; Trojnarski, J. Investigations of Externally Pressurized Gas Bearings with Different Feeding Systems. *J. Lubr. Technol.* **1980**, *102*, 59–64. [CrossRef]
6. Boffey, D.A.; Duncan, A.E.; Dearden, J.K. An Experimental Investigation of the Effect of Orifice Restrictor Size on the Stiffness of an Industrial Air Lubricated Thrust Bearing. *Tribol. Int.* **1981**, *14*, 287–291. [CrossRef]
7. Colombo, F.; Lentini, L.; Raparelli, T.; Trivella, A.; Viktorov, V. Dynamic Characterisation of Rectangular Aerostatic Pads with Multiple Inherent Orifices. *Tribol. Lett.* **2018**, *66*, 133. [CrossRef]
8. Arghir, M.; Hassini, M.-A.; Balducchi, F.; Gauthier, R. Synthesis of Experimental and Theoretical Analysis of Pneumatic Hammer Instability in an Aerostatic Bearing. *J. Eng. Gas Turbines Power* **2016**, *138*, 021602. [CrossRef]
9. Talukder, H.M.; Stowell, T.B. Pneumatic Hammer in an Externally Pressurized Orifice-Compensated Air Journal Bearing. *Tribol. Int.* **2003**, *36*, 585–591. [CrossRef]
10. Stowell, T.B. Pneumatic Hammer in a Gas Lubricated Externally Pressurized Annular Thrust Bearing. *J. Tribol.* **1971**, *93*, 498–503. [CrossRef]
11. Fourka, M.; Bonis, M. Comparison between Externally Pressurized Gas Thrust Bearings with Different Orifice and Porous Feeding Systems. *Wear* **1997**, *210*, 311–317. [CrossRef]
12. Kwan, Y.B.P.; Corbett, J. Porous Aerostatic Bearings: An Updated Review. *Wear* **1998**, *222*, 69–73. [CrossRef]
13. Raparelli, T.; Viktorov, V.; Colombo, F.; Lentini, L. Aerostatic Thrust Bearings Active Compensation: Critical Review. *Precis. Eng.* **2016**, *44*, 1–12. [CrossRef]
14. Al-Bender, F. On the Modelling of the Dynamic Characteristics of Aerostatic Bearing Films: From Stability Analysis to Active Compensation. *Precis. Eng.* **2009**, *33*, 117–126. [CrossRef]
15. Aguirre, G.; Al-Bender, F.; van Brussel, H. A Multiphysics Coupled Model for Active Aerostatic Thrust Bearings. In Proceedings of the 2008 IEEE/ASME International Conference on Advanced Intelligent Mechatronics, Xi'an, China, 2–5 July 2008; pp. 710–715.
16. Maamari, N.; Krebs, A.; Weikert, S.; Wegener, K. Centrally Fed Orifice Based Active Aerostatic Bearing with Quasi-Infinite Static Stiffness and High Servo Compliance. *Tribol. Int.* **2009**, *129*, 297–313. [CrossRef]

17. Maamari, N.; Krebs, A.; Weikert, S.; Wild, H.; Wegener, K. Stability and Dynamics of an Orifice Based Aerostatic Bearing with a Compliant Back Plate. *Tribol. Int.* **2009**, *138*, 279–296. [CrossRef]
18. Colombo, F.; Lentini, L.; Raparelli, T.; Viktorov, V. Actively Compensated Aerostatic Thrust Bearing: Design, Modelling and Experimental Validation. *Meccanica* **2017**, *52*, 3645–3660. [CrossRef]
19. Matsumoto, H.; Yamaguchi, J.; Aoyama, H.; Shimokohbe, A. An Ultra Precision Straight Motion System (2nd Report). *J. Jpn. Soc. Precis. Eng.* **1988**, *54*, 1945–1950. [CrossRef]
20. Aoyama, H.; Watanabe, I.; Akutsu, K.; Shimokohbe, A. An Ultra Precision Straight Motion System (1st Report). *J. Jpn. Soc. Precis. Eng.* **1988**, *54*, 558–563. [CrossRef]
21. Morosi, S.; Santos, I.F. Active Lubrication Applied to Radial Gas Journal Bearings. Part 1: Modeling. *Tribol. Int.* **2011**, *44*, 1949–1958. [CrossRef]
22. Pierart, F.G.; Santos, I.F. Active Lubrication Applied to Radial Gas Journal Bearings. Part 2: Modelling Improvement and Experimental Validation. *Tribol. Int.* **2016**, *96*, 237–246. [CrossRef]
23. Mizumoto, H.; Matsubara, T.; Yamamoto, H.; Okuno, K.; Yabuya, M. An Infinite-Stiffness Aerostatic Bearing with an Exhaust-Control Restrictor. In *Progress in Precision Engineering*; Seyfried, P., Kunzmann, H., McKeown, P., Weck, M., Eds.; Springer: Berlin/Heidelberg, Germany, 1991; pp. 315–316.
24. Bryant, M.R.; Velinsky, S.A.; Beachley, N.H.; Fronczak, F.J. A Design Methodology for Obtaining Infinite Stiffness in an Aerostatic Thrust Bearing. *J. Mech. Transm. Autom. Des.* **1986**, *108*, 448–453. [CrossRef]
25. Newgard, P.M.; Kiang, R.L. Elastic Orifices for Pressurized Gas Bearings. *ASLE Trans.* **1966**, *9*, 311–317. [CrossRef]
26. Blondeel, E.; Snoeys, R.; Devrieze, L. Externally Pressurized Bearings with Variable Gap Geometries. In Proceedings of the 7th International Gas Bearing Symposium, Cambridge, UK, 13–15 July 1976.
27. Snoeys, R.; Al-Bender, F. Development of Improved Externally Pressurized Gas Bearings. *KSME J.* **1987**, *1*, 81–88. [CrossRef]
28. Chen, M.-F.; Lin, Y.-T. Dynamic Analysis of the X-Shaped Groove Aerostatic Bearings with Disk-Spring Compensator. *JSME Int. J. Ser. C* **2002**, *45*, 492–501. [CrossRef]
29. Lentini, L.; Colombo, F.; Trivella, A.; Raparelli, T.; Viktorov, V. On the Design of a Diaphragm Valve for Aerostatic Bearings. *E3S Web Conf.* **2020**, *197*, 07006. [CrossRef]
30. Colombo, F.; Lentini, L.; Raparelli, T.; Trivella, A.; Viktorov, V. Air Pad Controlled by Means of a Diaphragm-Valve: Static and Dynamic Behaviour. In *Advances in Italian Mechanism Science*; Niola, V., Gasparetto, A., Eds.; Springer International Publishing: Cham, Switzerland, 2021; pp. 699–710.
31. Ghodsiyeh, D.; Colombo, F.; Lentini, L.; Raparelli, T.; Trivella, A.; Viktorov, V. An Infinite Stiffness Aerostatic Pad with a Diaphragm Valve. *Tribol. Int.* **2020**, *141*, 105964. [CrossRef]
32. Ghodsiyeh, D.; Colombo, F.; Raparelli, T.; Trivella, A.; Viktorov, V. Diaphragm Valve-Controlled Air Thrust Bearing. *Tribol. Int.* **2017**, *109*, 328–335. [CrossRef]
33. Lentini, L.; Colombo, F.; Raparelli, T.; Trivella, A.; Viktorov, V. An Aerostatic Pad with an Internal Pressure Control. *E3S Web Conf.* **2020**, *197*, 07002. [CrossRef]
34. Belforte, G.; Raparelli, T.; Viktorov, V.; Trivella, A. Discharge Coefficients of Orifice-Type Restrictor for Aerostatic Bearings. *Tribol. Int.* **2007**, *40*, 512–521. [CrossRef]
35. Colombo, F.; Lentini, L.; Raparelli, T.; Trivella, A.; Viktorov, V. An Identification Method for Orifice-Type Restrictors Based on the Closed-Form Solution of Reynolds Equation. *Lubricants* **2021**. under review.
36. Wardle, F. *Ultra-Precision Bearings*; Elsevier: Amsterdam, The Netherlands, 2015.

 *lubricants*

Article

Study of the Influence of Temperature on Contact Pressures and Resource of Metal-Polymer Plain Bearings with Filled Polyamide PA6 Bushing

Myron Chernets [1,*], Mykhaylo Pashechko [2], Anatolii Kornienko [1] and Andriy Buketov [3]

[1] Department of Applied Mechanics and Materials Engineering, National Aviation University, 03058 Kyiv, Ukraine; anatoliykor80@gmail.com
[2] Department of Fundamentals of Technology, Lublin University of Technology, 20-618 Lublin, Poland; mpashechko@hotmail.com
[3] Department of Transport Technologies, Kherson State Maritime Academy, 73000 Kherson, Ukraine; buketov@tntu.edu.ua
* Correspondence: myron.czerniec@gmail.com

Abstract: It is known that the elastic characteristics of polyamides change with increasing temperature; in particular, the Young's modulus decreases significantly. This fact is practically not taken into account in design calculations of metal-polymer plain (MP) bearings, operating under conditions of the boundary and dry friction. The purpose of the study was the analysis of the effect of temperature on the change of the Young's modulus and, accordingly, the contact strength and triboresource according to the developed method of calculating MP bearings. MP bearings with a bushing made of polyamide PA6 reinforced with glass or carbon-dispersed fibers were investigated. Quantitative and qualitative regularities of change of the maximum contact pressures and resource of the bearings at temperature increase under conditions of boundary and dry friction are established. The pressures in the bearing bushing made of PA6 + 30GF will be lower than for the bushing made of PA6 + 30CF. The resource of the bushing made of PA6 + 30CF will be significantly greater than for PA6 + 30GF. For thermoplastic polymers, the increase in temperature will have a useful practical effect due to the decrease in the rigidity of the polymer composites of the bearing bushing.

Keywords: metal-polymer plain bearings; PA6-based polyamide composites; dispersed glass and carbon fibers; temperature; Young's modulus; maximum contact pressures and resource; boundary and dry friction

Citation: Chernets, M.; Pashechko, M.; Kornienko, A.; Buketov, A. Study of the Influence of Temperature on Contact Pressures and Resource of Metal-Polymer Plain Bearings with Filled Polyamide PA6 Bushing. *Lubricants* **2022**, *10*, 13. https://doi.org/10.3390/lubricants10010013

Received: 22 September 2021
Accepted: 6 December 2021
Published: 17 January 2022

Publisher's Note: MDPI stays neutral with regard to jurisdictional claims in published maps and institutional affiliations.

1. Introduction

Metal-polymer (MP) plain bearings, which have been known since the 1930s, are quite common in modern conditions due to their characteristic positive qualities. These include sufficiently large load capacity, the possibility of use in dry friction and in a variety of special technological and operational conditions and under shock loads, small axial dimensions in comparison with their diameter, low noise level, damping ability, etc. They are able to operate in a wide range of temperatures, both low and high.

Various types of thermoplastic polymers and composites based on them are used as the material of MP bearing bushings, in particular, polyamides PA6, PA46, and PA66. To increase the strength and wear resistance of polymers, fillers of various types and structures (glass and carbon fiber, bronze powder, molybdenum disulfide, graphite, PTFE, polyethylene, and even semiconductor materials) are used [1–7]. As a result of modification by one or more fillers, significantly higher wear resistance, reduction of the sliding friction coefficient, and increase of strength characteristics in comparison with basic polymers are achieved. It should be noted that these performance characteristics of both unmodified and modified polymers are influenced by temperature. In addition, the temperature affects the elastic characteristics of polymeric materials. It is known that as the temperature decreases, the polymer materials become harder due to an increase in the Young's modulus and a decrease in the Poisson's ratio. As it increases, the Young's modulus will decrease and the Poisson's ratio will increase, i.e., the polymers become more deformable. However, despite the rather large area of distribution and scope of MP bearing, the question of the impact of changes in these elastic characteristics of polymer composites, in particular polyamide, under the influence of temperature on their life and the bearing capacity remains poorly understood. It is known [1] that the temperature significantly affects the decrease in the Young's modulus of polyamides.

A priori, it can be argued that with decreasing Young's modulus and increasing Poisson's ratio due to increasing temperature, the contact pressures in the MP bearing will be lower because these mechanical characteristics of the polymer bushing are in the expressions for their calculation in contact mechanics. It is logical to assume that the MP bearing life will increase at a temperature higher than normal. However, this type of research for MP bearings is absent in the literature as well as the corresponding effective analytical methods for their implementation. It should be noted that the computational [8–15] or numerical methods for the study of metal plain bearings [1,16,17] have not been used in the calculation of metal-polymer plain bearings. In numerical methods of research of metal-polymer plain bearings [18–20], the estimation of contact pressures is carried out; however, there is no possibility of their triboresource determination.

It should be noted that in the experimental study of MP bearings, in addition to the analysis of polymer bushing wear and sliding friction coefficient, some authors also assess the temperature in the bearing and its effect on these tribotechnical characteristics of various polymer composites [2–5], etc., in particular, PA6-based composites [2,3].

In [2], the tribological behavior of different bearings with polymer bushings was studied in dry friction conditions, where it was shown that friction and wear depend on speed, load, temperature, and duration of operation. In [3], the influence of sliding speeds, pressure, and temperature on friction and wear of metal-polymer bearings made of PA66, PA66 + 18% PTFE, and PA66 + 20% GF + 25% PTFE composites was described. The article [4] was devoted to the study of the influence of the radial clearance of composite plain bearings on the coefficient of friction, temperature, and wear. In [5], the wear under dry friction of bearings with bushings made of modified PEEK + 30% wt. GF, +10% wt. CF, graphite, and carbon fiber was investigated. The tribological properties of three different, highly efficient polymers and their composites were studied at different temperatures in [6].

Since the problem of estimating the influence of the temperature factor on the bearing capacity and triboresource of MP bearings is of great practical importance, the developed analytical method [21–25] for studying the contact strength and frictional stability of this type of sliding tribosystems was used. It is based on the methodology of research of wear kinetics at sliding friction and the authors' methods of calculation of sliding bearings with metal elements [22,25–27]. The article investigated MP bearings with bushings made of composites PA6 + 30%GF and PA6 + 30%CF, modified with glass and carbon fibers, correspondingly.

2. Materials and Methods

2.1. Influence of Temperature

The influence of temperature on the contact pressures and the resource of MP bearings is based on changes in the elastic characteristics of polymer materials. Such changes for the modulus of elasticity E and Poisson's ratio μ of polyamide PA6 and PA6-based composites PA6 + 30%GF and PA6 + 30%CF at $T = 25$–55 °C and relative humidity of 50% are given in Table 1 ([1,28,29] and our own research results obtained by the Oliver–Farr method).

The operating temperature in the metal-polymer plain bearings should not exceed 60 °C, as this can lead to their thermal aging. The decrease in the modulus of elasticity, i.e., the decrease in rigidity, is more than 30% for PA6 and PA6 + 30%GF, while for PA6 + 30%CF this decrease is smaller. As for the increase in the Poisson's ratio, i.e., the increase in plasticity, it is insignificant.

Table 1. Influence of temperature on the modulus of elasticity and Poisson's ratio.

Materials/Temperature, T °C	25	35	45	55	Decrease E, Times	Poisson's Ratio [1], μ	Increase μ, Times
PA6	2700	2550	2400	2050	1.32	0.39/0.416	1.067
PA6 + 30%GF	3900	3750	3500	3000	1.3	0.42/0.438	1.043
PA6 + 30%CF	5400	5270	5150	4450	1.21	0.4/0.418	1.045

[1] Poisson's ratio μ at 25 and 55 °C.

2.2. Calculation of Contact Pressures

To conduct a study on the calculation of maximum contact pressures, the authors' method of contact mechanics [21–25] on the internal contact of close radius cylindrical bodies was used. This contact occurs in the plain bearing. The view and diagram of the MP bearing is shown in Figure 1.

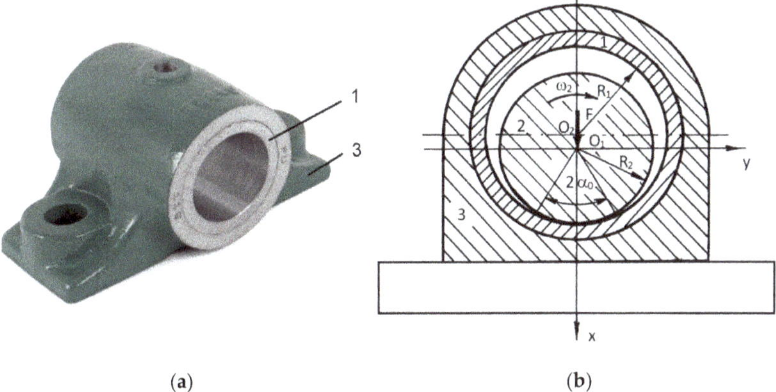

Figure 1. Metal-polymer plain bearing: (**a**) General view; (**b**) Diagram. 1, composite bushing; 2, shaft; 3, housing.

A static load N acts on the shaft 2 of the MP bearing. Since the plane contact problem of the elasticity theory is considered, the total load N is reduced to a single $F = N/l$, where l is the length of the shaft journal. The bearing has a guaranteed initial radial clearance $\varepsilon = R_1 - R_2 \geq 0$. Accordingly, R_1 is the inner radius of the bushing 1 and R_2 is the radius of the shaft 2. In MP bearings, the shafts are metal and the bushings are non-metallic. It should be noted that the strength characteristics of the shaft and the bushing materials differ significantly (8–10 times) as well as their modulus of elasticity (40–100 times), which has a fundamental effect on the contact parameters. It is also known that in metal-polymer tribocouples, polymer composites have 3–4 orders of magnitude lower wear resistance than steel. The external load leads to the contact interaction of the shaft elements, resulting in the appearance of the contact pressure $p(\alpha)$, unknown by magnitude and distribution acting on the contact arc $2R_2\alpha_0$. That is, the task is to determine the specified contact parameters.

In the general equation for determining the contact pressure in the contact area [22,25], we used the following function:

$$p(\alpha) \approx E_0\varepsilon\sqrt{\tan^2\frac{\alpha_0}{2} - \tan^2\frac{\alpha}{2}},\qquad(1)$$

where $E_0 = (e/R_2)\cos^2(\alpha_0/4)$, $e = 4E_1E_2/Z$, $Z = (1 + \kappa_1)(1 + \nu_1)E_2 + (1 + \kappa_2)(1 + \nu_2)E_1$, α is the polar angle, E_1, E_2 is the modulus of elasticity, ν_1, ν_2 are Poisson's ratios, and $\kappa = 3 - 4\nu$.

The bearing capacity of the bearing is characterized by the level of maximum contact pressures $p(0)$, and they occur at $\alpha = 0$. That is,

$$p(0) \approx E_0 \varepsilon \tan(\alpha_0/2), \tag{2}$$

The contact semiangle α_0 is determined from the equilibrium condition of external load F and contact pressures $p(\alpha)$. Taking into account expression (2), it is:

$$F = R_2 \int_{-\alpha_0}^{\alpha_0} p(\alpha)\cos\alpha d\alpha = 4\pi R_2 E_0 \varepsilon \sin^2(\alpha_0/4). \tag{3}$$

When the shaft rotates at a constant angular velocity, ω_2, the parameters of the static contact, i.e., the contact pressures $p(\alpha)$ and the contact angle, $2R_2\alpha_0$, will change due to wear of the composite bushing. Accordingly, there will be a decrease in $p(\alpha)$ and an increase in the contact area, $2R_2\alpha_0$. According to [22–26], wear contact pressures is

$$p(\alpha, t, h) = p(\alpha) + p(h), \tag{4}$$

where $p(h)$ is the change in initial pressures due to wear.

The law of change $p(h)$ is chosen as follows

$$p(h) = E_h \varepsilon_h \sqrt{\tan^2 \frac{\alpha_{0h}}{2} - \tan^2 \frac{\alpha}{2}}, \tag{5}$$

where $E_h = c_h(e/R_2)\cos^2(\alpha_{0h}/4)$ and $c_h > 0$ is the wear rate indicator.

The tribocontact semiangle α_{0h} characterizing the wear zone during wear is determined by condition (6)

$$F = 4\pi R_2 E_0(\varepsilon + c_{\alpha h}\varepsilon_h)\sin^2(\alpha_{0h}/4), \tag{6}$$

where $\varepsilon_h = h_{kmax}\left(-K_t^{(k)} + h'_k\right)$, $h'_1 = h_2/h_1$, $h'_2 = h_1/h_2$ is the relative wear in the tribosystem, $K_t^{(1)} = 1, K_t^{(2)} = \alpha_0/\pi$–coefficients of mutual overlap of bearing elements, $c_{\alpha h}$ is the indicator of the tribocontact angle growth rate, and h_{kmax} is the acceptable wear of elements 1 and 2 of the bearing:

$$h'_1 = \frac{B_1 \tau_{10}^{m_1}(\tau - \tau_{20})^{m_2}}{B_2 \tau_{20}^{m_2}(\tau - \tau_{10})^{m_1}} K_t^{(2)}, h'_2 = \frac{B_2 \tau_{20}^{m_2}(\tau - \tau_{10})^{m_1}}{B_1 \tau_{10}^{m_1}(\tau - \tau_{20})^{m_2}} K_t^{(1)}, \tag{7}$$

where $\tau = fp(0)$ is the Coulomb specific friction force, f is the sliding friction coefficient, B_k, m_k, τ_{k0} is the wear resistance characteristics of the tribocouple materials under the accepted external conditions of triboexperimental research, and k is the numbering of the bearing elements (Figure 1b).

2.3. Calculation of Resource

The calculation of the MP bearing service life when the bushing wears to $h_1 = h_{kmax}$ is carried out according to the formula:

$$t = \frac{-B_k \tau_{k0}^{m_k}}{v c_h \tau(h)\Sigma_k(1 - m_k)K_t^{(k)}} \left\{ [\tau - \tau_{k0}]^{1-m_k} - [(\tau - \tau_{k0}) + c_h h_{kmax}\Sigma_k \tau_h(h)]^{1-m_k} \right\}, \tag{8}$$

where $v = \omega_2 R$ is the sliding speed, c_h is the coefficient of wear rate, $\Sigma_1 = \left(-K_t^{(1)} + h'_1\right)$, $\Sigma_2 = \left(K_t^{(2)} - h'_2\right)$, and $\tau(h) = fp(0, t, h) = fE_h\tan(\alpha_{0h}/2)$ is the maximum specific friction force in tribocontact.

2.4. Conditional Calculation of Bearing Capacity

In modern mechanical engineering, in the design of metal plain bearings, there are two main criteria. The average pressure p (as its bearing capacity) and Zeiner's pv are used. The average contact pressure p is considered to be evenly distributed over the contact area $2R_2l''$

$$p = \frac{N}{2R_2l} = \frac{F}{D_2} \leq [p], \tag{9}$$

where $[p]$ is the allowable value of the contact pressure of less durable material (reference parameter).

From Formula (9), it follows that the contact arc of length $2R_2 = D_2$ corresponds to the contact angle $2\alpha_0 = 360°/\pi \approx 114.6° = $ const, which does not depend on the properties of the bearing materials. It should be noted that such a significant contact angle is achieved at significant bearing loads and minimal radial clearances. Accordingly, this is confirmed in the works available in the literature [1,8–20] and the authors' works [22–27], based on the methods of contact elasticity theory. It is also known that in plain bearings the magnitude and distribution of contact pressures are significantly influenced by the radial clearance and the elastic characteristics of the shaft and bushing materials. Actually, these significant factors of influence are absent in Formula (9).

Additionally, also known is a modified formula [30] for determining the maximum pressure p_{max} in the bearing:

$$p_{max} = \frac{4}{\pi}\frac{N}{D_2l} = \frac{4}{\pi}\frac{F}{D_2}, \tag{10}$$

Here it is assumed that the pressure varies according to the law of cosine and the contact arc $2\alpha_0 = 180°$. The structure of Formula (10) is the same as Formula (9). The maximum pressure p_{max} according to Formula (10) will be 1273 times higher than the average pressure p. It was also established [11] that at zero clearance in the cylindrical conjugation at static contact, $2\alpha_0 \approx 170°$. However, there must be a certain radial clearance in the plain bearing because without it, its reliable operation is impossible.

That is, the simplified conditional methods for calculating plain bearings according to the criterion of contact pressure p do not allow objectively assessing the actual pressure level. Their use in the case of MP bearings seems unreasonable. For this purpose, it is reasonable to use the above, presented in Section 2.2, the classical analytical method of the contact mechanics of close radii cylindrical bodies with internal contact.

3. Solution, Results, and Discussion

Data for the calculation of MP bearings were N = 500, 1000, and 1500 N; $F = N/l$; D_2 = 30 mm; $l = D_2$; ε = 0.2 mm; n_2 = 60 rpm is the shaft rotational speed; $f_{GF} = f_{CF}$ = 0.3 under dry friction; and h_{1max} = 1.0 mm is the acceptable bushing wear.

Materials of the metal-polymeric bearings were shaft-steel 45 (0.45% C) normalized; grinding; E_2 = 210 GPa, μ_2 = 0.3; $B_2 = 10^{13}$; m_2 = 2; and τ_{20} = 0.1 MPa. The bushings were (1) carbon-filled polyamide PA6 + 30%CF, E_{CF} = 5.40 GPa, μ_{CF} = 0.4, $B_{1CF} = 24 \times 10^{10}$, m_{1CF} = 1.9, and τ_{10} = 0.05 MPa and (2) glass-filled polyamide PA6 + 30%GF, E_{GF} = 3.90 GPa, μ_{GF} = 0.42, $B_{1GF} = 6.67 \times 10^{10}$, m_{1GF} = 1.9, and τ_{10} = 0.05 MPa [29,30]. Volume filler content was 30%.

The calculation was performed according to the given flow diagram (Figure 2).

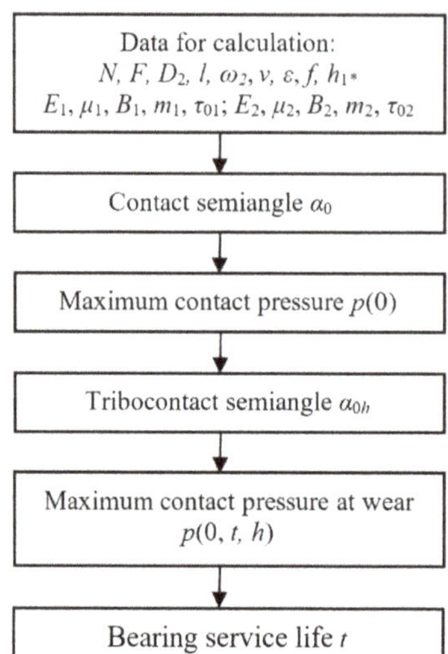

Figure 2. The flow diagram.

The example of the calculation examines friction both with greasing (boundary) and without greasing (dry). The mode of dry friction is widespread and dominant in MP bearings due to self-lubrication of polymeric materials [1–7,18,19]. It is possible, if necessary, to use in this type of bearings greasing with the implementation of boundary friction. In this case, for these hybrid combinations, a steel-polymer composite in the calculated ratios of the presented method, you only need to use the sliding friction coefficients corresponding to this type of lubrication. Instead, the established wear resistance characteristics, B, m, τ_0, of these polymer composites remained the same because they were set for the range of change of 0.2–4 MPa.

Figures 3–11 present the results of calculations. Accordingly, Figure 3 shows the effect of temperature on the change of the maximum contact pressures at different levels of external load.

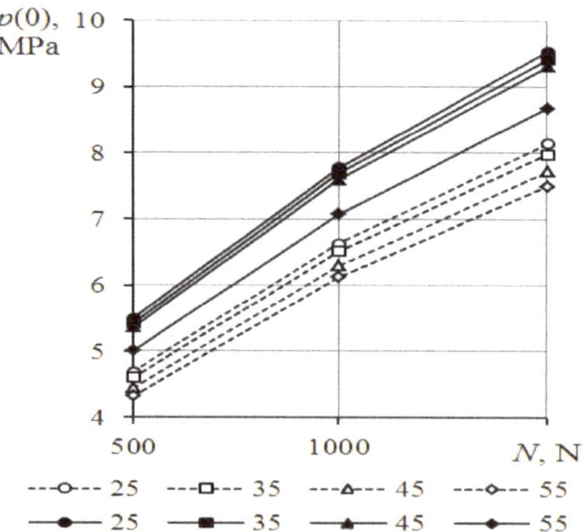

Figure 3. Influence of ambient temperature on maximum contact pressures: PA6 + 30%CF in solid lines, PA6 + 30%GF in dashed lines.

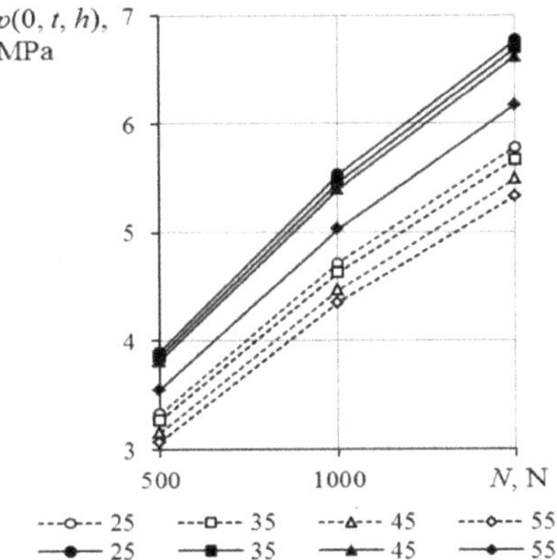

Figure 4. Influence of ambient temperature on maximum wear contact pressures.

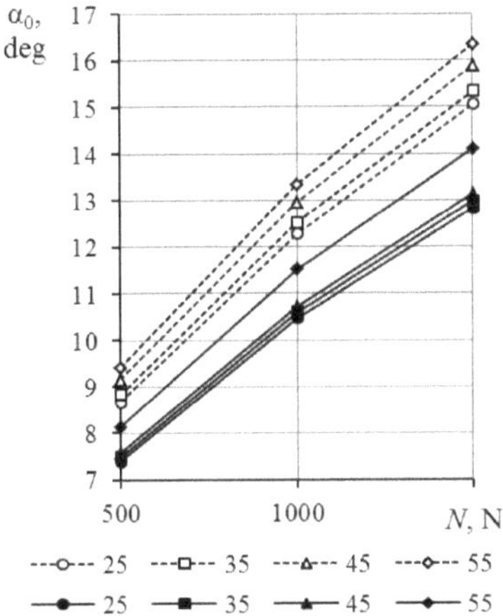

Figure 5. Influence of ambient temperature on the contact semiangle.

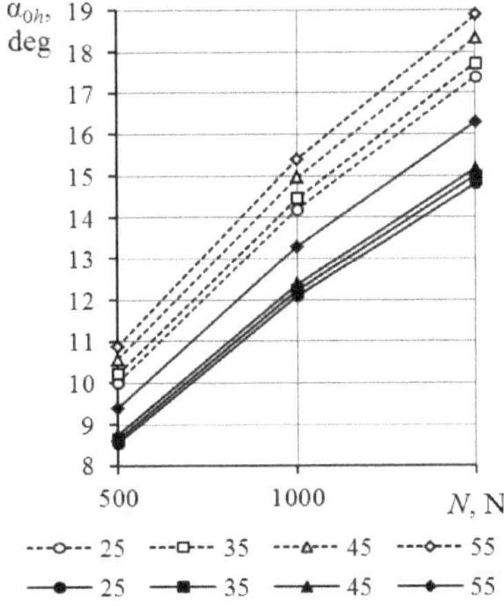

Figure 6. Influence of ambient temperature on change of contact semiangle due to bushing wear.

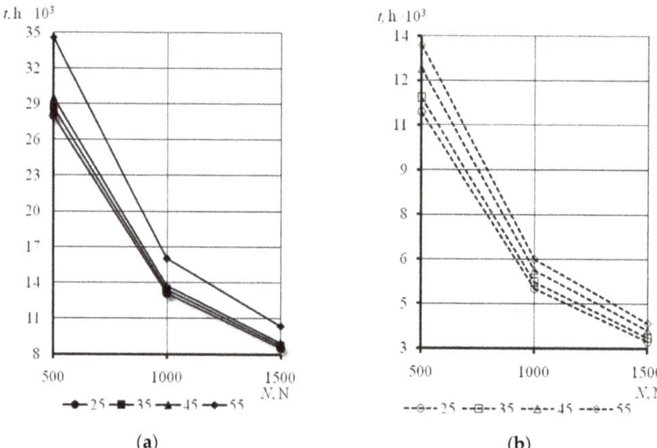

Figure 7. Influence of ambient temperature on the bearing resource (boundary friction): (**a**) PA6 + 30%CF; (**b**) PA6 + 30%GF.

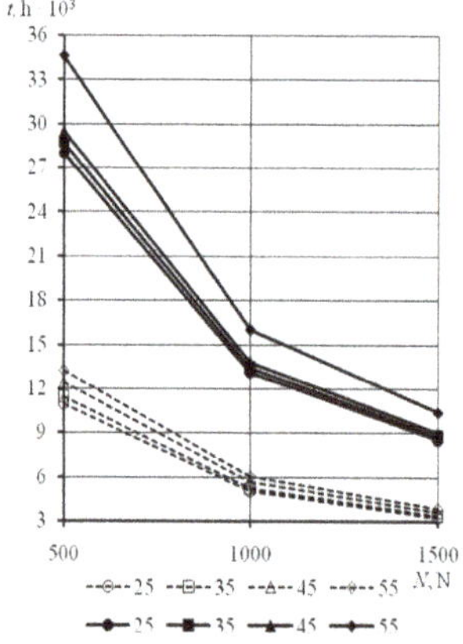

Figure 8. Comparison of the bearing resources under boundary friction.

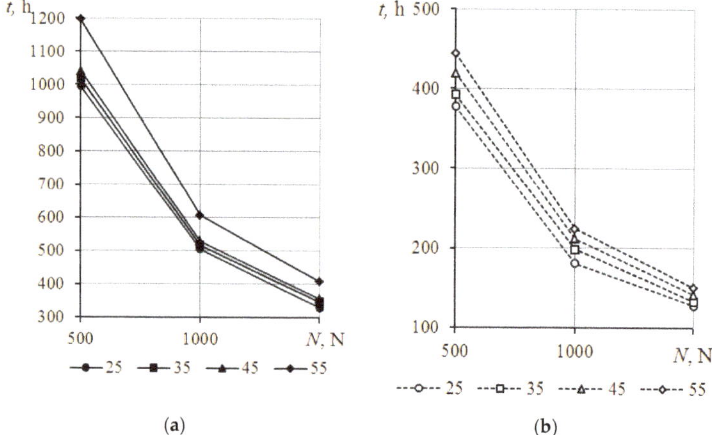

Figure 9. Influence of temperature on the bearing resource (dry friction): (**a**) PA6 + 30%CF; (**b**) PA6 + 30%GF.

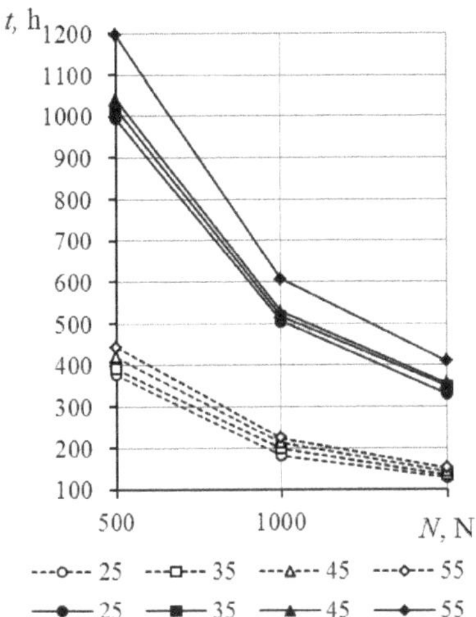

Figure 10. Comparison of the bearing resources under dry friction.

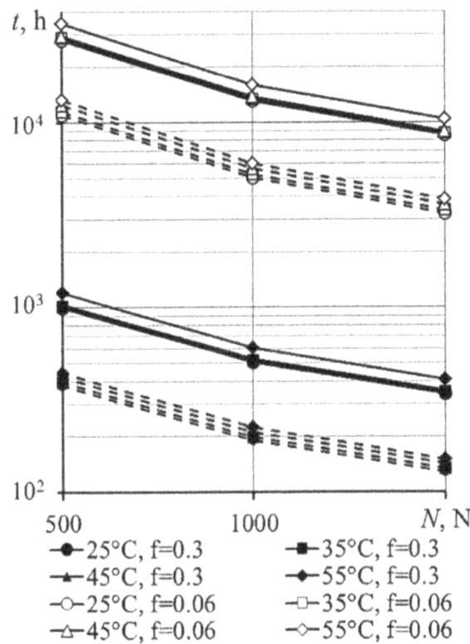

Figure 11. Resource of MP bearings under dry and boundary friction.

Quantitative growth of $p(0)$ is almost the same at different temperature levels. It is established that, in the case of MP bearing with a bushing made of carbon composite PA6 + 30%CF, the pressure at temperature $T = 25\ °C$ will be 1.1 times lower than at $T = 55\ °C$. For fiberglass bushings, this difference is almost the same, as it is 1085 times. In the case of a bushing made of carbon composite, the maximum contact pressures in the bearing will be 1.17–1.18 times higher than for a glass composite bushing at all investigated operating loads. The maximum contact pressures $p(0)$ in the MP bearings depend nonlinearly on the load N. An increase in load N by three times causes an increase in $p(0)$by 1736 times, i.e., $\approx \sqrt{3}$.

When the bushing wears during the operation, the initial pressures $p(0)$ will decrease. Accordingly, Figure 4 shows the tribocontact pressures when reaching the accepted allowable bushing wear h_{1max}.

Regarding the patterns of maximum wear contact pressures, $p(0, t, h)$ changes due to polymer bushing wear or with increasing load. They will be similar both qualitatively and quantitatively, as in the case of the initial maximum contact pressures $p(0)$ (Figure 3).

The initial semiangles α_0 were also calculated (Figure 5). It was established that, in a bearing with a glass-composite bushing, they will be, accordingly, up to 1.18 times smaller than in the presence of a carbon-composite bushing. With increasing load, α_0 increases as $p(0)$, i.e., by 1736 times.

Wear of the composite bushing leads to an increase in the initial contact semiangles α_0 (Figure 6). Qualitative and quantitative patterns of their increase are preserved and inverted to those that occur when the wear contact pressures $p(0, t, h)$ decrease.

As a result of solving the problem by Equation (8), a predictive estimation of the resource of the MP bearing with bushings PA6 + 30%CF and PA6 + 30%GF for both types of friction was performed.

Figures 7 and 8 show the dependence of the bearing resource on the temperature at the acceptable wear $h_{1max} = 1.0$ mm for the boundary friction at different load levels.

Quantitative growth of the resource with increasing ambient temperature will be 22% in the bearing with a carbon composite bushing and 20% with a glass composite bushing. At all loads, such dependences are practically identical.

Figure 8 shows the resource of both MP bearings. The resource of a bearing with a carbon composite bushing is much longer.

The ratio of resources of both MP bearings under boundary friction will be approximately the same for all loads at $T = 25$–$45\,^\circ$C: At $N = 500$ N it is 240–256%; at $N = 1000$ N it is 245–261%; and at $N = 1500$ N it is 247–264%. At $T = 55\,^\circ$C for all loads N, this ratio is higher, up to 270%.

Accordingly, Figures 9 and 10 show the bearing resource under dry friction.

When the acceptable wear $h_{1max} = 1.0$ mm is achieved, there is an increase in the resource with increasing ambient temperature: by 20% in the bearing with a carbon-composite bushing and by 17% in the bearing with a glass-composite bushing. At all loads, such dependence is practically identical.

The resource of a bearing with a carbon composite bushing is significantly greater than with a glass composite (Figure 10).

The ratio of resources of both MP bearings under dry friction will be approximately the same for all loads at $T = 25$–$45\,^\circ$C: At $N = 500$ N it is 248–264%; at $N = 1000$ N it is 250–264%; and at $N = 1500$ N it is 250–260%. At $T = 55\,^\circ$C for all loads N, this ratio is higher, up to 271%.

Figure 11 shows the calculated resources of MP bearings for both types of friction.

A decrease in the coefficient of friction five times under boundary friction compared to dry friction causes an increase in the resource of the MP bearings by several times. In particular, at $N = 1500$ N it is 25–25.6 times with increasing temperature, at $N = 1000$ N it is 25.9–26.8 times, and at $N = 500$ N it is 28.1–29.8 times.

4. Conclusions

1. An increase in the ambient temperature causes a decrease in the value of the Young's modulus of polyamide composites PA6 + 30%CF and PA6 + 30%GF, resulting in a decrease in the level of the initial maximum contact pressures $p(0)$. That is, the increase in temperature causes a decrease in the rigidity of polymer composites, which has a positive effect on the bearing capacity of MP bearings (up to 10%). When the bushing wears, the patterns of further reduction of the initial maximum contact pressures $p(0)$ remain similar to the initial pressures.

2. With increasing ambient temperature, there is an increase in the resource of MP bearings. That is, the higher temperature has a positive effect on the resource of MP bearings (up to 22%). The calculated resource under boundary friction of the bearing with the carbon composite bushing will be higher, by 240–270%, than with the glass composite bushing. Under dry friction, the similar correlation of resources takes place.

3. It is established that the resource of MP bearings increases approximately on a square dependence on decrease in coefficient of sliding friction

4. The developed analytical method of calculation of MP sliding bearings provides at a design stage an effective predictive estimation of their contact parameters and the resource, taking into account influence of ambient temperature. The presented method will be used for research of MP bearings made of other types of polymeric materials.

Author Contributions: Conceptualization, M.C.; methodology, M.C.; investigation, M.C. and A.K.; formal analysis, M.P.; funding acquisition, M.P.; writing—review and editing, A.K.; software, A.K.; validation, A.B. All authors have read and agreed to the published version of the manuscript.

Funding: This research received no external funding.

Conflicts of Interest: The authors declare no conflict of interest.

References

1. Wielieba, W. *Maintenance-Free Plain Bearings Made of Thermoplastic Polymers*; Wyd. Wrocław University of Science and Technology: Wrocław, Poland, 2013.
2. Feyzullahoglu, E.; Saffak, Z. The tribological behavior of different engineering plastics under dry friction conditions. *Mater. Des.* **2008**, *29*, 205–211. [CrossRef]
3. Demirci, M.T.; Düzcükoğlu, H. Wear behaviors of Polytetrafluoroethylene and glass fiber reinforced Polyamide 66 journal bearings. *Mater. Des.* **2014**, *57*, 560–567. [CrossRef]
4. Miler, D.; Škec, S.; Katana, B.; Žeželj, D. An Experimental Study of Composite Plain Bearings: The Influence of Clearance on Friction Coefficient and Temperature. *Stroj. Vestn. J. Mech. Eng.* **2019**, *65*, 547–556. [CrossRef]
5. Zhu, J.; Xie, F.; Dwyer-Joyce, R.S. PEEK Composites as Self-Lubricating Bush Materials for Articulating Revolute Pin Joints. *Polymers* **2020**, *12*, 665. [CrossRef]
6. Kurdi, A.; Kan, W.H.; Chang, L. Tribological behaviour of high performance polymers and polymer composites at elevated temperature. *Tribol. Ind.* **2019**, *130*, 94–105. [CrossRef]
7. Kindrachuk, M.; Volchenko, A.; Volchenko, D.; Volchenko, N.; Poliakov, P.; Tisov, O.; Kornienko, A. Polymers with enhanced energy capacity modified by semiconductor materials. *Funct. Mater.* **2019**, *26*, 629–634. [CrossRef]
8. Goryacheva, I.G. *Mechanics of Frictional Interaction*; Nauka: Moscow, Russia, 2001.
9. Kragelsky, I.V.; Dobychin, N.M.; Kombalov, V.S. *Fundamentals of Friction and Wear Calculations*; Mashinostroenie: Moscow, Russia, 1977.
10. Kuzmenko, A.G. *Development of Methods of Contact Tribomechanics*; KhNU: Khmelnytsky, Ukraine, 2010.
11. Teplyy, M.I. Determination of contact parameters and wear in cylindrical sliding bearings. *Frict. Wear* **1987**, *6*, 895–902.
12. Usov, P.P. Internal contact of cylindrical bodies of close radii during wear of their surfaces. *Frict. Wear* **1985**, *3*, 404–414.
13. Dykha, A.; Marchenko, D. Prediction the wear of sliding bearings. *Int. J. Eng. Technol.* **2018**, *7*, 4–8. [CrossRef]
14. Sorokatyi, R.; Chernets, M.; Dykha, A.; Mikosyanchyk, O. Phenomenological model of accumulation of fatigue tribological damage in the surface layer of materials. *Mech. Mach. Sci.* **2019**, *73*, 3761–3769. [CrossRef]
15. Zwieżycki, W. *Forecasting the Reliability of the Wearing Parts of Machines*; ITE: Radom, Poland, 1999.
16. Sorokatyi, R.V. Modeling the behavior of tribosystems using the method of triboelements. *J. Frict. Wear* **2002**, *23*, 16–22.
17. Sorokatyi, R.V. Solution of the problem of wear of a fine elastic layer with a rigid bearing mounted on a rigid shaft using the method of triboelements. *J. Frict. Wear* **2003**, *24*, 35–41.
18. Rezaei, A.; Ost, W.; Van Paepegem, W.; De Baets, P.; Degrieck, J. Experimental study and numerical simulation of the large-scale testing of polymeric composite journal bearings: Three-dimensional and dynamic modeling. *Wear* **2011**, *270*, 431–438. [CrossRef]
19. Rezaei, A.; Ost, W.; Van Paepegem, W.; De Baets, P.; Degrieck, J. A study on the effect of the clearance on the contact stresses and kinematics of polymeric composite journal bearings under reciprocating sliding conditions. *Tribol. Int.* **2012**, *48*, 8–14. [CrossRef]
20. Rezaei, A.; Van Paepegem, W.; De Baets, P.; Ost, W.; Degrieck, J. Adaptive finite element simulation of wear evolution in radial sliding bearing. *Wear* **2012**, *296*, 660–671. [CrossRef]
21. Andreikiv, M.V.; Chernets, M.V. *Evaluation of the Contact Interaction of Rubbing Machine Elements*; Naukova Dumka: Kiev, Ukraine, 1991.
22. Chernets, M.; Pashechko, M.; Nevchas, A. Methods of Forecasting and Increasing the Wear Resistance of Tribotechnical Sliding Systems. In *Vol.1. Research and Calculation of Sliding Tribosystems, Methods to Increase Durability and Wear Resistance*; KOLO: Drogobich, Ukraine, 2001.
23. Chernets, M.; Pashechko, M.; Nevchas, A. Methods of forecasting and increasing the wear resistance of tribotechnical sliding systems. In *Vol. 2. Surface Reinforcement of Structural Materials of Sliding Tribosystems*; KOLO: Drogobich, Ukraine, 2001.
24. Chernets, M.V.; Andreikiv, O.E.; Liebiedieva, N.M.; Zhydyk, V.B. A model for evaluation of wear and durability of plain bearing with small out-of-roundness. *Mater. Sci.* **2009**, *2*, 279–290. [CrossRef]
25. Chernets, M.V. Prediction of the life of a sliding bearing based on a cumulative wear model taking into account the lobing of shaft contour. *J. Frict. Wear* **2015**, *36*, 163–169. [CrossRef]

26. Chernets, M.V. Contact problem for a cylindrical joint with technological faceting of the contours of its parts. *Mater. Sci.* **2009**, *6*, 859–868. [CrossRef]
27. Chernets, M.; Chernets, J. Generalized method for calculating the durability of sliding bearings with technological out-of-roundness of details. *Proc. Inst. Mech. Eng. Part J J. Eng. Tribol.* **2015**, *229*, 216–226. [CrossRef]
28. Chernets, M.V.; Shil'ko, S.V.; Pashechko, M.I.; Barshch, M. Wear resistance of glass- and carbon-filled polyamide composites for metal-polymer gears. *J. Frict. Wear* **2018**, *39*, 361–364. [CrossRef]
29. Chernets, M.; Chernets, J.; Kindrachuk, M.; Kornienko, A. Methodology of calculation of metal-polymer sliding bearings for contact strength, durability and wear. *Tribol. Ind.* **2020**, *42*, 572–584. [CrossRef]
30. Budynas, R.G.; Nisbett, J.K. *Shigley's Mechanical Engineering Design*, 11th ed.; McGraw-Hill: New York, NY, USA, 2019.

lubricants

Article

Investigation of the Wettability Properties of Different Textured Lead/Lead-Free Bronze Coatings

Amani Khaskhoussi, Giacomo Risitano, Luigi Calabrese and Danilo D'Andrea *

Department of Engineering, University of Messina, Contrada di Dio, 98166 Messina, Italy;
amani.khaskhoussi@unime.it (A.K.); grisitano@unime.it (G.R.); luigi.calabrese@unime.it (L.C.)
* Correspondence: dandread@unime.it; Tel.: +39-39-3020-9246

Abstract: Hydraulic components are often subjected to sliding contacts under starved or mixed lubrication. The condition of starved lubrication occurs during the start-up phase of the hydraulic machines or at low working temperature, causing friction and wear of components such as the cylinder block or the valve plate. The aim of this paper was to evaluate the hydrophobicity and oleophilic behavior of lead/lead-free bronze coatings under different texture conditions obtained by varying the diameter and the density of the dimples. The wettability tests were performed using sessile drop tests with oil and water liquids. The dimple parameters were analyzed using confocal microscopy, while the XRF analyses were performed to evaluate the composition of the bronze coatings. Based on the wettability measurements using oil and water, it was possible to assess that the porous surface acted as oil reservoirs that could prolong the life of lubricating oil layer, and may have resulted in a superior wear resistance. Furthermore, a relevant hydrophobicity was highlighted, suggesting that the surface texturing promoted the water-repellent barrier action on the surface. The experimental results showed that the discrepancy in surface properties in oil and water was raised when using the lead bronze coating. These coupled oleophilic and hydrophobic behaviors could play a beneficial role in sustaining the durability of a lubricating oil layer under a condition of continuous water-droplet impact.

Keywords: wettability; lubrification; oleophilicity; hydrophobicity; laser surface texturing

Citation: Khaskhoussi, A.; Risitano, G.; Calabrese, L.; D'Andrea, D. Investigation of the Wettability Properties of Different Textured Lead/Lead-Free Bronze Coatings. *Lubricants* **2022**, *10*, 82. https://doi.org/10.3390/lubricants10050082

Received: 25 March 2022
Accepted: 29 April 2022
Published: 3 May 2022

Publisher's Note: MDPI stays neutral with regard to jurisdictional claims in published maps and institutional affiliations.

1. Introduction

Friction and wear processes are generated in all parts subjected to a sliding contact. The study of the tribological behavior of systems in reciprocating contact is a topic of fundamental importance in many industrial fields, such as mechanics [1–3], hydraulics [4], and prosthetics [5,6], in order to reduce the deterioration of components and improve their working life. Ruggiero et al. [7] investigated the tribological performances of tooth-to-tooth contact and material-to-natural tooth contact (zirconia vs. zirconia and natural tooth vs. zirconia) using a reciprocating tribometer under lubricated conditions (artificial saliva). Wang et al. [8] analyzed the effects of textured surfaces on the friction performance of a low-speed and high-torque water hydraulic motor. The experimental results showed that about 62.6% of the wear-loss reduction could be reached using an ellipsoidal pit surface, and the wear loss mainly occurred on the edges of pits.

Typically, the problem of friction and wear is addressed by using lubricants, but in some working conditions, the lubrication may be insufficient. For this reason, in the last few decades, different types of coatings have been studied to guarantee lubrication even in the most severe conditions, such as coatings based on solid lubricants (lead and bismuth) [9,10], functionalized surfaces to create superhydrophobic/superoleophobic coatings [11,12], self-lubricating multilayered coatings [13], or textured surfaces. Surface texturing has been known for a long time, as shown by the numerous studies reported in the scientific literature [14–17], but the aim of this study was to evaluate how the surface texturing affected the oleophilic and hydrophobic properties of the coating. Yang et al. [18] studied the effects of micro/nano hierarchical structures on the surface of a titanium alloy (Ti-6Al-4V), and observed that the contact angle of the drop increased as the density of the microtextured surface increased, and the wetting state of the textured surfaces conformed to the Cassie model. Volpe et al. [19] analyzed the effects of three different surface-texture geometries on aluminum alloy surfaces. It was shown that by improving the laser texture strategy, it was possible to reduce the laser processing time to produce superhydrophobic surfaces.

In recent times, several surface-texturing techniques have been developed, including micromilling [20], hot embossing [21], electrochemical machining [22], wire EDM machining [23], and surface laser texturing [24–27], which is the most applied technique for obtaining various micropatterns on a material surface due to its high precision, lower environmental impact, good controllability, and flexibility without any chemical treatment. The most widely used surface laser texturing taxonomies include the nanosecond (ns), picosecond (ps), and femtosecond (fs), as they have better controllability, accuracy, and complexity [28]. Due to these advantages, the surface laser texturing technique has been extensively used in numerous applications to improve the tribological performance of coatings, such as lubrication in bearing applications; to improve adhesion bond strength in various coating applications; and fabrication of structures for hydrophobic/superhydrophobic surfaces [29–31].

The laser surface texturing has several advantages in terms of tribo-mechanical behavior. Indeed, the presence of micro-dimples, could allow an enhancement of the tribological behavior optimizing the surface lubrication, thanks to a reservoir action supplied by the tailored surface dimples [32]. Besides, in hydrodynamic or mixed lubrication, these surface microcavities acts as hydrodynamic bearings [33]. A further relevant feature of the laser surface texturing is the capacity to decrease the abrasive wear, since the cavities of the profile behave like trap for the debris that are formed during the relative motions with other bodies improving the tribological properties of the coated surface [27].

Although the tribological behavior of textured coatings has been analyzed in many scientific studies, the focus of the authors is not to evaluate the effect of the texture on the friction and wear of the coating, but rather how the wettability of the coating varies and in particular the hydrophobicity and oleophilic, and therefore lubrication, changing the geometric parameters of the dimples and the composition of the bronze alloy.

The possibility of correlating the microstructural properties with the surface interaction with polar and non-polar liquids, such as water and oil, represents an increase in knowledge for this class of materials, providing further stimuli for their development and industrial applicability of this technique in this field [34,35].

In this paper the wettability measurements with pump oil and water were performed on different types of lead/lead-free textured coating, varying the diameter and density of the dimples. The surface and dimples parameters were analyzed using confocal microscopy, while XRF analyses were performed to evaluate the composition of the bronze coatings.

2. Materials and Methods

2.1. Materials

All tests were carried out on two different bronze alloys (the specimens are shown in Figure 1): one called EN CC480K, which did not contain lead; and one called EN CC496K, with a percentage of lead of about 12%.

These materials are typically used for antifriction coatings in hydraulic pumps and motors, such as on valve plates [36], cylinder blocks [4], and slippers [30], or in the automotive field [37,38]. As is known, in the bronze alloys used for components that can work in conditions of starved lubrication and high friction, lead is used as a solid lubricant. However, lead is a toxic element, and therefore manufacturers are attempting to eliminate or replace it (with bismuth). For the above reasons, we decided to evaluate the behavior of two bronze alloys, one containing lead and one free of lead.

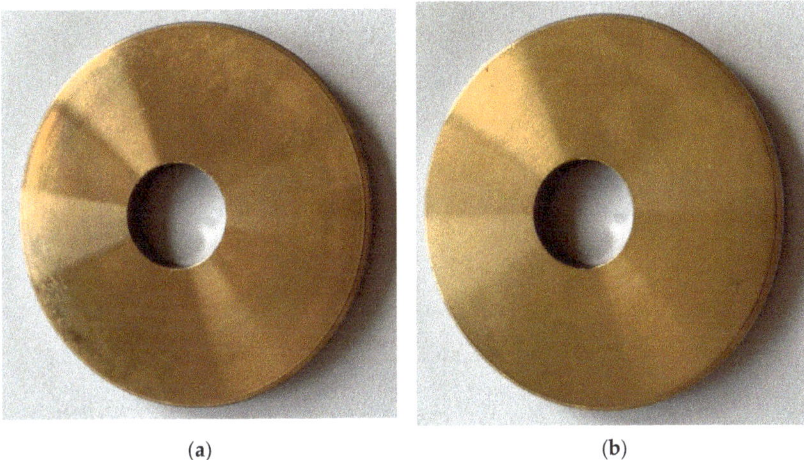

(**a**)   (**b**)

Figure 1. (**a**) EN CC480K bimetal disk (lead-free bronze); (**b**) EN CC496K bimetal disk.

Table 1 shows the results of the XRF analyses of the two different bronze alloys. As shown in the table, the EN CC480K alloy contained 90% copper and a trace amount of Pb, while the EN CC496 K alloy had a high lead content, ranging between 10% and 15%. The effects of the different compositions on the wettabilities of the coatings were analyzed through the use of an EDS analysis.

Table 1. Compositions of bronze alloys.

Bronze Alloys	Sn%	Pb%	Ni%	Si%	Cu%
EN CC480K	7.94	0.06	1.08	0.31	90.5
EN CC496K	10.86	12.43	0.03	0.15	76

The textured microstructures on the bimetal coated cylinder disk samples were obtained by using a surface laser texturing technique (Laser P 400, GF Machining Solution, Schaffhausen, Switzerland), with maximum power of 40 W and a spot size of 15 μm. In particular, the surface texture was tailored by varying the diameter and the density of the dimples on the surface, for both the lead-based and lead-free bronze coatings. The codes and surface-texture parameters for all batches are reported in Table 2.

Table 2. Codes and surface-texture parameters for all batches.

Sample	Typology of Texture	Dimple Diameter (μm)	Density Area (%)
AR	As received	/	/
B100-D10		100	10
B150-D10		150	10
B200-D10		200	10

Table 2. *Cont.*

Sample	Typology of Texture	Dimple Diameter (μm)	Density Area (%)
B100-D20	10μm ⌀ 100μm	100	20
B100-D30	10μm ⌀ 100μm	100	30
B100-D40	10μm ⌀ 100μm	100	40

Samples were investigated with three laser-beam diameters (100, 150, and 200 μm). Furthermore, for a laser beam of 100 μm, four dimple surface densities were analyzed, in the range of 10–40%. For each batch, 3 replicas were created.

A total of 42 samples were created: they were classified based on their textured surface and coating characteristics. In particular, each batch was coded with an acronym using a prefix that referred to the coating characteristic (PbB and PbF, for the Pb-based and Pb-free surfaces, respectively. A second suffix, "B" coupled with a number, referred to the beam diameter (expressed in microns). The last suffix, "C" coupled with a number, referred to the density of the surface area of the laser-melted surface. As an example, the code PbF-B100-D20 indicates a sample with a Pb-free surface, textured with a beam diameter of 100 μm and a surface density of 20%. PbB-AR and PbF-AR codes were used for the as-received samples of Pb-based and Pb-free surfaces, respectively.

2.2. Wettability Measurments

Water-contact-angle measurements were performed by using a tensiometer instrument (AttensionTheta by Biolin Scientific, Gothenburg, Sweden). The test was performed with bidistilled water and pump oil. The oil lubricant used was LI-HIV 46 (viscosity index 175). It is commonly applied in pumps and motors. Table 3 summarizes the main physical parameters of the lubricating oil.

Table 3. LI-HIV 46 oil parameters.

Parameters	Value
Density at 20 °C	873 kg/m^3
Viscosity at 40 °C	46 cSt
Viscosity at 100 °C	9 cSt
Viscosity index	175
Freezing °C	−35
Flammability °C	210

A droplet of the liquid (volume of 3 μL) was softly sited on the coating surfaces in conditions that were open to air and at room temperature (25 °C). After the deposition, the droplet profile was recorded by a microcamera and automatically analyzed by the software supplied with the tensiometer instrument. For each sample, 10 replicas (randomly located on the surface) were performed for all batches.

Morphological analyses of the textured surface and related surface features were carried out by means a 3D confocal optical microscope (Leica DCM 3D, Wetzlar, Germany).

3. Results and Discussion

3.1. Surface Morphology

To better evaluate the surface characteristics of the coatings and the geometry of the dimples, confocal microscopies were carried out on both the EN CC480K alloy coating and the EN CC496K alloy coating. Figure 2 shows the 3D scans of the AR samples (Figure 2a,b) and those of two B150-D10 textured samples (Figure 2c,d). As reported in Table 4, the surface of the PbB-AR sample had a higher average roughness than the PbF-AR sample, measuring approximately 100 nm. This may have been because lead bronze alloys exploit the insolubility of lead in copper to create lead-free globules in a copper–tin matrix. The soft lead phase deformed easily and was smeared on the surface to form a solid lubricant, leaving empty pockets that could have been the cause of the increased roughness.

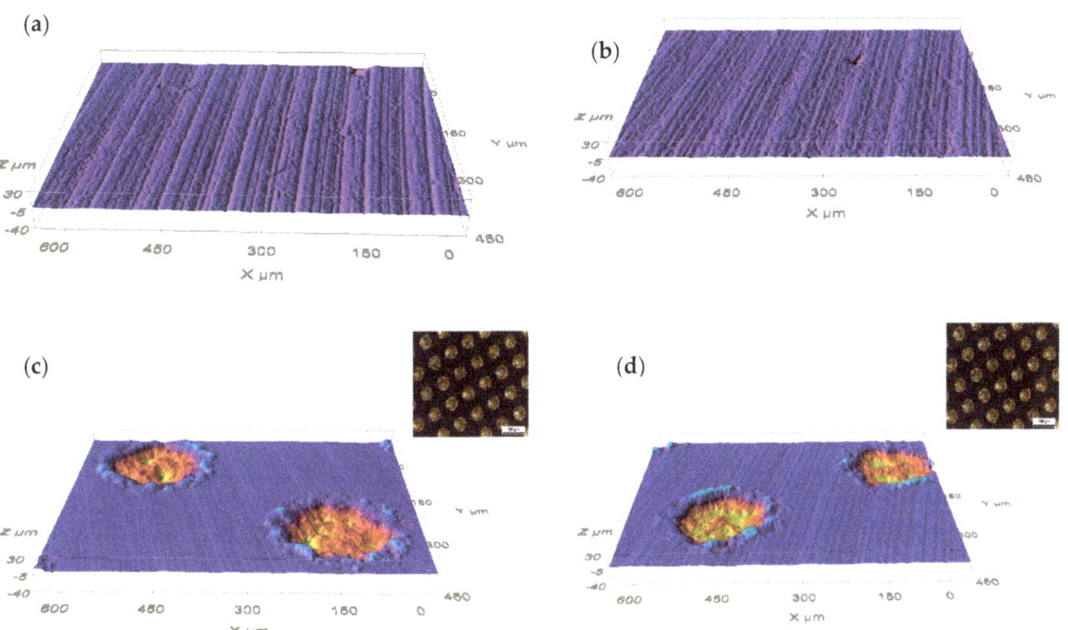

Figure 2. Comparison of 3D confocal microscopies: (**a**) PbF-AR sample; (**b**) PbB-AR sample; (**c**) PbF-B150-D10; (**d**) PbB-B150-D10.

Table 4. Roughness parameters of untextured specimens in an area of 636.61 × 477.25 (μm²).

Roughness	PbB-AR	PbF-AR
Max (μm)	1.728	2.495
Min (μm)	−6.220	−6.181
Mean (nm)	474.79	583.25

As shown in Figure 2c,d, the dimensions of the dimples were comparable with the project dimension, although the surface laser texturing technique did not allow them to have a perfect hemisphere. However, it should be noted that the surface laser texturing process caused asperities called pile-ups, which increased the surface roughness by involving an increase in friction and possible debris formation [27,39].

To evaluate the effects of the surface laser texturing on the surface of the bronze coatings, the roughness profiles of the textured samples obtained by varying the diameter of the dimples (Figure 3) and the profiles of the textured samples obtained by varying the density of the dimples were compared (Figure 4). The roughness profiles were obtained while taking into consideration the portion of the specimen with the maximum density of dimples. When examining the graphs in Figure 3, it is possible to notice some of the negative aspects related to the laser beam texturing technique. In fact, by increasing the diameter of the dimples, the depth of the dimples also increased, reaching peaks of 30 μm (Figure 3c). This was due to the inability of the diameter of the laser beam to draw a perfect hemisphere. Furthermore, as shown in Table 5, as the diameter of the dimples increased, the average surface roughness also increased. This aspect was probably due to the formation of pile-ups (more visible in Figure 3a), which were much greater than the quantity of material removed.

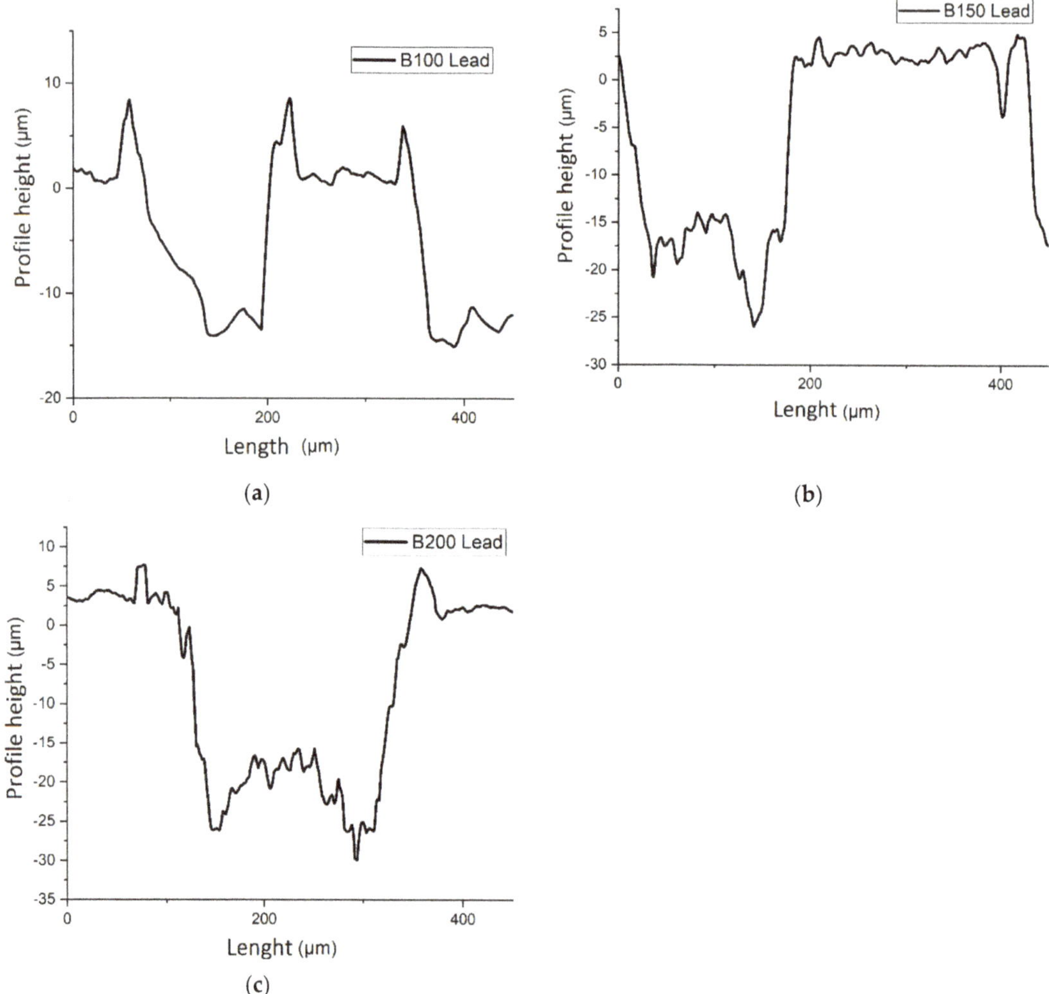

Figure 3. Comparison of roughness profiles when varying dimple diameter: (**a**) PbB-B100-D10; (**b**) PbB-B150-D10; (**c**) PbB-B200-D10.

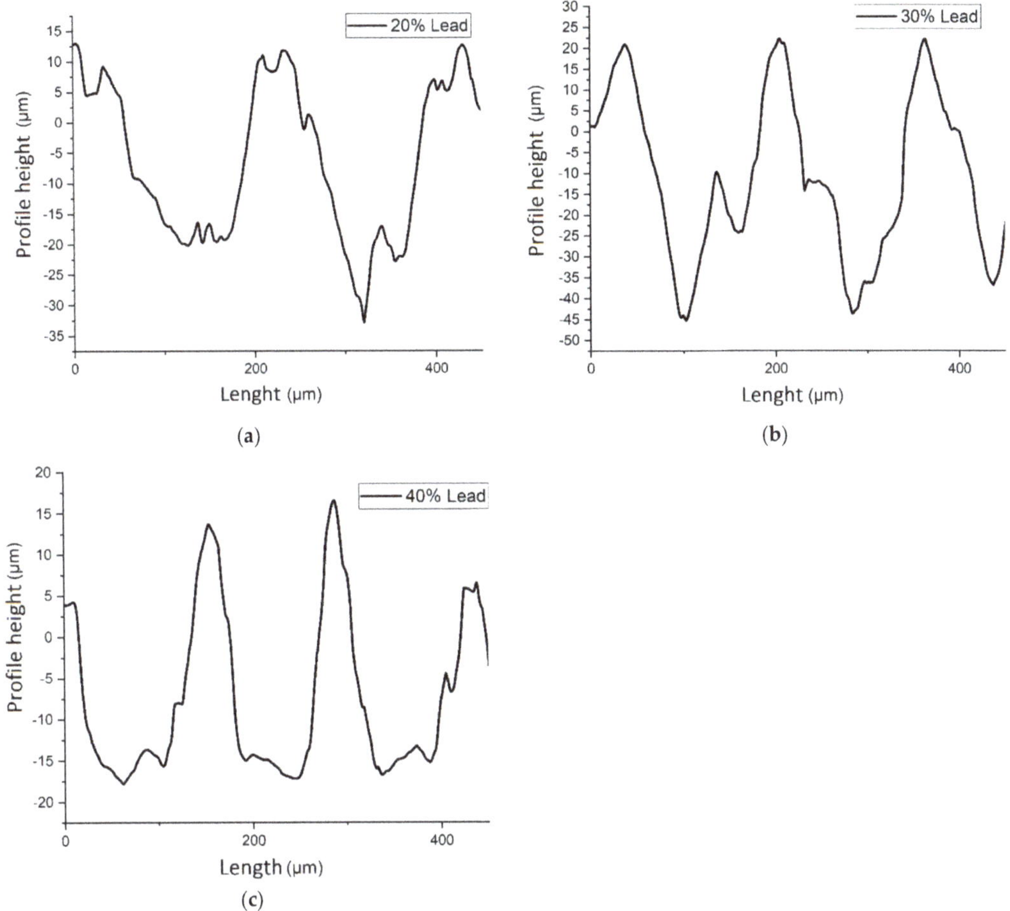

Figure 4. Comparison of roughness profiles when varying dimple density: (**a**) PbB-B100-D20; (**b**) PbB-B100-D30; (**c**) PbB-B100-D40.

Table 5. Roughness parameters of PbB specimens with different dimple diameters in an area of 636.61 × 477.25 (μm^2).

Roughness	PbB-B100-D10	PbB-B150-D10	PbB-B200-D10
Max (μm)	17.409	11.174	8.866
Min (μm)	−16.384	−31.646	−34.295
Mean (μm)	3.330	4.161	4.584

Figure 4 shows a comparison of the roughness profiles as the density varied. What should be noted when analyzing the curves is that the greater the density of the dimples on the surface, the smaller the portion of material that separated one dimple from another, until it became real roughness in the case of the PbB-B100-D40 sample. While the increase in the density of the dimples ensured a more effective lubrication due to the greater presence of "oil reservoirs" on the surface, on the other hand, it also involved a substantial increase in the surface roughness, approximately 3–4 times greater than that obtained by varying the diameter of the dimples. Furthermore, as shown in Table 6, in this case the depth values for the dimples that were obtained were even higher than in the previous case, up to 40–45 μm in depth. However, as will be analyzed in the next chapter, a greater increase in roughness led to an increase in the hydrophobic characteristics of the coatings.

Table 6. Roughness parameters of PbB specimens with different dimple densities in an area of 636.61×477.25 (μm^2).

Roughness	PbB-B100-D20	PbB-B100-D30	PbB-B100-D40
Max (μm)	14.472	28.936	27.515
Min (μm)	-45.045	-49.443	-18.330
Mean (μm)	8.1984	12.579	10.467

Figure 5 shows a comparison of the roughness profiles of the lead-free bronze and lead bronze samples. As shown in Figure 5a,b, in the cases of both the PbB-B100-D10 and PbB-B100-D40 samples, the profile drawn by the dimples had a more regular trend, and the lower surface of the dimples had a less-indented appearance. This more regular trend was justified by the lower hardness of the lead bronze coating compared to the free-lead bronze. In fact, for both the PbF-B100-D10 and PbB-B100-D40 samples, the depths of the dimples were much greater than that defined in Table 2, and this was justified by the longer time taken by the laser to melt the coating.

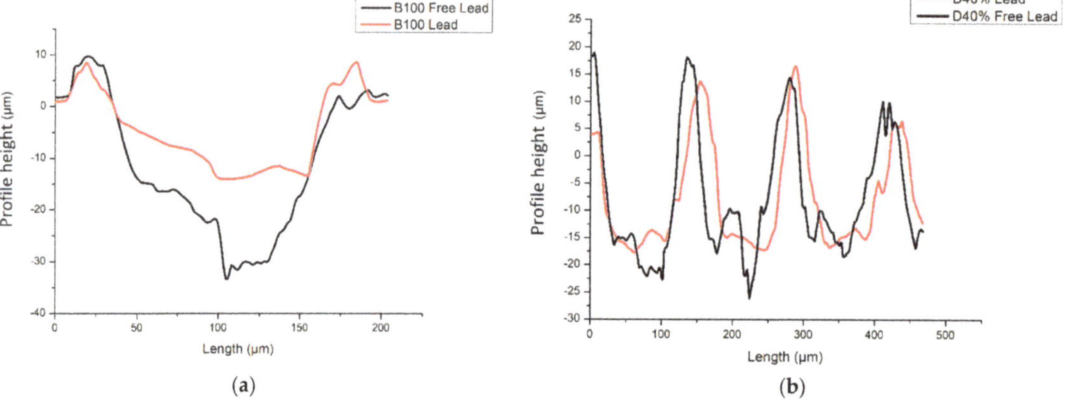

Figure 5. (**a**) Comparison of the roughness profiles of the PbB-B100-D10 and PbF-B100-D10 samples; (**b**) comparison of the roughness profiles of the PbB-B100-D40 and PbF-B100-D40 samples.

Figure 6 shows the statistical parameters of the surface profile used to analyze the correspondence between the geometric design parameters of the texture and the real geometric parameters averaged as a function of 10 measurements for each type of texture. Therefore, by analyzing the statistical parameters of the texture, the numerical values of which are shown in Table 7, it was possible to confirm that the surface laser texturing process generated a variability of a few tens of microns with respect to the design values, and this was due to the diameter of the laser beam of the machine used, as well as the type of material. It is important to specify that the data were obtained while considering both the PbF and PbB specimens to have broader statistical data. However, as already mentioned, and as shown in Figure 5, there was a difference in workability between the two types of coatings that certainly depended on the presence of lead.

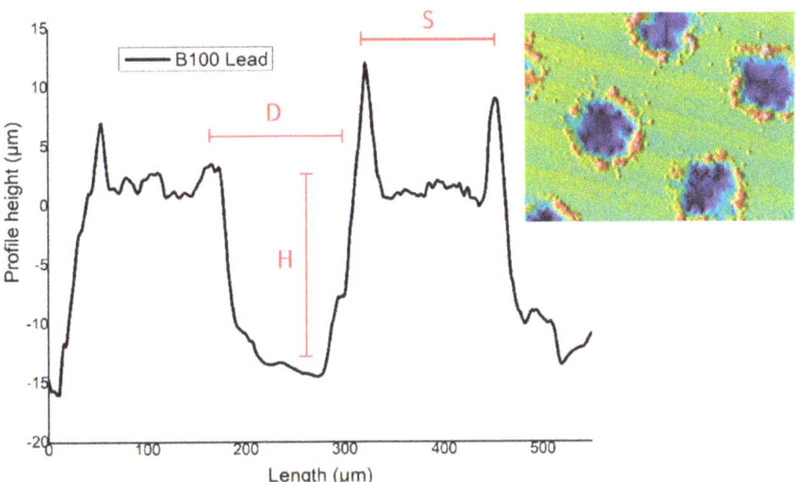

Figure 6. Representation of the geometric parameters used for statistical analysis.

Table 7. Variation of surface geometric parameters.

Parameters	B100-D10	B150-D10	B200-D10	B100-D20	B100-D30	B100-D40
S (µm)	119.50 ± 8	243 ± 20	411 ± 22	38.2 ± 5	16 ± 6	15.7 ± 2
D (µm)	113 ± 10	165.4 ± 15	221.6 ± 20	119.2 ± 7	138 ± 8	154.3 ± 4
H (µm)	14.1 ± 3	22.7 ± 5	27 ± 2	26.4 ± 7	37.3 ± 10	20.4 ± 4

3.2. Surface Wettability

In order to better correlate the morphologies of the textured surfaces with the surface properties of the bronze coatings, wettability measurements were carried out. Some representative images of 3 mL droplets deposited on the textured surfaces, with varying surface texturing, are shown in Figures 7 and 8 for both the Pb-based and Pb-free coated surfaces. In particular, the wettability measurements were carried out with water (polar liquid—Figure 7) and pumping oil (nonpolar liquid 002D—Figure 8).

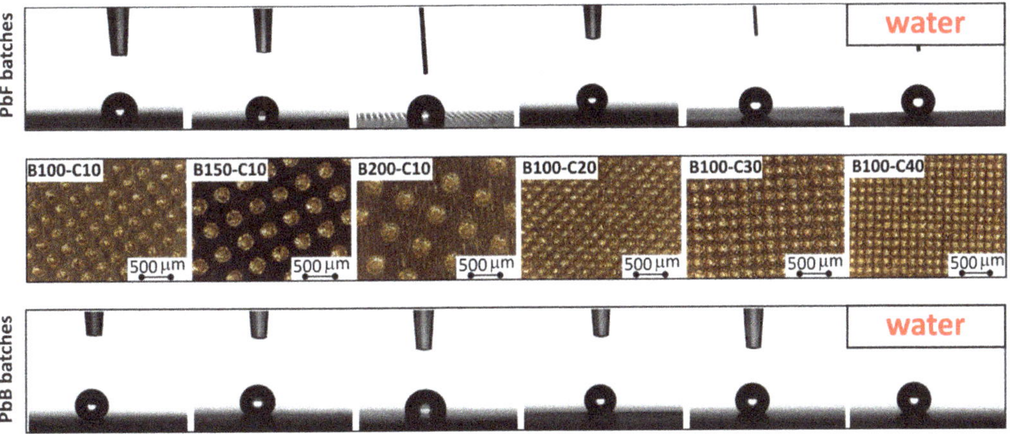

Figure 7. Water droplets for all Pb-based and Pb-free textured surfaces.

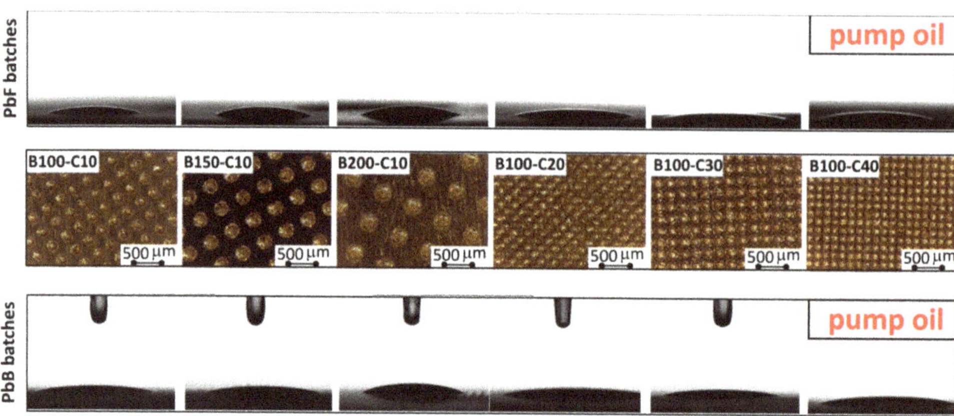

Figure 8. Oil droplets for all Pb-based and Pb-free textured surfaces.

When evaluating the wettability results, an evident correlation of the surface properties with the polar and nonpolar characteristics of the liquid was identified. All the specimens in both the PbB and PbF batches showed a predominantly hydrophobic behavior with water and an oleophilic one with pump oil. In particular, the surfaces showed a strong interaction with the oil, which spread easily in the interstices of the textured surface, which led to very low contact angles.

Moreover, it is worth noting that the texturing procedure contributed significantly to modifying the surface properties of the bronze coatings. The PbB-AR and PbF-AR specimens showed a more evident surface hydrophilicity and a less marked oleophilic one compared to the surface-treated ones. This qualitatively suggested that the surface without laser treatment did not allow selective operation toward liquids, potentially offering a less-effective lubricating capacity.

In order to be able to better quantitatively correlate the surface wettability performances with the surface morphologies, Table 8 summarizes the water contact angle (WCA) and pump oil contact angle (OCA) for all the investigated batches.

Table 8. Water and pump oil contact angles for PbF and PbB batches.

PbF Batches	WCA (°)	OCA (°)	PbB Batches	WCA (°)	OCA (°)
PbF-AR	93.0 ± 3.7	16.5 ± 1.6	PbB-AR	100.0 ± 3.7	29.3 ± 2.9
PbF-B100-D10	109.6 ± 3.6	17.2 ± 1.7	PbB-B100-D10	149.9 ± 4.2	12.2 ± 1.2
PbF-B150-D10	122.1 ± 3.2	20.2 ± 2.0	PbB-B150-D10	133.5 ± 4.8	11.6 ± 1.2
PbF-B200-D10	100.1 ± 2.5	20.1 ± 2.0	PbB-B200-D10	105.3 ± 4.0	22.0 ± 2.2
PbF-B100-D20	134.9 ± 7.0	14.0 ± 1.4	PbB-B100-D20	140.8 ± 3.6	10.5 ± 1.0
PbF-B100-D30	149.5 ± 4.2	9.3 ± 0.9	PbB-B100-D30	141.5 ± 2.1	11.2 ± 1.1
PbF-B100-D40	144.3 ± 4.8	15.6 ± 1.6	PbB-B100-D40	145.5 ± 3.0	13.0 ± 1.3

Concerning the PbF batch, the laser surface texturing induced an increase in the contact angle in a range of about 15°–50°. The higher result was observed for PbF-B100-D30, which showed a WCA of 149.5°, close to the superhydrophobic threshold [31].

When evaluating the evolution of the WCA values with varied surface texturing, it was seen that the dimple size played a less-relevant role than the dimple density. Indeed, the specimens characterized by the same density and a growing dimple size showed slight similar water contact angle values. Conversely, with increasing dimple density, a significant increase in the hydrophobic surface properties occurred.

In addition, when assessing the evolution of the OCA trend when varying the surface texturing, the oleophilic properties were quite similar, if not worse. Only the PbF-B100-D30 batch showed an evident increase in the interfacial affinity with nonpolar liquids, as shown by an average OCA value of 9.3°.

Different considerations could be addressed by evaluating the contact angle values for the PbB batches. For samples with the lead-based bronze coating, the laser surface texturing had an evident effect on the surface wettability performances. A coupled increase in the hydrophobic and oleophilic behaviors took place. All samples characterized by a laser beam diameter of 100 µm exhibited a WCA close to 150°. Furthermore, the OCA experienced a reduction of up to 20°.

Not only the liquid/solid contact angles were evaluated, but also the sliding angles, in order to understand the surface wettability behavior [40]. All the samples before and after texturing showed a high liquid sliding angle (90°), indicating a high liquid adhesion (data not reported in the table). Thus, creating a textured structure did not change the water adhesion on the sample surface, regardless of the coating nature.

The marked hydrophobic and oleophilic behaviors found in the laser-textured samples could be related to the intrinsic surface morphology acquired by the coatings at the end of the surface treatment. The resulting surface profile was constituted by a larger number of peaks and valleys, making the surface regularly jagged and rough.

This was in accordance with Wenzel's theory, which relates the surface roughness to the liquid/solid contact angle [41,42]. Indeed, Wenzel proposed a relationship between the surface-roughness ratio (*R*: ratio of the rough surface area to the smooth surface one) and the contact angles on smooth and rough surfaces:

$$cos(\theta w) = R\ cos(\theta) \tag{1}$$

where θw is the Wenzel contact angle (contact angle on rough surface) and θ is the ideal contact angle (contact angle on smooth surface). According to this equation, by increasing the surface roughness, the hydrophobic surface becomes more hydrophobic, and the hydrophilic surface becomes more hydrophilic [43]. In fact, as shown in Table 4, both PbR_AR and PbB-AR were hydrophobic and oleophilic before texturing, and they become more hydrophobic and oleophilic after texturing. In this Wenzel state, the liquid penetrated the rough surface cavities; such behavior is known as the homogenous wetting mode, in which the interaction and the adhesion between the liquid and the solid surface are high [44].

In order to better investigate the correlation between the wettability and the characteristics of the surface profile of the specimens, an index related to the dimples' surface morphologies, named the DS index, was calculated according to the following expression:

$$\text{DS index} = (d * h)/s \tag{2}$$

where *d* and *h* are the diameter and height of the dimple, respectively. Consequently, $d * h$ is the dimple area, and *S* is the length of the peak between two dimples. All the measurements were defined in µm, thus the DS index was expressed in µm. The greater this index, the greater the contribution of the cavities to the surface peaks.

Figure 9 shows the evolution of the water contact angle (WCA) and oil contact angle (OCA) with the DS index for Pb-based and Pb-free textured samples.

For PbF batches (blue diamond marker in Figure 9), the WCA and OCA had a quite good linear relationship with the DS index. The slope increased for the WCA values, and vice versa (the slope decreased for the OCA). This result suggested that an increase in the size of the dimples and their depth played a key role in enhancing the hydrophobic and oleophilic behavior of the surface.

For the PbB batches (orange circle marker in Figure 9), a clear bilinear trend in the WCA and OCA vs. DS index could instead be identified. The trend initially showed a strong change in the contact angles due to a slight change in the DS index. The slope of the fitting line (positive for WCA and negative for OCA) was very high. Subsequently, for values of DS > 1.4, the trend showed a plateau zone. An increase in the dimple shape (diameter and depth) did not provide a statistically significant change in the contact angle. The fitting line had a very low slope, indicating a trend almost parallel to the x axis.

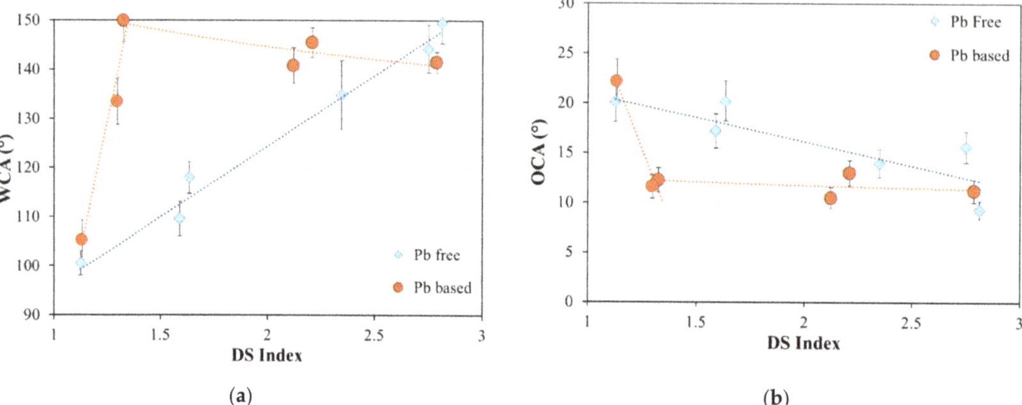

Figure 9. Evolution of (**a**) water contact angle (WCA) and (**b**) oil contact angle (OCA) with the DS index for Pb-based and Pb-free textured samples.

This behavior showed that for the PbB batch, a very extensive and invasive laser texturing was not necessary to induce a surface-modification effect in the coating. A DS index close to 4 was sufficient to guarantee an effective surface hydrophilicity and hydrophobicity.

Based on wettability measurements with oil and water, it was possible to assess that the porous textured surface was tailored to act an oil reservoir thanks to its good oleophilic behavior. This can prolong the life of lubricating oil layer and may result in a superior wear resistance. Furthermore, a relevant hydrophobicity was highlighted, suggesting that the surface texturing promoted the water-repellent barrier action on the surface.

4. Conclusions

The experimental studies in this paper were conducted to understand the influence of surface laser texturing and roughness on the wettability of lead-free bronze and lead bronze coatings. In order to increase the efficiency of components that work under reciprocating sliding, classes of textured specimens were obtained by varying the diameter of the dimples in a range of values between 100 and 200 μm and varying the surface density of the dimples between 10 and 40%. By using a low-economic-impact technique, such as surface laser texturing, it is possible to decrease the friction coefficient by increasing the hydrophobicity and oleophilic, and therefore improve lubrication, especially in lubrication-starved conditions. The results obtained on the basis of the wettability measurements with oil and water highlighted that the porous textured surface was adapted to serve as an oil reservoir thanks to its good oleophilic behavior. This can extend the life of the lubricating oil layer and ensure better lubrication, even in hard-working conditions, resulting in superior wear resistance. Furthermore, a significant hydrophobicity was highlighted, suggesting that the surface texturing favored the water-repellent barrier action on the surface. For the tests carried out on the PbF specimens, it was experimentally shown that an increase in the size of the dimples and their depths played a key role in enhancing the hydrophobic and oleophilic behavior of the surface, while regarding the PbB specimens, an increase in the dimple shape (diameter and depth) did not provide a statistically significant change in the contact angle, and therefore it was sufficient to guarantee a DS equal to 4, meaning a coating with good hydrophobicity and oleophilic. Future work will concern the correlation of the wettability tests with the tribological behavior of the coatings (both PbF and PbB) in order to evaluate the extent to which the application of the surface texture affects the friction coefficient of the coatings and the wear rate.

Author Contributions: Conceptualization, G.R. and L.C.; methodology, D.D. and A.K.; validation, G.R. and L.C.; formal analysis, D.D. and A.K.; investigation, D.D. and A.K.; data curation, D.D., L.C., and A.K.; writing—original draft preparation, D.D. and L.C.; writing—review and editing, D.D., L.C., and A.K.; visualization, D.D. and A.K.; supervision, G.R. and L.C. All authors have read and agreed to the published version of the manuscript.

Funding: The authors received no financial support for the research, authorship, and/or publication of this article.

Institutional Review Board Statement: Not applicable.

Informed Consent Statement: Not applicable.

Data Availability Statement: The data presented in this study are available on request from the corresponding author.

Conflicts of Interest: The authors declare no potential conflict of interest with respect to the research, authorship, and/or publication of this article.

References

1. Tang, Z.; Liu, X.; Liu, K. Effect of surface texture on the frictional properties of grease lubricated spherical plain bearings under reciprocating swing conditions. *Proc. Inst. Mech. Eng. Part J J. Eng. Tribol.* **2017**, *231*, 125–135. [CrossRef]
2. Ali, M.K.A.; Xianjun, H.; Essa, F.A.; Abdelkareem, M.A.A.; Elagouz, A.; Sharshir, S.W. Friction and wear reduction mechanisms of the reciprocating contact interfaces using nanolubricant under different loads and speeds. *J. Tribol.* **2018**, *140*, 051606. [CrossRef]
3. Hua, D.; Wang, W.; Luo, D.; Zhou, Q.; Li, S.; Shi, J.; Fu, M.; Wang, H. Molecular dynamics simulation of the tribological performance of amorphous/amorphous nano-laminates. *J. Mater. Sci. Technol.* **2022**, *105*, 226–236. [CrossRef]
4. D'Andrea, D.; Epasto, G.; Bonanno, A.; Guglielmino, E.; Benazzi, G. Failure analysis of anti-friction coating for cylinder blocks in axial piston pumps. *Eng. Fail. Anal.* **2019**, *104*, 126–138. [CrossRef]
5. Liu, X.; Tang, J.; Li, X.; Li, W. Study on friction behavior at the interface between prosthetic socket and liner. *Acta Bioeng. Biomech.* **2021**, *23*, 83–93. [CrossRef]
6. D'Andrea, D.; Pistone, A.; Risitano, G.; Santonocito, D.; Scappaticci, L.; Alberti, F. Tribological characterization of a hip prosthesis in Si3N4-TiN ceramic composite made with Electrical Discharge Machining (EDM). *Procedia Struct. Integr.* **2021**, *33*, 469–481. [CrossRef]
7. Ruggiero, A.; D'Amato, R.; Sbordone, L.; Haro, F.B.; Lanza, A. Experimental comparison on dental BioTribological pairs zirconia/zirconia and zirconia/natural tooth by using a reciprocating tribometer. *J. Med. Syst.* **2019**, *43*, 97. [CrossRef]
8. Wang, Z.; Xiang, J.; Fu, Q.; Wood, R.J.K.; Wang, S. Study on the friction performance of textured surface on water hydraulic motor piston pair. *Tribol. Trans.* **2022**, *65*, 308–320. [CrossRef]
9. Yang, D.; Li, W.; Wang, H.; Gao, G.; Cheng, S.; Ren, J.; Tian, S. Effect of Counterpart Ring Surface Roughness on Wear Process of Bismuth Bronze. *Chin. J. Mater. Res.* **2021**, *35*, 732–740.
10. Kestursatya, M.; Kim, J.K.; Rohatgi, P.K. Wear performance of copper-graphite composite and a leaded copper alloy. *Mater. Sci. Eng. A* **2003**, *339*, 150–158. [CrossRef]
11. Dyett, B.; Lamb, R. Correlating Material Properties with the Wear Behavior of Sol–Gel Derived Superhydrophobic Films. *Adv. Mater. Interfaces* **2016**, *3*, 1500680. [CrossRef]
12. Calabrese, L.; Khaskhoussi, A.; Patane, S.; Proverbio, E. Assessment of Super-Hydrophobic Textured Coatings on AA6082 Aluminum Alloy. *Coatings* **2019**, *9*, 352. [CrossRef]
13. Luo, D.; Zhou, Q.; Ye, W.; Ren, Y.; Greiner, C.; He, Y.; Wang, H. Design and Characterization of Self-Lubricating Refractory High Entropy Alloy-Based Multilayered Films. *ACS Appl. Mater. Interfaces* **2021**, *13*, 55712–55725. [CrossRef] [PubMed]
14. Kovalchenko, A.; Ajayi, O.; Erdemir, A.; Fenske, G.; Etsion, I. The effect of laser texturing of steel surfaces and speed-load parameters on the transition of lubrication regime from boundary to hydrodynamic. *Tribol. Trans.* **2004**, *47*, 299–307. [CrossRef]
15. Etsion, I. Improving tribological performance of mechanical components by laser surface texturing. *Tribol. Lett.* **2004**, *17*, 733–737. [CrossRef]
16. Etsion, I.; Halperin, G. A laser surface textured hydrostatic mechanical seal. *Seal. Technol.* **2003**, *45*, 430–434. [CrossRef]
17. Shum, P.W.; Zhou, Z.F.; Li, K.Y. Investigation of the tribological properties of the different textured DLC coatings under reciprocating lubricated conditions. *Tribol. Int.* **2013**, *65*, 259–264. [CrossRef]
18. Yang, Z.; Zhu, C.; Zheng, N.; Le, D.; Zhou, J. Superhydrophobic surface preparation and wettability transition of titanium alloy with micro/nano hierarchical texture. *Materials* **2018**, *11*, 2210. [CrossRef]
19. Volpe, A.; Covella, S.; Gaudiuso, C.; Ancona, A. Improving the laser texture strategy to get superhydrophobic aluminum alloy surfaces. *Coatings* **2021**, *11*, 369. [CrossRef]
20. Chen, L.; Liu, Z.; Shen, Q. Enhancing tribological performance by anodizing micro-textured surfaces with nano-MoS2 coatings prepared on aluminum-silicon alloys. *Tribol. Int.* **2018**, *122*, 84–95. [CrossRef]

21. Gao, P.; Liang, Z.; Wang, X.; Zhou, T.; Xie, J.; Li, S.; Shen, W. Fabrication of a micro-lens array mold by micro ball end-milling and its hot embossing. *Micromachines* **2018**, *9*, 96. [CrossRef] [PubMed]
22. Chen, X.L.; Dong, B.Y.; Zhang, C.Y.; Wu, M.; Guo, Z.N. Jet electrochemical machining of micro dimples with conductive mask. *J. Mater. Process. Technol.* **2018**, *257*, 101–111. [CrossRef]
23. Holmberg, J.; Berglund, J.; Wretland, A.; Beno, T. Evaluation of surface integrity after high energy machining with EDM, laser beam machining and abrasive water jet machining of alloy 718. *Int. J. Adv. Manuf. Technol.* **2019**, *100*, 1575–1591. [CrossRef]
24. Despres, L.; Costil, S.; Cormier, J.; Villechaise, P.; Cariou, R. Impact of Laser Texturing on Ni-Based Single Crystal Superalloys. *Metals* **2021**, *11*, 1737. [CrossRef]
25. Conradi, M.; Kocijan, A.; Klobčar, D.; Godec, M. Influence of laser texturing on microstructure, surface and corrosion properties of Ti-6Al-4V. *Metals* **2020**, *10*, 1504. [CrossRef]
26. Travessa, D.N.; Guedes, G.V.B.; de Oliveira, A.C.; Cardoso, K.R.; Roche, V.; Jorge, A.M., Jr. The effect of surface laser texturing on the corrosion performance of the biocompatible β-Ti12Mo6Zr2Fe alloy. *Surf. Coat. Technol.* **2021**, *405*, 126628. [CrossRef]
27. Senatore, A.; Risitano, G.; Scappaticci, L.; D'andrea, D. Investigation of the tribological properties of different textured lead bronze coatings under severe load conditions. *Lubricants* **2021**, *9*, 34. [CrossRef]
28. Rajab, F.H.; Liu, Z.; Li, L. Long term superhydrophobic and hybrid superhydrophobic/superhydrophilic surfaces produced by laser surface micro/nano surface structuring. *Appl. Surf. Sci.* **2019**, *466*, 808–821. [CrossRef]
29. Li, M.; Li, Y.; Xue, F.; Jing, X. A robust and versatile superhydrophobic coating: Wear-resistance study upon sandpaper abrasion. *Appl. Surf. Sci.* **2019**, *480*, 738–748. [CrossRef]
30. Rizzo, G.; Massarotti, G.P.; Bonanno, A.; Paoluzzi, R.; Raimondom, M.; Blosi, M.; Veronesi, F.; Caldarelli, A.; Guarini, G. Axial piston pumps slippers with nanocoated surfaces to reduce friction. *Int. J. Fluid Power* **2015**, *16*, 1–10. [CrossRef]
31. Khaskhoussi, A.; Calabrese, L.; Proverbio, E. An Easy Approach for Obtaining Superhydrophobic Surfaces and their Applications. *Key Eng. Mater.* **2019**, *813*, 37–42. [CrossRef]
32. Rosén, B.-G.; Nilsson, B.; Thomas, T.R.; Wiklund, D.; Xiao, L. Oil pockets and surface topography: Mechanisms of friction reduction. In Proceedings of the XI. International Colloquium on Surfaces, Chemnitz, Germany, 24–28 September 2004.
33. Costa, H.L.; Hutchings, I.M. Hydrodynamic lubrication of textured steel surfaces under reciprocating sliding conditions. *Tribol. Int.* **2007**, *40*, 1227–1238. [CrossRef]
34. Bharatish, A.; Rajkumar, G.R.; Gurav, P.; Satheesh Babu, G.; Narasimha Murthy, H.N.; Roy, M. Optimization of laser texture geometry and resulting functionality of nickel aluminium bronze for landing gear applications. *Int. J. Light. Mater. Manuf.* **2021**, *4*, 346–357. [CrossRef]
35. Li, J.; Liu, S.; Yu, A.; Xiang, S. Effect of laser surface texture on CuSn6 bronze sliding against PTFE material under dry friction. *Tribol. Int.* **2018**, *118*, 37–45. [CrossRef]
36. Wang, Z.; Gu, L.; Li, L. Experimental studies on the overall efficiency performance of axial piston motor with a laser surface textured valve plate. *Proc. Inst. Mech. Eng. Part B J. Eng. Manuf.* **2013**, *227*, 1049–1056. [CrossRef]
37. Ryk, G.; Kligerman, Y.; Etsion, I. Experimental investigation of laser surface texturing for reciprocating automotive components. *Tribol. Trans.* **2002**, *45*, 444–449. [CrossRef]
38. Guarnaccio, A.; Belviso, C.; Montano, P.; Toschi, F.; Orlando, S.; Ciaccio, G.; Ferreri, S.; Trevisan, D.; Mollica, D.; Parisi, G.P.; et al. Femtosecond laser surface texturing of polypropylene copolymer for automotive paint applications. *Surf. Coat. Technol.* **2021**, *406*, 126727. [CrossRef]
39. Kümmel, D.; Hamann-Schroer, M.; Hetzner, H.; Schneider, J. Tribological behavior of nanosecond-laser surface textured Ti6Al4V. *Wear* **2019**, *422*, 261–268. [CrossRef]
40. Khaskhoussi, A.; Calabrese, L.; Proverbio, E. Superhydrophobic Self-Assembled Silane Monolayers on Hierarchical 6082 Aluminum Alloy for Anti-Corrosion Applications. *Appl. Sci.* **2020**, *10*, 2656. [CrossRef]
41. Wenzel, R.N. Resistance of solid surfaces to wetting by water. *Ind. Eng. Chem.* **1936**, *28*, 988–994. [CrossRef]
42. Khaskhoussi, A.; Calabrese, L.; Proverbio, E. Effect of the Cassie Baxter-Wenzel behaviour transitions on the corrosion performances of AA6082 superhydrophobic surfaces. *Metall. Ital.* **2021**, *5*, 15–21.
43. Jiang, G.; Hu, J.; Chen, L. Preparation of a Flexible Superhydrophobic Surface and Its Wetting Mechanism Based on Fractal Theory. *Langmuir* **2020**, *36*, 8435–8443. [CrossRef] [PubMed]
44. Khaskhoussi, A.; Calabrese, L.; Patané, S.; Proverbio, E. Effect of Chemical Surface Texturing on the Superhydrophobic Behavior of Micro-Nano-Roughened AA6082 Surfaces. *Materials* **2021**, *14*, 7161. [CrossRef] [PubMed]

 *lubricants*

Article

Impact of Graphene Nano-Additives to Lithium Grease on the Dynamic and Tribological Behavior of Rolling Bearings

Mohamed G. A. Nassef [1,2,*], Mina Soliman [3], Belal Galal Nassef [2,4], Mohamed A. Daha [2] and Galal A. Nassef [2]

[1] Industrial and Manufacturing Engineering Department, Egypt-Japan University of Science and Technology, New Borg El Arab City 21934, Egypt

[2] Production Engineering Department, Alexandria University, Bab Sharqi 21544, Egypt; belal.nassef@ejust.edu.eg (B.G.N.); m.daha@alexu.edu.eg (M.A.D.); galalnassef@alexu.edu.eg (G.A.N.)

[3] Atlasco Egypt Company, 1st Industrial Zone, El-Obour 11828, Egypt; mina@atlascoegypt.com

[4] Materials Science and Engineering Department, Egypt-Japan University of Science and Technology, New Borg El Arab City 21934, Egypt

* Correspondence: mohamed.nassef@ejust.edu.eg

Abstract: In recent years, reduced graphene oxide (rGO) received considerable interest as a lubricant nano-additive for enhancing sliding and rolling contacts. This paper investigates the tribological and dynamic behavior of ball bearings lubricated by lithium grease at different weight percentages of rGO. Full bearing tests were conducted for experimental modal analysis, vibration analysis, ultrasonic analysis, and infrared thermography. Modal analysis indicated considerable improvements of the damping ratio values up to 50% for the bearings with rGO nano-additives. These findings were confirmed by the corresponding reductions in vibrations and ultrasound levels. The steady-state temperatures of bearings running with lithium grease reached 64 °C, whereas the temperature of bearings lubricated by grease with 2 wt.% rGO measured only 27 °C. A Timken Load test was conducted on grease samples with and without rGO additives. Grease samples having 2, 3.5, and 5 wt.% rGO showed the highest OK load with an increase of 25%, 50%, and 100% as compared to values of lithium grease. For comparison, all tests were conducted on samples of the same grease blended with graphite and MWCNTs' nano-additives. The results proved the superiority of graphene in enhancing the load-carrying capacity and damping of grease in rolling bearings.

Keywords: lithium grease; graphene; load-carrying capacity; wear scar; infrared thermography; vibration analysis; damping ratio

Citation: Nassef, M.G.A.; Soliman, M.; Nassef, B.G.; Daha, M.A.; Nassef, G.A. Impact of Graphene Nano-Additives to Lithium Grease on the Dynamic and Tribological Behavior of Rolling Bearings. *Lubricants* **2022**, *10*, 29. https://doi.org/10.3390/lubricants10020029

Received: 15 January 2022
Accepted: 15 February 2022
Published: 18 February 2022

Publisher's Note: MDPI stays neutral with regard to jurisdictional claims in published maps and institutional affiliations.

1. Introduction

Rolling element bearings (REBs) are critical components in rotating machinery for various mechanical systems in wind turbines, aerospace, and automotive industries. It is reported that around 10 billion bearings are manufactured worldwide every year [1], and this bearing market is expected to grow in size from about USD 118.7 billion in 2020 with a compound annual growth rate (CAGR) of 8.5% from 2021 to 2028 [2].

Unexpected failures of REBs present a crucial impact on the performance of machinery and may cause a catastrophic failure of the assets and losses of lives [3]. A review analysis by Vencl et al. [4] classified the most frequent bearing failures according to wear mechanisms while reporting that bearings fail prematurely due to improper bearing selection, incorrect bearing handling and mounting, and inappropriate operating conditions. More than one-third of the failed bearings are attributed to inadequate lubrication, including improper lubricant viscosity, overdose/underdose lubrication, and wrong lubrication intervals. By combining lubricant contamination from the harsh surrounding environment and worn-out sealing, this ratio is increased to about 80%. Harris [5] classified lubricant deficiency as one of the main failure mechanisms of rolling bearings. For example, if the lubricant is insufficient at the Hertzian contact zone (i.e., lubricant starvation case), an adhesion wear mechanism takes place where the contacting surfaces weld together; then, they get torn apart as the rolling action follows [5,6].

During operation, lubricant migration from the contact zone in bearings is a common cause of lubricant starvation [7,8]. Different theoretical and experimental approaches for controlling oil migration considering the thermal gradient factor, surface texture, and chemical surface properties were discussed by Grützmacher et al. [9]. One advanced approach presented multi-scale surface patterning using two techniques in journal bearings as a countermeasure for oil migration [10]. It proved highly efficient in reducing lubricant migration and wear reduction. A further study in the same direction considered the effect of thermal gradient on the oil spreading (migration) on single and multi-scale surface patterned stainless steel samples [11].

Oil and grease are mainly employed in REBs to prevent metal-to-metal contact between the rubbing surfaces of the inner race, rolling elements, and outer race [12]. Their purpose is to minimize friction and wear in REBs during operation; hence, they significantly contribute to reducing energy losses and maintaining high reliability of power transmission systems [13]. Grease is preferred over oil lubricants for REBs as it shows superior performance in resisting the washing action of water, preventing any leakage, operating without frequent relubrication, showing good vibration damping capacities, and increasing the sealing efficiency against any contaminants [14,15].

Grease lubrication consists of a thickener (such as lithium soap, calcium soap, polyurea), 70–97% base oil (synthetic or mineral oil), and 3% additives. The prominent role of the additives is to introduce or enhance specific properties, such as extreme pressure (EP) additives, anti-wear (AW) agents, oxidation inhibitors, and lubricity additives [16].

The current increasing usage of heavy machinery and advanced mechanical systems in sectors of construction, mining, railway and aerospace, and electric vehicles' industries requires a corresponding significant improvement in grease lubrication additives to achieve higher load-carrying capacities, better vibration damping, and minimum energy losses. Advances in synthesis and characterization of two-dimensional (2-D) materials such as molybdenum disulfide (MoS2), tungsten disulfide (WS2), hexagonal boron nitride, graphene, and other 2-D nano-materials have opened the door for their application as solid lubricants or lubricant additives in rotating machinery. A review by Rosenkranz et al. [17] summarized the directions of research work conducted on 2-D nano-materials, including their chemical properties, tribological behavior, and their potential to be applied as solid lubricants well as lubricant additives. The review also highlighted the current issues related to large-scale production and the obtained quality of each 2-D nano-material to be reasonably applied in the industry as solid lubricants. In order to efficiently incorporate 2-D nano-materials to reach a superlubricity regime on the nano- and macro-scales, different mechanisms of energy dissipation due to friction between sliding surfaces were studied and discussed in [18]. The review in [19] discussed different theoretical and experimental models describing physical mechanisms such as the interlayer sliding between single or multilayered 2-D nano-materials such as graphene and graphene oxide in comparison to graphite and how this significantly contributed to reducing the friction coefficient to the superlubricity zone. The review recommended that more studies should address the practical application of these 2-D materials in controlling friction and wear in macro- and meso-mechanical systems.

Adding carbonaceous nanomaterials such as CNTs, graphene, and reduced graphene oxide (GO) to the lubricating grease has been proposed as a popular trend to improve lubricant properties [20–22] significantly. Carbonaceous nanomaterials enhance the load-carrying capacity, decrease the coefficient of friction (COF) and wear scar diameter (WSD) between the mating parts [23,24], and impart new chemical, mechanical, and physical properties to grease [25]. Due to their strong in-plane bonding, weak interlayer interactions, and dangling bonds, their application in dry and boundary lubrication of machine elements such as gears and bearings was highly recommended by Berman et al. [26].

Graphene has a two-dimensional structure and small particle size, providing a large contact area sufficient to resist oxidation [27,28]. Its unique nano-scaled structure introduced high thermal and electrical conductivities, promoting it to be one of the best choices in supercapacitors and energy harvesting [29]. Due to its outstanding physical properties and self-lubricity characteristic, researchers reported that graphene enhanced the lubricity of grease and thermal conductivity as well [30,31].

Zhang et al. [20] investigated the addition of different graphene weight fractions with lithium-based grease. The results showed about 15% improvement of the COF when adding 2 wt.% graphene. Jayant et al. [30] conducted experimental trials by mixing reduced graphene oxide (rGO) with lithium-based grease. The rGO concentration of 0.4 wt.% reduced the COF by about 20% and 30% for sliding-induced rolling contact and rolling contact compared to base grease. Fu et al. [32] tried different graphene additions with calcium grease, and they found that 2 wt.% was the optimum percentage regarding the tribological properties as this percentage reduced the COF by about 15% compared to base grease. Furthermore, a reduction in the friction coefficient by up to 60% was obtained after adding 1 wt.% graphene to a semi-solid lubricant [33]. Nan et al. [34] investigated the effect of mixing different graphene percentages with attapulgite-based grease. They observed that graphene additions improve the tribological behavior of the base grease, especially at high loading conditions. It was also found that the addition of 1 wt.% graphene reduces the friction coefficient by 67% as compared to the base grease.

From the above review of literature, it was concluded that complex lithium grease is commonly used as a lubricant in REBs due to its good tribological behavior. However, there is still a need to improve the tribological properties of the REBs in rotating machinery by reducing generated friction forces and energy losses. According to a previous investigation, achieving 1% reduction in the frictional power concerning the automotive industry can save approximately billions of fuel liters [35].

Most of the research work highlighted the importance of nano-additives in general and graphene in particular in enhancing the tribological performance of grease lubricant. Extensive research work focused on determining the COF and WSD of the developed grease using standard tests, such as the four-ball wear test and ball-on-disk test. Despite the promising tribological results of graphene as a nano-additive in grease, its application in the lubrication of essential mechanical components, such as rolling bearings, remains relatively unexplored.

The literature lacks any attempts to provide an insight into the influence of adding graphene on the damping capacity of grease lubricant and how this would enhance the dynamic response of REBs during highly dynamic operations. An individual attempt in [36] found that have rolling bearings experienced the lowest frictional torque when adding 1 wt.% GNPs with the thickness of 6–8 nm as an additive to grease. It reduced the frictional torque inside the bearing by more than 60% as compared to base grease, but the investigation was limited to only 1 wt.% of graphene as an additive. A more recent investigation studied the effect of adding different rGO percentages and several other carbonaceous additives such as MWCNTs to the bearing lubrication and reported maximum power saving when 2 wt.% rGO was added to lithium grease [37].

The present research had the objective to focus on investigating important aspects of graphene grease lubricant that are still unexplored and have not received considerable research attention on rolling machinery components such as rolling bearings. Vibration damping properties, heat chilling potential, and load-carrying capacity are among the important aspects of this investigation. Full bearing tests, vibration modal analysis tests, and standard lubricant tests were conducted in this research investigation. Other prominent carbonaceous additives, namely, graphite and MWCNTs, were considered for comparison to stand upon the level of improvement when applying graphene to grease lubricant.

2. Materials and Methods

2.1. Test Setup

A customized test rig was designed and manufactured to conduct a full bearing test, as shown in Figure 1. It was used to evaluate the effect of each nano-additive in lithium grease lubricant on the vibrational response, ultrasound emissions, and heat generation of rolling element bearings (REBs) [37]. The test rig previously designed by Nassef [37] consisted of seven main components: (1) the electric motor, (2) coupling, (3) base, (4) shaft, (5) two support bearings, (6) test bearing, and (7) the applied radial load. The specifications of each component are listed in Table 1, whereas the detailed design of the rig is described elsewhere [37].

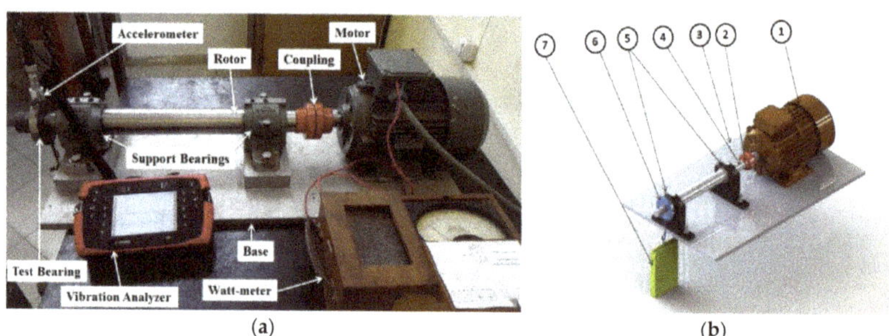

Figure 1. (**a**) Test rig setup, and (**b**) Assembly model of the test showing: (1) an electric motor with (2) a suitable coupling, (3) base, (4) shaft, (5) two support bearings and their housings, (6) test bearing with a specially designed housing, and (7) the applied radial load.

Table 1. Specifications of test rig components.

Component	Specifications
Electric Motor	GAMAK (3 hp, and 1400 rpm)
Base	C45 Carbon Steel
Shaft	SUS 420 Stainless Steel
Two Support Bearings	NU1011M Roller Bearing

The test bearing was a deep-groove ball bearing 6006 zz, whose specifications are summarized in Table 2 as provided by NSK bearing manufacturer. The test bearing was fitted to its special housing, which was assembled in an interference to the rotor. The motor ran at 1400 RPM while a set of dead weights between 2050 N and 2250 N were separately attached to the bearing housing through a suitable hook.

Table 2. Specifications of the test bearing (6006 zz) provided by NSK manufacturer.

	D	D	B	r	
Dimensions (mm)	30	55	13	1	
Mass (kg)		0.116			
Dynamic load rating, CD (N)		13,200			
Static load rating, Co (N)		8300			
	C2	CN	C3	C4	C5
Clearances (μm)	1–11	5–20	13–28	23–41	30–53

2.2. Material Preparation and Characterization

The nano-additive material under investigation in this work was graphene nano-sheets supplied, by Nanogate Company, Cairo, Egypt, in the form of reduced graphene oxide (rGO) from the local market, which was prepared via the Modified Hummer's Method [38–40]. X-ray diffraction (XRD) (Empyrean Malver panalytical, Almelo city, The Netherlands) was used to obtain the average grain size, % crystallinity, and d-spacing of graphene nano-sheets. XRD analysis operated at 30 mA and 40 keV with an angular range of 0.08–80 degrees and Cu-Kα1 radiation of 0.154-nm wavelength.

A transmission electron microscope (TEM) (JEOL JEM-2100, JOEL Ltd., Tokyo, Japan) was used to obtain the shape and number of layers. The sample was first immersed in ethanol, and the suspended solution was then sonicated for around 5 min. Then, a 5-μm sample from the suspension was drop cast on a carbon-coated copper grid. The TEM results were further confirmed using Raman spectroscopy (WITec alpha 300 R confocal Raman microscope, WITec Company, Ulm, Germany) at room temperature through the exposure of the rGO sample to a 532-nm laser. The Raman spectrum was recorded using a WITec alpha 300 R confocal Raman microscope (Germany) with 50× magnification objective. The sample was excited using a 532-nm laser line and 3-mW power impinging on the sample. Finally, the Raman spectrum was collected with 20-s detector time and five accumulations.

Moreover, Fourier transform infrared (FTIR) spectroscopy (Bruker Vertex 70 with a wave-number range of 400–4000 cm^{-1}, Bruker Company, Billerica, MA, USA) was used to obtain information about the functional groups attached to the graphene structure to assess its quality. MWCNTs were characterized using TEM to confirm their shape and size, whereas a scanning electron microscope (SEM) (JSM-IT200) was used to investigate the graphite structure.

The rGO weight percentages of 0.5, 1, 2, 3.5, and 5 wt.% were separately mixed with the appropriate quantity of commercial, heavy-duty, lithium-based grease (4 g). The blending process was performed using a commercial mixer of 150 kW. The stirring process was completed after the grease color became uniformly dark. The exact mixing conditions were applied for mixing constant MWCNTs' percentage of 1 wt.% and graphite percentages of 2, 3.5, and 5 wt.% with grease to compare their tribological behavior with rGO. The blended grease and nano-additives were then used to lubricate the test bearings according to the SKF formula shown in Equation (1).

$$Gq = 0.114(D)(B) \tag{1}$$

where Gq is the amount of grease used to re-lubricate the bearing in ounces, (D) is the outer diameter of the bearing, and (B) is the bearing width.

2.3. Experimental Modal Analysis

Experimental modal analysis technique was conducted on the test bearings to obtain their dynamic characteristics in terms of natural frequencies and damping, according to ISO 7626-5:2019 [41]. The purpose of this test was to evaluate the effect of added nano-additives with a lithium grease lubricant on the dynamic response of a test bearing. A customized setup was used for supporting each rolling bearing. Each test bearing was suspended from the support using elastic cords to reach approximate free boundary conditions. This was an essential step to obtain the flexible body modes of the bearing structure while ignoring the effect of the rigid body modes.

An impact hammer was used to induce a low to moderate force impulse in a very short time in the time domain, which excited a wide (theoretically infinite) range of frequencies, including the natural frequencies of the test bearing itself. The transient impulsive signals were measured using a piezoelectric force transducer attached to the hammer. Meanwhile, the vibrational response of the bearing to the impulsive shock was measured using a piezoelectric triaxial accelerometer mounted using wax on the outer ring of the bearing (in the opposite direction to the force signal). The accelerometer was carefully selected so that its weight was less than one-tenth the weight of the test bearing.

Both force and response signals were detected and recorded by a multi-channel signal analyzer unit. The signals were then conditioned and processed in PULSE labshop© software to calculate the frequency response function (FRF) of each bearing as the ratio between response to force signal. An arithmetic average of four impulse signals was determined for each bearing to ensure the high repeatability of results. The FRF and coherence function for each test bearing were determined. The first fundamental mode of vibration (first resonant mode) was selected for further analysis. Hence, the natural frequency (fn), damping ratio (ξ), FRF amplitude (H), and corresponding coherence value at the first resonant mode of vibration were recorded. Only resonant modes with high coherence values were considered during the selection of bearing modes. This was essential to ensure high autocorrelation and cross-correlation between response and force signals.

For proper signal conditioning, transient and exponential windowing functions were selected for excitation and response signals, respectively. The frequency range of interest for the tested bearing fell between 0–4000 Hz. An anti-aliasing filter and proper least-squares curve were used to avoid aliasing and truncations of signals. To ensure the consistency of results, the same locations were considered for the impact force, vibration response, and fixation points of the elastic cords during testing for all test bearings.

2.4. Vibration Analysis

Vibration measurements and analysis in the time domain and spectrum were conducted to investigate the influence of rGO nano-additives on the vibration performance of each bearing during operation. The vibration measurements were performed on each test bearing mounted to the test rig under a radial load of 2250 N. SPM Lenova Diamond vibration analyzer (SPM, Strängnäs, Sweden) was used for condition monitoring of the test REBs according to ISO 10816 part 7 [42].

Test bearing vibrations were measured using an accelerometer attached to the housing of the test bearing in the radial vertical and horizontal directions via magnet. The measured signals were directly fed into the data analyzer, which conducted signal conditioning and processing using anti-aliasing filtering, amplifiers, and integrators. Vibration measurements were taken in terms of acceleration (g) and velocity (mm/s). In the analysis part, the root mean square (RMS) was selected as a vibration amplitude indicator and was calculated for all signals to detect any abnormal behavior of rolling bearings due to localized faults or improper lubrication. RMS results were used for comparison between grease types as it was directly proportional to the energy consumption in bearings due to vibration.

2.5. Ultrasound Analysis

Ultrasonic testing is a commonly applied technique for condition monitoring and early detection of localized faults in rolling bearings. It is effectively used to measure the ultrasonic energy generated by friction forces and identify the lubrication status, whether it is normal or improper lubrication, such as over lubrication or lack of lubrication. Ultrasound technique operates in the spectrum range between 20 kHz to 100 kHz. In this work, an SDT 340 ultrasound device (STD Ultrasound Solutions, Brussels, Belgium) was employed to measure and analyze dynamic signals in time and frequency domains for each test bearing. The aim was to detect changes in the high-frequency ultrasonic signatures of the test bearings caused by changing the tested lubrication type according to ISO 29821:2018 [43].

The ultrasound detector consisted of a transducer mounted on the bearing support, which dynamically detected the signal, i.e., each measurement on each test bearing was recorded as a signal for 10 s. The data were then converted by a heterodyne circuit board into an audible signal (<20 kHz). This processed signal was furtherly conditioned by an audio amplifier to be heard using standard headphones. The heterodyned signals were analyzed in the time domain and frequency domain. The root mean square (RMS) was used for analyzing each measured signal in the time domain.

2.6. Infrared Thermographic Analysis

Infrared thermography (IRT) was conducted in this work using a FLIR E8 Thermal Camera (Teledyne FLIR LCC, Wilsonville, OR, USA) as a non-contact and non-invasive condition monitoring technique to measure the surface temperature and its distribution on test bearing according to the ISO18434-1:2008 [44]. It is commonly used in maintenance activities to record infrared radiation emitted by the bearing and indicate any abnormal friction due to improper lubrication during operation [45]. Considering the fact that the intensity of the emitted infrared radiation from the bearing is proportional to its temperature, it presents a relevant indicator for the impact of lubricant additives on the lubricity and generated rolling friction during operation and on the resultant temperature [46,47].

Proper calibration of the test setup of the IRT camera is essential to obtain accurate temperature recordings [48]. Since the test bearing is the main source of heat during operation, the passive IRT technique was adopted, i.e., no external heat source was used for each test. It should be noted that IRT measurements were conducted in a controlled environment. Therefore, the effect of humidity, wind, and solar radiations were ignored. Two main parameters were considered during IRT calibration: emissivity and reflected temperature measurement. The test bearing material had low emissivity; therefore, a high emissivity electrical tape was placed on the outer ring of the bearing to measure the apparent temperature accurately. The tape had high emissivity, which was set to be 0.95. The IR camera was located at a distance of 500 mm from the test bearing to fit within the region of interest (ROI).

The bearing surface temperature, i.e., the environmental temperature, was recorded at the beginning of each test. Then, the test bearing ran at 1400 rpm supporting a 2250-N radial load. The transient temperature was monitored every 2 min until reaching a steady state after 10 min. After being transformed into equivalent electric signals, the measured infrared radiations were revealed as visible images (thermograms) of different colored energy levels.

2.7. Timken Load Test

The load-carrying capacity of each test grease aggregate was determined using the Timken load test method, according to ASTM D 2509 [49]. The Timken OK load is described as the maximum load that can be applied before the lubricant film's breakdown and the roller element's seizure. It is used to characterize the extreme pressure (EP) of each test grease. The duration of the test is around 10 min. The test procedure starts by installing a bearing roller element with a 15-mm width and 11-mm diameter in contact with the inner race. The material of both the inner race and the roller element was made from GCr15 bearing steel (SAE 52100). A sample of 10 gm from the test grease was applied to the inner race and the roller element.

During Timken load tests, the temperature was maintained at 25 °C before the start of the test. The load was increased by a constant rate until the load broke the lubricant film, separating a rolling element and the inner raceway, leading to scoring and motor seizure. The number of used unit loads was recorded in each test (each unit load weighed 497 g). After each test, the Timken OK load was recorded and the corresponding roller element scar area was inspected and measured for each test grease using the Stereo microscope (Zeiss Company, Jena, Germany).

3. Results and Discussion

3.1. Nano-Additive Material Characterization

Figure 2a shows the XRD pattern of graphene, which reveals a broad (near sharp) characteristic peak at 2θ = 24.65 °C, referring to the basal plane (002) with about 54% crystallinity. The broad peak indicates the randomly arranged crystal phase due to the formation of a few layers of rGO after the reduction process [50–52]. Additionally, the average grain size was estimated from Scherrer's formula to be 1.66 nm, and the interlayer spacing

(d) (FWHM = 5.11) accounted for 0.37 nm according to Bragg's law, which confirmed the removal of oxygen functional groups during the fabrication process [53,54].

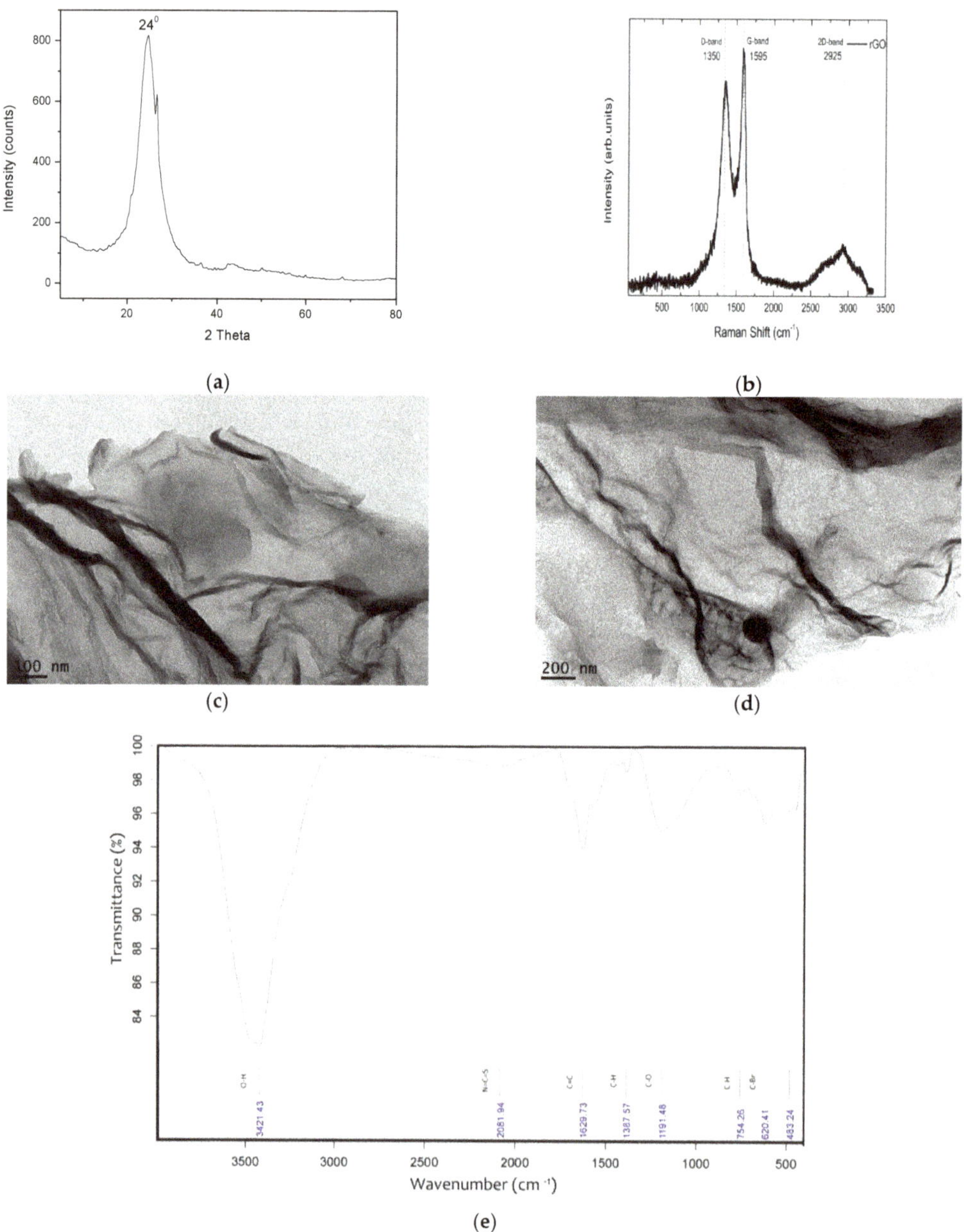

Figure 2. (**a**) XRD pattern of reduced graphene oxide, (**b**) Raman spectrum of rGO sample, (**c**) TEM analysis of rGO structure with a resolution of 100 nm, (**d**) TEM analysis of rGO structure with a resolution of 200 nm, and (**e**) FTIR analysis results of rGO sample.

From Raman spectroscopy results, three characteristic peaks attributed to rGO, are shown in Figure 2b. The first one, which is the D-band, appeared at 1350 cm^{-1} and this peak was mainly responsible for the structural disorders and defects [55], while the second (G-band) and third (2-D-band) peaks were observed at 1595 and 2925 cm^{-1}, respectively. The G-band characterized the graphene symmetry and order [56], while the stacking of carbon atoms was deduced from the 2-D-band [57]. The intensity ratio of 2-D/G was estimated to be 0.91, and this ratio was essential to evaluate the number of graphene layers, which accounted for four layers, confirming the TEM results [58]. Turning to the D-band, it strongly appeared in the spectrum, reflecting a significant distortion in the symmetry of the graphite structure [59,60]. The intensity ratio of D/G showed a great importance in the detection of the structural defects and disorders. By calculating the ratio, it was found that ID/IG was 0.98, which indicated a remarkable density of defects represented in edge defects obtained from the reduction of the C=C bond [61].

In Figure 2c,d, TEM images reveal a nanosheet-layered structure of graphene with four stacked layers and dimensions of up to 3 μm in length and 5 nm in thickness, respectively. Additionally, the graphene structure had fewer wrinkles and folds at different TEM resolutions. Figure 2e reveals the spectrum obtained from FTIR analysis of rGO. It was found that there was a strong and broad stretching band at 3421.43 cm^{-1}, which was attributed to the intermolecular bonded O-H function group. The weak presence of the C-O function group at 1191.48 cm^{-1} indicated the insufficiency of oxidizing substances to break covalent bonds at the basal plane, reducing structural defects and distortions. Another medium stretching band of C=C appeared at 1629.73 cm^{-1}, indicating the stability and integrity of the graphene structure. Moreover, weak peaks were observed at 1387.57 and 754.26 cm^{-1}, which belonged to C-H bending, revealing the small amounts of hydrogen species, which were functionalized with carbon atoms [61].

Both graphite and MWCNTs were used in this study and were characterized using SEM for the former and TEM for the latter. The SEM micrograph of the graphite structure revealed a sheet-like structure with a flake size of about 27 μm, as shown in Figure 3a, while for MWCNTs, TEM images confirmed the nanotube structure with an average external diameter of about 17.5 nm, as revealed in Figure 3b.

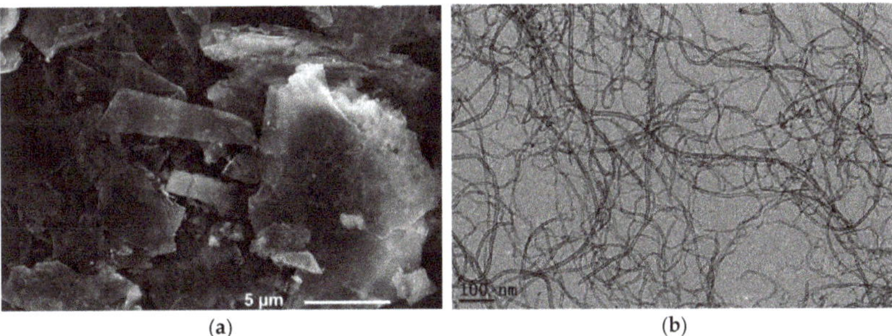

(a)   (b)

Figure 3. (**a**) Microstructural analysis of graphite using SEM with 5-μm resolution and (**b**) microstructural analysis of MWCNTs using SEM with 100-nm resolution.

3.2. Modal Analysis Results

Figure 4 shows an example of the FRF plot, including the first mode of vibration for bearing from impact excitation of bearing and the corresponding coherence function to the first dominant mode. The damping ratio values for the test rolling bearings were calculated from their individual FRF plots and are illustrated as shown in Figure 5.

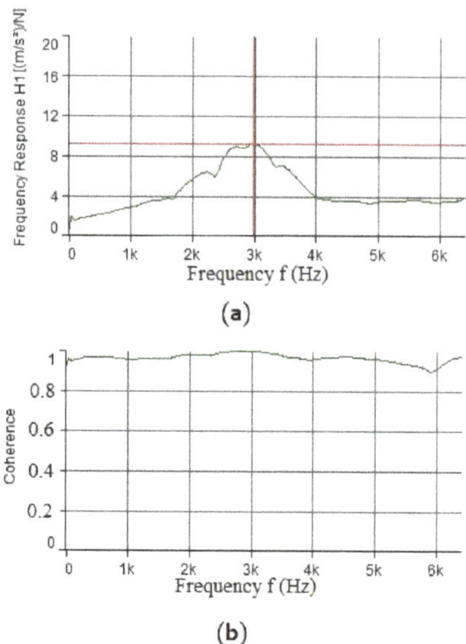

Figure 4. Modal test results of bearing with base grease lubrication in the form of: (**a**) frequency response function (FRF) and (**b**) coherence function.

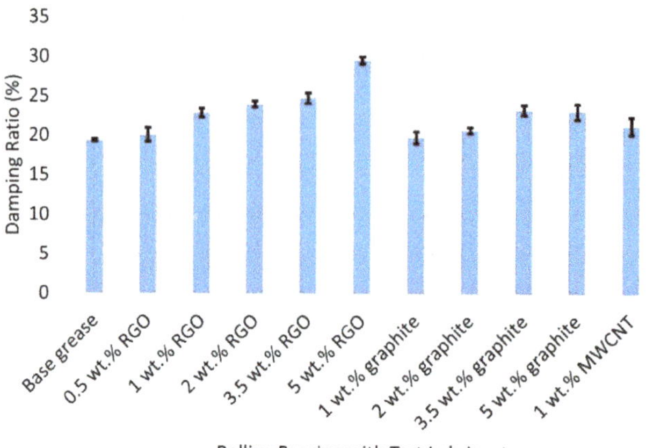

Figure 5. Damping ratio values for rolling bearings with each test lubricant.

Test rolling bearings of the same type and size, same material, same manufacturer, and same interference fit with the shaft showed various dynamic characteristics. This was potentially attributed to variations in the lubricant characteristics of the tested rolling bearings. The damping ratio values for bearings lubricated by the addition of 1, 2, 3.5, and 5 wt.% rGO were increased by 18, 24, 28, and 52% when compared to bearings with base grease lubrication. While the damping ratio values of bearings with 1 wt.% rGO grease fell in the same range as those for 1 wt.% graphite and 1 wt.% MWCNTs, increasing the contribution of rGO to the grease to 2 wt.% and up to 5 wt.% improved the damping ratios by 15% and 28% when compared to 2 wt.% and 5 wt.% graphite.

For a typical ball bearing, the damping capacity originates from three mechanisms [62]. The first damping mechanism is the material damping due to internal friction within the material structure during the material dynamic deformation, while the second damping contributor is the dry friction at the contact points between rolling elements and raceways. The third and the most effective mechanism of damping is the viscous damping originated from the oil full film lubricant at the elastohydrodynamic lubrication zone (EHL). Other possible sources for damping are the contact surfaces between the outer ring to the housing and the inner ring to the shaft. Since bearings are assembled with the shaft in interference, the contact areas between the bearing rings and the housing possess high stiffness and barely contribute to the damping mechanisms of REBs. This is because damping can only be effective in the case of relative motion between REB elements [62,63]. As REBs suffer from low damping characteristics compared to other bearing types, enhancement of oil film damping characteristics contributes to a corresponding improvement in reliability of bearings during service [63].

On the nanoscale level, interfaces between stacked layers of graphene are significantly weaker than grease–graphene interfaces [64]. The addition of graphene to grease increases the viscosity of grease, which in turn increases its shear strength, while weak Van der Waal bonding between graphene layers reduces their interface shear strength [65]. Thus, lower shear forces are sufficient to cause slippage of graphene layers with respect to each other during operation. This latter action would contribute to damping enhancement. The larger surface area of 2-D graphene nanosheets makes it more compliant than 1-D CNTs, especially when subjected to small shear forces resulting in easier activation of damping mechanisms.

3.3. Vibration Analysis Results

For each test bearing, vibrations were measured and analyzed in radial vertical and horizontal directions [66]. Time waveform and frequency spectrum for test bearings lubricated with base grease are shown in Figure 6. Among all results, test bearings lubricated with base grease showed the maximum vibration velocity levels, reaching 3.31 mm/s and 3.93 mm/s in the vertical and horizontal directions, respectively. These levels comply with the maximum permissible levels in ISO 10816-7:2009 [67].

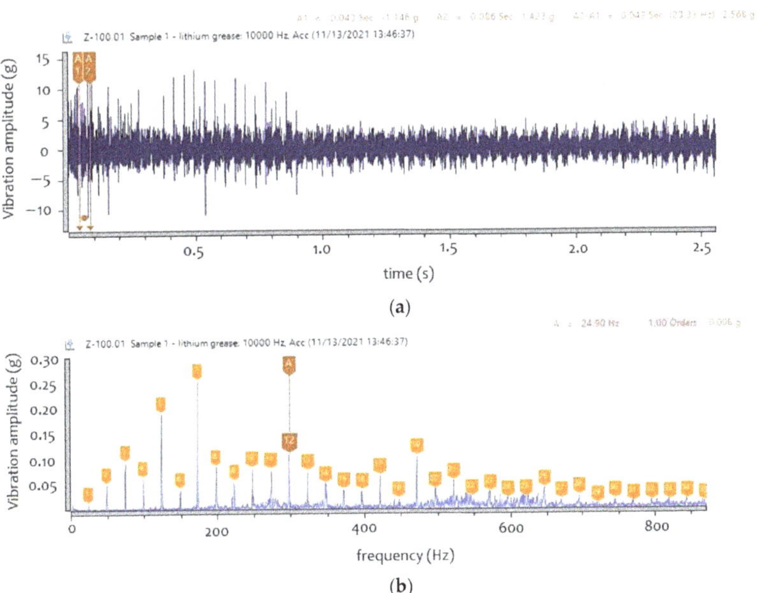

Figure 6. (**a**) Time waveform and (**b**) frequency spectrum of rolling bearings lubricated with base grease.

Vibration acceleration values (RMS) were calculated from the frequency domain for all test bearings, as shown in Figure 7. It can be seen that test bearings with nano-additives' lubricant demonstrated significantly lower vibration values than a base lithium grease condition. Furthermore, increasing the weight percentage of rGO nano-additive in grease resulted in a corresponding reduction in the calculated vibration RMS levels until reaching its lowest at 0.71 g and 0.88 g in the horizontal and vertical directions, respectively, which was lower by around 70% and 34% when compared to base grease results. Increasing the aggregate rGO in grease to 3.5 and 5 wt.% caused an increase in vibration levels by 40% and 80%, respectively. This is attributed to the agglomeration effect of graphene in the grease, which resulted in more resistance to the flow of lubricant between the rolling element and raceway, causing more friction and vibrations. A similar trend took place in the case of graphite nano-additives. Grease with 1 wt.% MWCNTs' nano-additives showed similar vibration levels to the 1 wt.% rGO case and is lower by 38% than the 1 wt.% graphite results.

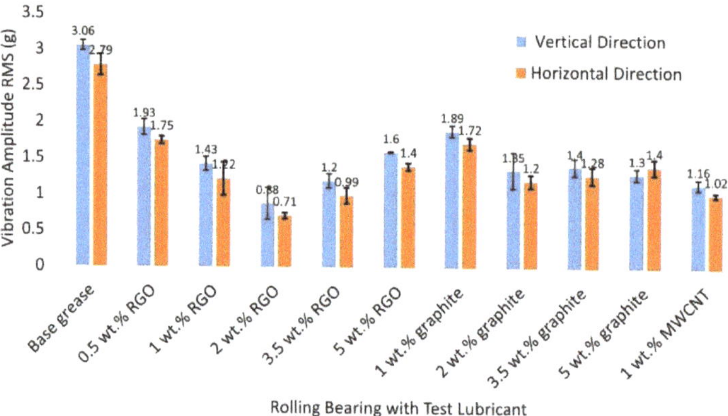

Figure 7. Vibration levels (RMS) measured in radial vertical and horizontal directions on each test bearing.

The vibration levels in the case of test bearings with rGO, MWCNTs, and graphite nano-additive lubricants proved that these additives contributed to decreasing the transmitted vibrations inside the bearing structure [68]. Adding these nano-additives to grease produced a stable tribo-film between the rolling elements, inner ring, and outer ring, which acted as spring and damper connectors. The carbonaceous nano-additives possessed high damping and elasticity properties that significantly reduced vibration signal transmissibility between bearing elements. Furthermore, the 2-D structure of rGO had a large area that covered the asperities between the mating surfaces and inhibited metal-to-metal contact in the elastohydrodynamic zone [64]. Hence, it reduced the vibration levels due to rolling friction in the ball bearings during operation.

Another aspect is that adding graphene stabilizes the fibrous structure of grease thickener to a large extent [69]. However, it weakens the cross-linking in the thickener fibrous network, which facilitates the discharge of base oil from voids to formulate full film lubrication [70]. Adding rGO also contributed to widening the voids in the fibrous structure of the lithium grease thickener. This, in turn, increased the amount of discharged base oil to the clearance gap between the rolling elements and raceways. Hence, the supplied nano-additive lubricant between the bearing elements eliminated metal-to-metal contact and acted as a good spring-damper system that reduced transmitted vibration in the bearing structure. Adding more rGO wt.% beyond 2 wt.% did not cause any further reduction in vibrations levels as the grease structure became more saturated with the graphene nano-additives.

3.4. Ultrasound Results

Figure 8 shows the ultrasound signal measured in the time domain and spectrum for test bearings with base grease lubrication. It can be seen that the test bearing was in a normal condition, showing ultrasound RMS value of 64.18 dBμV. The main source of this value was friction forces during operation. In comparison, test bearings with rGO, graphite, and MWCNTs' nano-additive cases showed a lower ultrasound RMS value by around 15%–20%, as shown in Figure 9.

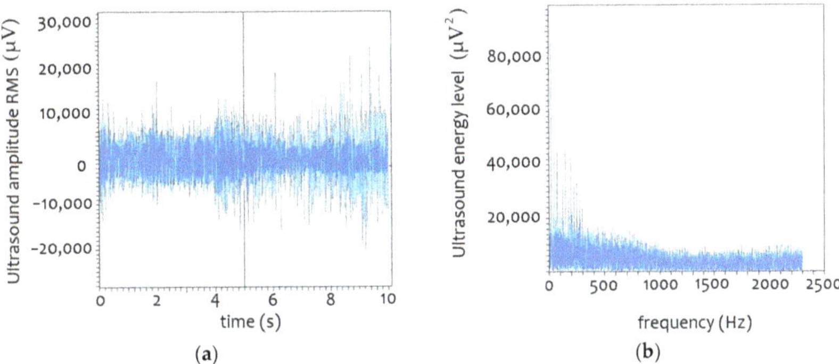

Figure 8. Ultrasound signal for test bearing with base grease lubrication in (**a**) time domain and (**b**) frequency domain.

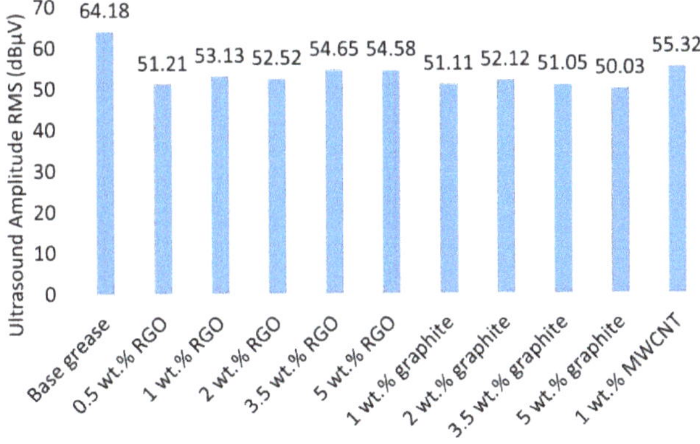

Figure 9. Ultrasound amplitude (RMS) values for test bearings with test lubricants.

The results indicated the efficiency of rGO nano-additives in reducing frictional forces generated during operation while providing quieter operation. The obtained ultrasound values confirmed the vibration analysis results in the fact that nano-additive lubricants showed significant reduction in friction forces of rolling bearings during operation in comparison with base grease lubricant.

3.5. IRT Thermographic Results

In this section, thermograms were obtained from the IRT measurement for each test bearing. Figure 10 shows the IRT measurement and corresponding thermogram for the test bearing with base grease lubricant. The thermograms were post-processed, and the average temperature at the test bearing hotspot was detected and considered for analysis, as shown in Figure 11. A noticeable variation of average temperature was detected between bearings at steady-state operating conditions, i.e., after running the test rig for 10 min.

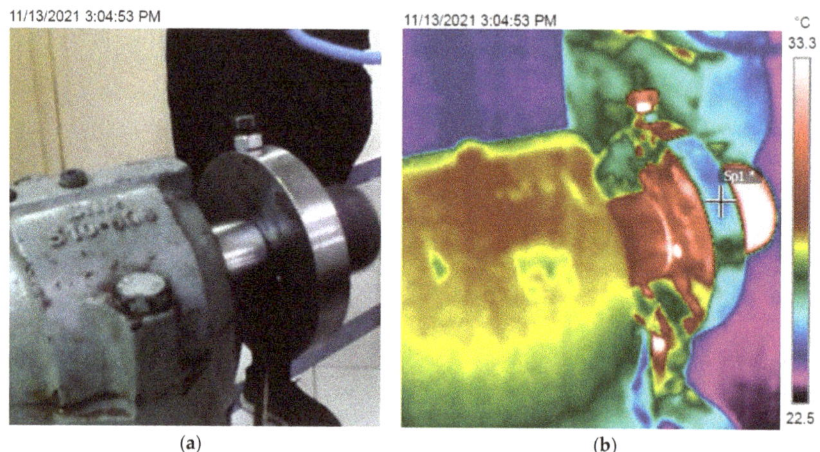

Figure 10. (**a**) The test bearing mounted on the test rig and (**b**) Infrared thermal image (thermogram) of the test bearing.

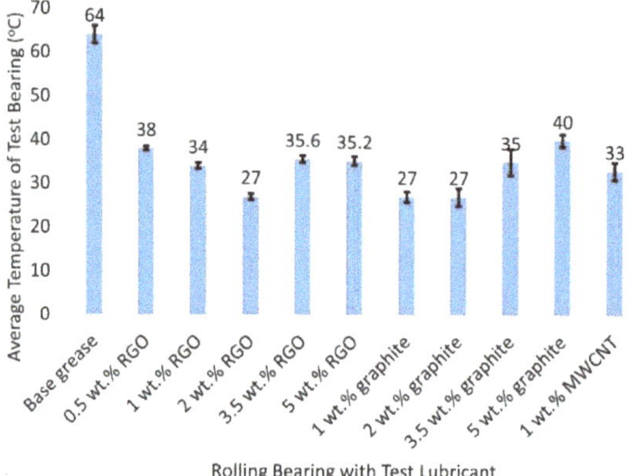

Figure 11. Average temperatures of bearings under selected test lubricants after running for 10 min.

The thermal images and the bar chart results indicated that the test bearing with base grease lubrication exhibited the highest temperature during operation, peaking at 64 °C. This was confirmed by thermogram (A), which demonstrated the distribution of heat and maximum temperature at the tested bearing. All test bearings lubricated with nano-additives showed temperature below 40 °C. Adding rGO additives gradually reduced the temperature until reaching the lowest temperature, which accounted for 27 °C in the case of the 2 wt.% rGO addition. This means that the heat energy generated on the test bearing decreased by 57% due to adding 2 wt.% rGO compared to the base grease case. Additionally, this value was repeated when adding 1 wt.% graphite and 2 wt.% graphite to the grease. However, the MWCNTs' addition reached only 33 °C, which surprisingly denoted the superiority of small graphite additions over MWCNTs in reducing the heat inside rolling bearings.

It was also noticed that the effect of the agglomeration in the case of 3.5 and 5 wt.% rGO and graphite nano-additives gave rise to more friction between the rolling elements and lubricant nano-additives themselves; hence, this led to more energy losses in the form of heat. IRT results were consistent with the vibrations' measurements obtained for the same test bearings.

3.6. Timken Load Test Results

Table 3 shows the Timken OK load values of each test grease. Commercial Mobil fibra x235 grease was tested in a Timken tester, and its results were used for comparison. The results revealed that lithium grease and Mobil fibra x235 grease showed the lowest load-carrying capacities among the test samples. However, samples with 1 wt.% rGO, 1 wt.% MWCNT, and 1 wt.% graphite showed an insignificant difference in Timken OK load values from base grease samples. Increasing rGO nano-additive percentages in grease to 2 wt.%, 3.5 wt.%, and 5 wt.% resulted in a remarkable increase of Timken OK load by 25%, 50%, and 100% compared to samples of base grease (without additives).

Table 3. Timken OK load and corresponding wear scar dimensions and area for each roller element.

Roller Number	Test Grease	Timken OK Load (N)	X (mm)	Y (mm)	Wear Scar Area (mm)
a	Base grease (Lithium Grease)	1988	5.3	9.3	154.771
b	1 wt.% rGO	1988	4.7	8.6	126.918
c	2 wt.% rGO	2485	4.7	8	118.064
d	3.5 wt.% rGO	2982	4	7.7	96.712
e	5 wt.% rGO	3976	4.5	7	98.911
f	1 wt.% graphite	1988	5	8.3	130.312
g	1 wt.% MWCNT	1988	4.7	8.5	125.443
h	Mobil fibra x235	1988	5.3	9.3	154.771

The rolling action between the loaded roller element and the ring component of the Timken test was accompanied by wear, which resulted in uniform elliptical-shaped surface scars. These wear scars on the test rollers were examined using brinelling microscopy and found to be smooth. The wear scar area for the roller elements under each lubricant was calculated using measurements of the minor (x) and major (y) axes [71]. The wear scar dimensions and area are summarized in Table 3.

It can be seen that the wear scar area was the largest in the case of lithium grease and Mobil fibra x235 grease, and it decreased gradually with the increase of wt.% rGO nano-additives in grease samples. Lithium grease with 3.5 wt.% and 5 wt.% rGO nano-additives produced around 50% smaller scar areas than those produced in the case of base grease. This revealed a positive influence of rGO nano-additives to improve anti-wear (AW) and extreme pressure (EP) properties.

4. Conclusions

The influence of adding reduced graphene oxide to commercial lithium grease on the dynamic and tribological behavior of a deep-groove ball bearing was investigated. Timken load test results revealed a considerable increase in test grease load-carrying capacities containing rGO nano-additives. Grease samples having 2, 3.5, and 5 weight percent rGO showed the highest load-carrying capacity (among other tested samples) with an increase of 25%, 50%, and 100% compared to base grease values. The wear scar dimensions measured on the rolling element after the rupture of the lubricant film between the rolling element and raceway indicated a remarkably high resistance of the formed graphene layer to wear. The results also proved the superiority of graphene as a lubricant nano-additive in enhancing the tribological properties in terms of load-carrying capacity and wear scar size.

After 10 min running under a radial test load of 2250 N and 1400 rpm, the steady-state temperature of the bearing lubricated by 2 wt.% rGO grease leveled up at 27 °C, whereas, for that lubricated by base grease, the temperature approached 65 °C. These findings implied a reduction in friction forces between contacting surfaces of bearing elements during operation, which also complied with the vibrations and ultrasound measurements. The vibration and ultrasound levels for 2 wt.% rGO grease were found to be 50% and 20% lower than those for bearings lubricated with base grease. Modal analysis results indicated considerable improvement of the damping ratio values of the bearings lubricated with rGO additives up to 50% as compared to base grease. Increasing the aggregate of rGO in grease to 2 wt.% and 5 wt.% resulted in higher damping ratios by 15% and 28% when compared to 2 wt.% and 5 wt.% graphite. The outcomes of this work represent a significant contribution towards improving the reliability and quiet running of rolling element bearings during their service time.

Author Contributions: Conceptualization, G.A.N. and M.G.A.N.; Methodology, M.S., B.G.N. and M.A.D.; Samples' Preparation, B.G.N. and M.G.A.N.; Validation, M.S., B.G.N. and M.G.A.N.; Formal Analysis, G.A.N., B.G.N. and M.A.D.; Investigation, M.S. and B.G.N.; Resources, M.S. and M.G.A.N.; Data Curation, M.G.A.N. and G.A.N.; Writing—Original Draft Preparation, B.G.N. and M.G.A.N.; Writing—Review and Editing, G.A.N. and M.A.D.; Visualization, M.G.A.N.; Project Administration, G.A.N. All authors have read and agreed to the published version of the manuscript.

Funding: This research received no external funding.

Data Availability Statement: Not applicable.

Conflicts of Interest: The authors declare no conflict of interest.

References

1. S. Group. *Bearing Damage and Failure Analysis*; Technical Report; SKF Group: Gothenburg, Sweden, 2017.
2. Market Analysis Report. Bearings Market Size, Share & Trends Analysis. Available online: https://www.grandviewresearch.com/industry-analysis/bearings-market (accessed on 8 November 2021).
3. Randall, R.B.; Antoni, J. Rolling element bearing diagnostics—A tutorial. *Mech. Syst. Signal Process.* **2011**, *25*, 485–520. [CrossRef]
4. Vencl, A.; Gašić1, V.; Stojanović, B. Fault tree analysis of most common rolling bearing tribological failures. *IOP Conf. Ser. Mater. Sci. Eng.* **2017**, *174*, 012048. [CrossRef]
5. Harris, T.A. *Rolling Bearing Analysis*, 4th ed.; Wiley: New York, NY, USA, 2001.
6. Howard, I.A. *Review of Rolling Element Bearing Vibration Detection, Diagnosis and Prognosis*; Technical Report; Defense Science and Technology Organization Canberra: Canberra, Australia, 1994.
7. Cann, P. Starved Grease Lubrication of Rolling Contacts. *Tribol. Trans.* **1999**, *42*, 867–873. [CrossRef]
8. Wandel, S.; Bader, N.; Schwack, F.; Glodowski, J.; Lehnhardt, B.; Poll, G. Starvation and relubrication mechanisms in grease lubricated oscillating bearings. *Tribol. Int.* **2021**, *165*, 107276. [CrossRef]
9. Grützmacher, P.; Jalikop, S.; Gachot, C.; Rosenkranz, A. Thermocapillary lubricant migration on textured surfaces—A review of theoretical and experimental insights. *Surf. Topogr. Metrol. Prop.* **2021**, *9*, 013001. [CrossRef]
10. Grützmacher, P.; Rosenkranz, A.; Szurdak, A.; Grüber, M.; Gachot, C.; Hirt, G.; Mücklich, F. Multi-scale surface patterning–an approach to control friction and lubricant migration in lubricated systems. *Ind. Lubr. Tribol.* **2019**, *71*, 1007–1016. [CrossRef]
11. Grützmacher, P.; Rosenkranz, A.; Szurdak, A.; Gachot, C.; Hirt, G.; Mücklich, F. Lubricant migration on stainless steel induced by bio-inspired multi-scale surface patterns. *Mater. Des.* **2018**, *150*, 55–63. [CrossRef]
12. S. Group. *FAG Lubrication of Rolling Bearings*; Technical Report; SKF: Bresler, Germany, 2013; p. 8.
13. Dresel, W.; Heckler, R.P. Lubricating Greases. In *Lubricants and Lubrication*, 2nd ed.; Mang, H., Dresel, W., Eds.; John Wiley & Sons: Hoboken, NJ, USA, 2006; pp. 648–693.
14. Lansdown, A.R. *Lubrication and Lubricant Selection: A Practical Guide*, 3rd ed.; ASME Press: New York, NY, USA, 2004.
15. Gow, G. *Lubricating Grease in Chemistry and Technology of Lubricants*, 3rd ed.; Mortier, R.M., FoxStefan, M.F., Orszulik, T., Eds.; Springer: Dordrecht, The Netherlands, 2010; pp. 411–432.
16. Lugt, P.M.; Pallister, D.M. Grease Composition and Properties. In *Grease Lubrication in Rolling Bearings*, 1st ed.; Lugt, P.M., Ed.; John Wiley & Sons, Ltd.: Hoboken, NJ, USA, 2012; pp. 23–69.
17. Rosenkranz, A.; Liu, Y.; Yang, L.; Chen, L. 2D nano-materials beyond graphene: From synthesis to tribological studies. *Appl. Nanosci.* **2020**, *10*, 3353–3388. [CrossRef]
18. Berman, D.; Erdemir, A.; Sumant, A. Approaches for Achieving Superlubricity in Two-Dimensional Materials. *ACS Nano* **2018**, *12*, 2122–2137. [CrossRef]
19. Zhang, S.; Ma, T.; Erdemir, A.; Li, Q. Tribology of two-dimensional materials: From mechanisms to modulating strategies. *Mater. Today* **2018**, *26*, 67–86. [CrossRef]
20. Zhang, J.; Li, J.; Wang, A.; Edwards, B.J.; Yin, H.; Li, Z.; Ding, Y. Improvement of the Tribological Properties of a Lithium-Based Grease by Addition of Graphene. *J. Nanosci. Nanotechnol.* **2018**, *18*, 7163–7169. [CrossRef] [PubMed]
21. Mohamed, A.; Osman, T.A.; Khattab, A.; Zaki, M. Tribological Behavior of Carbon Nanotubes as an Additive on Lithium Grease. *J. Tribol.* **2014**, *137*, 011801. [CrossRef]

22. Rawat, S.S.; Harsha, A.P.; Khatri, O.P.; Wäsche, R. Pristine, Reduced, and Alkylated Graphene Oxide as Additives to Paraffin Grease for Enhancement of Tribological Properties. *J. Tribol.* **2020**, *143*, 11. [CrossRef]

23. Akbulut, M.; Belman, N.; Golan, Y.; Israelachvili, J. Frictional Properties of Confined Nanorods. *Adv. Mater.* **2006**, *18*, 2589–2592. [CrossRef]

24. Rapoport, L.; Leshchinsky, V.; Lapsker, I.; Volovik, Y.; Nepomnyashchy, O.; Lvovsky, M.; Popovitz-Biro, R.; Feldman, Y.; Tenne, R. Tribological properties of WS2 nanoparticles under mixed lubrication. *Wear* **2003**, *255*, 785–793. [CrossRef]

25. Cameron, A. *The Principles of Lubrication*, 1st ed.; Longmans: London, UK, 1966.

26. Marian, M.; Berman, D.; Rota, A.; Jackson, R.; Rosenkranz, A. Layered 2D Nanomaterials to Tailor Friction and Wear in Machine Elements—A Review. *Adv. Mater.* **2022**, *9*, 2101622. [CrossRef]

27. Kamel, B.M.; Mohamed, A.; El Sherbiny, M.; Abed, K.A.; Abd-Rabou, M. Tribological properties of graphene nanosheets as an additive in calcium grease. *J. Dispers. Sci. Technol.* **2017**, *38*, 1495–1500. [CrossRef]

28. Tabandeh-Khorshid, M.; Omrani, E.; Menezes, P.L.; Rohatgi, P.K. Tribological performance of self-lubricating aluminum matrix nanocomposites: Role of graphene nanoplatelets. *Int. J. Eng. Sci. Technol.* **2016**, *19*, 463–469. [CrossRef]

29. Nassef, B.G.; Nassef, G.A.; Daha, M.A. Graphene and Its Industrial Applications—A Review. *Int. J. Mater. Eng.* **2020**, *10*, 1–12. [CrossRef]

30. Singh, J.; Anand, G.; Kumar, D.; Tandon, N. Graphene based composite grease for elastohydrodynamic lubricated point contact. *IOP Conf. Ser. Mater. Sci. Eng.* **2016**, *149*, 012195. [CrossRef]

31. Curà, F.; Mura, A.; Adamo, F. Experimental investigation about tribological performance of grapheme-nanoplatelets as additive for lubricants. *Procedia Struct. Integr.* **2018**, *12*, 44–51. [CrossRef]

32. Fu, H.; Yan, G.; Li, M.; Wang, H.; Chen, Y.; Yan, C.; Lin, C.-T.; Jiang, N.; Yu, J. Graphene as a nanofiller for enhancing the tribological properties and thermal conductivity of base grease. *RSC Adv.* **2019**, *9*, 42481–42488. [CrossRef]

33. Cheng, Z.L.; Qin, X.-X. Study on friction performance of graphene-based semi-solid grease. *Chin. Chem. Lett.* **2014**, *25*, 1305–1307. [CrossRef]

34. Nan, F.; Yin, Y. Improving of the tribological properties of attapulgite base grease with graphene. *Lubr. Sci.* **2021**, *33*, 380–393. [CrossRef]

35. Schwarz, U.D. Tracking antiwear film formation. *Science* **2015**, *348*, 40–41. [CrossRef]

36. Pape, F.; Poll, G. Investigations on Graphene platelets as dry lubricant and as grease additive for rolling and sliding contacts. *Lubricants* **2019**, *8*, 3. [CrossRef]

37. Nassef, B.G. Performance of Rolling Element Bearings Lubricated with Grease-Graphene Mixtures. Master's Thesis, Alexandria University, Alexandria, Egypt, 2021.

38. Brisebois, P.P.; Siaj, M. Harvesting graphene oxide—years 1859 to 2019: A review of its structure, synthesis, properties and exfoliation. *J. Mater. Chem. C* **2020**, *8*, 1517–1547. [CrossRef]

39. Shahriary, L.; Athawale, A. Graphene oxide synthesized by using modified Hummers approach. *Renew. Energy Environ. Eng.* **2014**, *2*, 58–63.

40. Kaur, M.; Kaur, H.; Kukkar, D. Synthesis and characterization of graphene oxide using modified Hummer's method. *AIP Conf. Proc.* **2018**, *1953*, 030180. [CrossRef]

41. ISO 7626-5:2019; Mechanical Vibration and Shock-Experimental Determination of Mechanical Mobility-5: Measurements Using Impact Excitation with an Exciter Which Is Not Attached to the Structure. International Organization for Standardization ISO: Geneva, Switzerland, 2019. Available online: https://www.iso.org/standard/68735.html (accessed on 17 February 2022).

42. ISO 10816-8:2014; Mechanical Vibration-Evaluation of Machine Vibration by Measurements on Non-Rotating Parts-Part 1: General Guidelines. International Organization for Standardization ISO: Geneva, Switzerland, 2014. Available online: https://www.iso.org/standard/56782.html (accessed on 17 February 2022).

43. ISO 29821:2018; Condition Monitoring and Diagnostics of Machines-Ultrasound-General Guidelines, Procedures and Validation. International Organization for Standardization ISO: Geneva, Switzerland, 2018. Available online: https://www.iso.org/standard/71196.html (accessed on 17 February 2022).

44. ISO 18434-1:2008; Condition Monitoring and Diagnostics of Machines-Thermography-Part 1: General Procedures. International Organization for Standardization ISO: Geneva, Switzerland, 2008. Available online: https://www.iso.org/standard/41648.html (accessed on 17 February 2022).

45. Luong, M.P. Fatigue limit evaluation of metals using an infrared thermographic technique. *Mech. Mater.* **1998**, *28*, 155–163. [CrossRef]

46. Maldague, X. *Theory and Practice of Infrared Technology for Nondestructive Testing*, 1st ed.; Wiley: New York, NY, USA, 2001.

47. La Rosa, G.; Risitano, A. Thermographic methodology for rapid determination of the fatigue limit of materials and mechanical components. *Int. J. Fatigue* **2000**, *22*, 65–73. [CrossRef]

48. ISO 18434-2:2019; Condition Monitoring and Diagnostics of Machine systems-Thermography-Part 2: Image Interpretation and Diagnostics. International Organization for Standardization ISO: Geneva, Switzerland, 2019. Available online: https://www.iso.org/standard/67617.html (accessed on 17 February 2022).

49. ASTM D2509-20ae1; Standard Test Method for Measurement of Load-Carrying Capacity of Lubricating Grease (Timken Method). American Society for Testing and Materials (ASTM International): West Conshohocken, PA, USA. Available online: https://www.astm.org/d2509-20ae01.html (accessed on 17 February 2022).

50. Krishnamoorthy, K.; Veerapandian, M.; Mohan, R.; Kim, S.-J. Investigation of Raman and photoluminescence studies of reduced graphene oxide sheets. *Appl. Phys. A* **2012**, *106*, 501–506. [CrossRef]
51. Thakur, S.; Karak, N. Green reduction of graphene oxide by aqueous phytoextracts. *Carbon* **2012**, *50*, 5331–5339. [CrossRef]
52. Tai, M.; Liu, W.; Khe3, C.; Hidayah, N.; Teoh, Y.; Voon, C.; Lee, H.; Adelyn, P. Green synthesis of reduced graphene oxide using green tea extract. *AIP Conf. Proc.* **2018**, *2045*, 020032. [CrossRef]
53. Song, P.; Cao, Z.; Cai, Y.; Zhao, L.; Fang, Z.; Fu, S. Fabrication of exfoliated graphene-based polypropylene nanocomposites with enhanced mechanical and thermal properties. *Polymer* **2011**, *52*, 4001–4010. [CrossRef]
54. Hidayah, N.; Liu, W.; Lai, C.; Noriman, N.; Khe, C.; Hashim, U.; Lee, H. Comparison on graphite, graphene oxide and reduced graphene oxide: Synthesis and characterization. *AIP Conf. Proc.* **2017**, *1892*, 150002. [CrossRef]
55. Stankovich, S.; Dikin, D.; Piner, R.; Kohlhaas, K.; Kleinhammes, A.; Jia, Y.; Wu, Y.; Nguyen, S.; Ruoff, R. Synthesis of graphene-based nanosheets via chemical reduction of exfoliated graphite oxide. *Carbon* **2007**, *45*, 1558–1565. [CrossRef]
56. Tuinstra, F.; Koenig, J. Raman spectrum of graphite. *J. Chem. Phys.* **1970**, *53*, 1126–1130. [CrossRef]
57. Ferrari, A.; Meyer, J.; Scardaci, V.; Casiraghi, C.; Lazzeri, M.; Mauri, F.; Piscanec, S.; Jiang, D.; Novoselov, K.; Roth, S.; et al. Raman Spectrum of Graphene and Graphene Layers. *Phys. Rev. Lett.* **2006**, *97*, 187401. [CrossRef]
58. Kumar, V.; Kumar, A.; Lee, D.; Park, S. Estimation of Number of Graphene Layers Using Different Methods: A Focused Review. *Materials* **2021**, *14*, 4590. [CrossRef]
59. Moon, I.K.; Lee, J.; Ruoff, R.; Lee, H. Reduced graphene oxide by chemical graphitization. *Nat. Commun.* **2010**, *1*, 73. [CrossRef] [PubMed]
60. Yasin, G.; Arif, M.; Shakeel, M.; Dun, Y.; Zuo, Y.; Khan, W.; Tang, Y.; Khan, A.; Nadeem, M. Exploring the Nickel–Graphene Nanocomposite Coatings for Superior Corrosion Resistance: Manipulating the Effect of Deposition Current Density on its Morphology, Mechanical Properties, and Erosion-Corrosion Performance. *Adv. Eng. Mater.* **2018**, *20*, 1701166. [CrossRef]
61. Gupta, B.; Kumar, N.; Panda, K.; Kanan, V.; Joshi, S.; Visoly-Fisher, I. Role of oxygen functional groups in reduced graphene oxide for lubrication. *Sci. Rep.* **2017**, *7*, 45030. [CrossRef]
62. Zeillinger, R.; Kottritsch, H. Damping in a rolling arrangement, Technical Report ed. *SKF Business and Technology Magazine (Evolution)*. 15 February 1996, pp. 27–30. Available online: https://evolution.skf.com/damping-in-a-rolling-bearing-arrangement/ (accessed on 14 January 2022).
63. Ali, N.J.; García, J.M. Experimental studies on the dynamic characteristics of rolling element bearings. *Proc. Inst. Mech. Eng. Part J. Eng. Tribol.* **2010**, *224*, 659–666. [CrossRef]
64. Singh, R.; Dixit, A.; Sharma, A.; Tiwari, A.; Mandal, V.; Pramanik, A. Influence of graphene and multi-walled carbon nanotube additives on tribological behaviour of lubricants. *Int. J. Surf. Sci. Eng.* **2018**, *12*, 207–227. [CrossRef]
65. Kamel, B.M.; Mohamed, A.; El-Sherbiny, M.; Abed, K.; Abd-Rabou, M. Rheological characteristics of modified calcium grease with graphene nanosheets Fuller Nanotube Carbon Nanostructures. *Fuller. Nanotub. Carbon Nanostructures* **2017**, *25*, 429–434. [CrossRef]
66. *ISO 13373-2:2016*; Condition Monitoring and Diagnostics of Machines—Vibration Condition Monitoring—Part 2: Processing, analysis and Presentation of Vibration Data. International Organization for Standardization ISO: Geneva, Switzerland, 2016. Available online: https://www.iso.org/standard/68128.html (accessed on 17 February 2022).
67. *ISO 10816-7:2009*; Mechanical Vibration—Evaluation of Machine Vibration by Measurements on Non-Rotating Parts—Part 7: Rotodynamic Pumps for Industrial Applications, Including Measurements on Rotating Shafts. International Organization for Standardization ISO: Geneva, Switzerland, 2009. Available online: https://www.iso.org/standard/41726.html (accessed on 17 February 2022).
68. Wu, C.; Yang, K.; Chen, Y.; Ni, J.; Yao, L.; Li, X. Investigation of friction and vibration performance of lithium complex grease containing nano-particles on rolling bearing. *Tribol. Int.* **2021**, *155*, 106761. [CrossRef]
69. Rawat, S.S.; Harsha, A.P.; Chouhan, A.; Khatri, O.P. Effect of Graphene-Based Nanoadditives on the Tribological and Rheological Performance of Paraffin Grease. *J. Mater. Eng. Perform.* **2020**, *29*, 2235–2247. [CrossRef]
70. Adhvaryu, A.; Sung, C.; Erhan, S.Z. Fatty acids and antioxidant effects on grease microstructures. *Ind. Crops Prod.* **2005**, *21*, 285–291. [CrossRef]
71. Tippayawong, N.; Sooksarn, P. Assessment of lubricating oil degradation in small motorcycle engine fueled with gasohol. *Maejo Int. J. Sci. Technol.* **2010**, *4*, 201–209.

 *lubricants*

Article

Stability Effects of Non-Circular Geometry in Floating Ring Bearings

Giovanni Adiletta

Dipartimento di Ingegneria Industriale, Università Degli Studi di Napoli "Federico II", 80125 Napoli, Italy; adiletta@unina.it

Received: 13 September 2020; Accepted: 12 November 2020; Published: 18 November 2020

Abstract: The present study theoretically evaluates the stability potential of noncircular geometries when they are adopted in the outer bearing of floating ring bearings (FRB). A numerical study is carried out to evaluate the stability about the static equilibrium position of a balanced, symmetrical, rigid rotor, horizontally placed, and supported at both ends by identical FRBs. In the analysis, the outer bearing of these FRBs is alternatively shaped with common circular bore (CB), two lobe-wave bore (2LWB) or lemon bore (LB), assuming a linearization of the film forces. A minor part of the study consists of partially supporting the results of the above study by means of a nonlinear, transient analysis. Despite limiting to the theoretical aspect, dealt with under several simplifying hypotheses, the investigation highlights the influence of the examined non-circular geometries on the stability of the static equilibrium position, when these geometries are adopted for shaping the outer housing of the FRB. The paper shows that contrasting effects are obtained, depending on the chosen geometrical parameters. In the paper, the acronyms CB, 2LWB, and LB are used to indicate the FRB layouts respectively equipped with outer circular, wave, and lemon bearing.

Keywords: journal bearing; rotor stability; floating ring bearing; lobe bearing; wave bearing; lemon bearing; Hopf bifurcation

---

1. Introduction

The present analysis theoretically investigates the effects of giving the outer bearing of a floating ring bearing (FRB) a non-circular geometry. The common FRB is a lubricated bearing that incorporates a thin bush or ring between the journal and the casing. This layout presents consequently an inner and an outer oil film, respectively delimited by the journal and the bush and by the bush and the casing. Besides floating, the bush is free to spin, so that the whole bearing practically doubles a simple sleeve bearing. The result is a device with remarkable capability to support ultra-high speed, high-temperature rotating machinery, owing to the significant damping accomplished by the two oil films operating in series. Reduced power loss and low costs are mainly the further key aspects that justify the wide recourse to such bearings, especially in the support of turbochargers. Instabilities are unavoidable effects of the rotor-support nonlinearities, to which the pronounced hydrodynamic concept of the FRB highly contributes. They are typically represented by subsynchronous, self-excited vibrations, which manifest in addition to the synchronous component excited by the rotor unbalance. Oil whirl and whip, bifurcations from one limit-cycle to a different one, with collapse of the limit-cycle itself or the possible evolution to a severe, total instability, are the ingredients of the dynamics of FRBs supported rotors [1,2]. The above remarks justify a bulk investigation carried out in decades, especially focused on the effects of the different parameters on the static and dynamic properties of the FRB [3–8], largely motivated by the need to guarantee higher specific power with lightweight assemblies and to overcome as far as possible the major drawbacks of non-linear instabilities. Shaw and Nussdorfer [3] mentioned the early use of such bearings in the connecting rods of the 1920–1930 Bristol aircraft engines.

They analytically evaluated the ring to journal speed ratio, taking account of both the clearances ratio and the radii ratio, as well as the heat generated in the FRB in comparison to that obtained in an equivalent journal bearing. An experimental measurement of the whirl to rotating speed ratio of the ring was also carried out in their study. Ng and Orcutt [4] calculated the stiffness and damping coefficients of the FRB around the static equilibrium position through linearization of the film forces. Fractional-frequency whirling was observed in the experimental tests, whereas the whirl amplitude was controlled through the oil feeding pressure, the oil viscosity, and the outer clearance. Tanaka and Hori [5] calculated the stability charts for a simple rotor on FRBs, evaluating the effects of the different FRB parameters and rotor flexibility as well as proving the superior stability of the FRB with respect to the conventional journal bearing. A good agreement between experimental and theoretically predicted stability limits was also obtained in the investigation. Boyaci et al. [6] theoretically investigated the Hopf-bifurcation of a rigid, balanced rotor on FRBs. The π-film, short bearing model was adopted together with the linearization of the film forces around the static equilibrium position. The sub-critical or super-critical nature of the bifurcations, occurring under different operating conditions, was also characterized with recourse to non-linear analysis. Hatakenaka et al. [7] set up a Reynolds equation with a suitable cavitation model that could take into account the centrifugal force. This way the FRB-model showed how the increase of the cavitation zone with the journal speed could influence the stiffness cross-coupling coefficients and justify the reduction of the bush-to-shaft speed ratio. Chasalevris [8] set up an analytical model of the finite-length FRB. Comparison to the numerical approach with finite difference method confirmed the capability of the method to give accurate predictions at reduced computational cost, when adopted within rotordynamic algorithms. Yet, a significant part of the literature was specifically addressed to the study of realistic turbocharger (TC) models equipped with FRBs [1,2,9–12]. These analyses turn out to be relatively involved. In order to guarantee an adequate predictive level, the implemented theoretical tools had to consider the mass and unbalance distributions and temperature gradients, in addition to the different parameters that characterize the operation of the FRBs. Gunther and Chen [9] and Inagaki et al. [10] investigated the onset of translational and conical unstable whirling modes, respectively through finite element analysis and flexible multibody model, with further use of experimental tests. To save computational time, the short bearing approximation was often preferred to the finite-length characterization of the oil films [10,11]. Tian et al. [11] and Schweizer [1,12] carried out thorough nonlinear analyses by means of run-up and coast-down simulations, putting in evidence the influence of the unbalance distribution within the TC rotor and giving particular emphasis to the character of the whirling modes and the phenomenon of total instability. The dynamical behavior of TC rotor on semi floating ring bearings, which differ from FRBs because the locking of the bush rotation was also studied by Bonello [13] and San Andrés et al. [14]. The effect of adopting a non-circular geometry for the bearing clearance, in place of the common circular one, remained though practically neglected in the research literature about FRBs. This circumstance contrasts with the widespread and consolidated recourse to lemon, multi-lobe, and pressure-dam shapes in journal bearings, as a mean to guarantee higher stability of the lubricated pair with respect to the simple circular geometry. To the best author's knowledge, only a few, recent papers have dealt with this topic. Eling et al. [15] theoretically analyzed a TC on FRBs with six-lobe-shaped bush. The study showed that, in spite of an increase in the friction losses due to the lobe geometry, a reduction in sub-synchronous vibrations, with respect to the use of common bush, could be obtained, as confirmed by experimental tests. Soni and Vakharia [16] investigated the steady-state characteristics of a FRB presenting an outer lemon bore. The theoretical analysis, which assumed a finite length bearing and solved the lubrication problem through the finite element method, made it possible to assess a better load capacity of the lobe geometry with respect to the reference circular one. Bernhauser et al. [17] presented a run-up analysis of a TC multibody model with FRBs similar to those adopted in [15]. Regarding the practical employment of TCs within combustion engines, the authors pointed out the importance of imbalance-excite and self-excited TC vibration as responsible of disturbing tonal noise during the engine operation. The study theoretically proved that an optimization of the TC dynamics, through reduction or suppression of the said vibrations, can be actually achieved by means of the assumed, uncommon geometry.

Previous studies carried out by the present author were focused on lubricated bearings equipped with non-circular casing, specifically shaped as a two-lobe wave bore. The highest degree of dissymmetry, in the class of multi-lobe geometries, showed by the 2LWB, makes it possible to test the effects of the casing orientation with respect to external loads presenting a fixed direction (e.g., gravity load). On the other hand, the wave amplitude represents a parameter apt to grade the preloading effect. The non-linear behavior of unbalanced rotors supported by these bearings, with special regard to the stability of periodic solution and bifurcations, was investigated in [18,19] with recourse to the continuation method. The theoretical outcomes from [18] showed that a suitable choice of the above geometric parameters is beneficial regarding the stability behavior of a rigid rotor on 2LWB. The present investigation shares with these past studies the consideration of a non-circular geometry, extending its use to an FRB, whose outer bearing, formed by the ring and the fixed casing, was made as a 2LWB or a lemon bearing (LB). Nevertheless, the dynamic aspects under examination limit here to the stability of the steady equilibrium point of a perfectly balanced rotor. The only bifurcation investigated in the following of the paper, by means of the classic linear stability analysis, is the Hopf-transition that gives rise to a limit cycle. How the said non-circular FRB geometry influences this dynamics, in comparison to the common circular geometry, represents the main topic of the paper. By way of example, a further inspection into the bifurcation scenario, with extension to a higher speed range, is carried out through the brute-force integration of the system's equations. This numerical, non-linear approach completes the investigation by enabling the dynamics survey through observation of the journal and ring orbits, the evaluation of the minimum film thickness, and that of a suitable orbit-based index.

2. Rotor Model and Aspects of the Linear Stability Analysis

2.1. Bearing Geometries

A horizontal, rigid rotor is supported with FRB placed at both ends. Owing to assumed symmetry with respect to the rotor middle plane, the analysis is restricted to a half rotor supported by one end FRB. The floating ring is rotating with ω_R speed around its geometrical axis and whirling in a precession motion within the fixed housing. No misalignment of the ring nor of the journal is hypothesized. When the shape of the outer bore is non-circular, with difference to the CB, the ring rotates inside of a clearance whose radial dimension is not more constant. In the present analysis two different non-circular shapes are adopted:

(1) Two-lobe wave bore (2LWB): the outer clearance $Cl_{o,WB}$ (the subscript "o" indicates the outer bearing) is expressed in dimensionless way and polar coordinates as

$$Cl_{o,WB}(\delta,\varphi) = 1 + B\cos(2\delta + \varphi) \tag{1}$$

The wave amplitude B represents the degree of deviation from the circular shape, while the angle φ indicates the angular displacement of the outer bearing with respect to the horizontal axis of a fixed frame of reference whose origin is in the bore center (Figure 1a). A null B value corresponds to the CB case.

(2) Lemon bore (LB): the clearance in this well-known geometry [20–22] is enclosed between two shells, each one presenting a circumferential extension that is not a complete half-circle (Figure 1b). It can be expressed in polar coordinates as

$$\overline{Cl}_{o,LB\_1,2}(\delta,\theta) = \overline{C}_H + (-1)^k \overline{d}\sin(\delta+\theta), \quad \delta \in [(k-1)\pi, k\pi], \; k = 1,2 \tag{2}$$

In the above expression, the k index refers to the k shell. The whole lengths are generally made dimensionless by division with $\overline{C}_H$. Differently, they are expressed here in the ratio to the dimension C_o

of the CB clearance, according to the plot in Figure 1b, where the orientation angle θ is also indicated. The ratio $p_E = \overline{d}/\overline{C}_H = d/C_H$ represents the ellipticity of the profile. It is worth to remark that, differently to the CB and accordingly to the geometry in Figure 1b, the journal (i.e., the ring, in the present case) is impeded to cover the whole clearance, owing to its radial dimension and to the shape discontinuity of the clearance wall. It must be also noticed, in Figure 1, the absence of holes or grooves for the oil circulation, according to the assumed hypothesis of a bearing model with side inlet-outlet at ambient pressure.

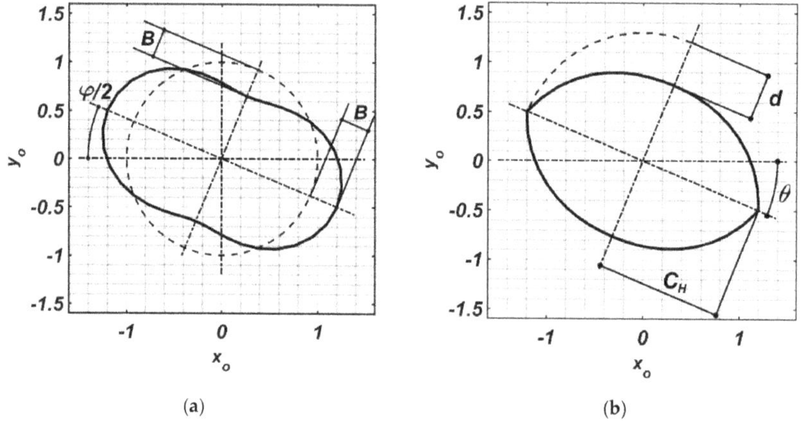

Figure 1. Geometry of the 2LWB and LB clearances. B, d and C_H are dimensionless quantities, obtained by division with the radius C_o of the CB clearance (represented by the dashed circle with dimensionless unit radius in the left figure). In figure $\theta = \varphi/2 = \varphi/8$. (**a**) 2LWB; (**b**) LB.

In order to carry out a comparison between the two shapes shown in Figure 1, regarding their use in the outer bearing, their geometrical parameters are to be suitably assigned in order to establish some mutual equivalence and further reference to the CB case. In this regard, in addition to setting the same angular displacement through the relationship $\theta = \varphi/2$, it seems quite reasonable to assume, on the one hand, that the dimensional clearances respectively pertaining to the two profiles, i.e., $\overline{Cl}_{o,WB}(\delta,\varphi) = C_o + C_o B \cos(2\delta + \varphi)$ and $\overline{Cl}_{o,LB\_1,2}(\delta,\theta) = \overline{C}_H \mp \overline{d}\sin(\delta + \theta)$, have equal mean values

$$\overline{Cl}_{o,WB}\Big|_M = C_o \text{ and } \overline{Cl}_{o,LB}\Big|_M = \overline{C}_H - 2\overline{d}/\pi, \tag{3}$$

On the other hand, imposing equal maxima

$$\overline{Cl}_{o,WB}\Big|_{MAX} = C_o(1 + B) \text{ and } \overline{Cl}_{o,LB}\Big|_{MAX} = \overline{C}_H \tag{4}$$

or, alternatively, equal minima

$$\overline{Cl}_{o,WB}\Big|_{MIN} = C_o(1 - B) \text{ and } \overline{Cl}_{o,LB}\Big|_{MIN} = \overline{C}_H - \overline{d} \tag{5}$$

is assumed to represents a plausible, further condition of equivalence. As a result, the ellipticity parameter and the dimensionless law of the LB clearance $Cl_{o,LB}(\delta,\theta)$, which comply with the requirement of equal mean values and maxima (LBmx), are

$$p_{E,LBmx} = \frac{\pi}{2}\frac{B}{1+B} \text{ and } Cl_{o,LBmx}(\delta,\varphi) = \frac{\overline{Cl}_{o,LB}(\delta,\theta)}{C_o} = 1 + B\left[1 + (-1)^k \frac{\pi}{2}\sin(\delta + \theta)\right], \; k = 1,2 \tag{6}$$

whereas equating mean values and minima it is alternatively obtained (LBmn)

$$p_{E,LBmn} = B\frac{\pi}{\pi - 2(1-B)} \text{ and } Cl_{o,LBmn}(\delta,\varphi) = \frac{1-B}{1-p_E}\left[1 + (-1)^k p_E \sin(\delta+\theta)\right], \ k = 1,2 \tag{7}$$

Figure 2a,b show the comparison of the two elliptical clearances LBmx and LBmn, respectively determined according to Equations (6) and (7).

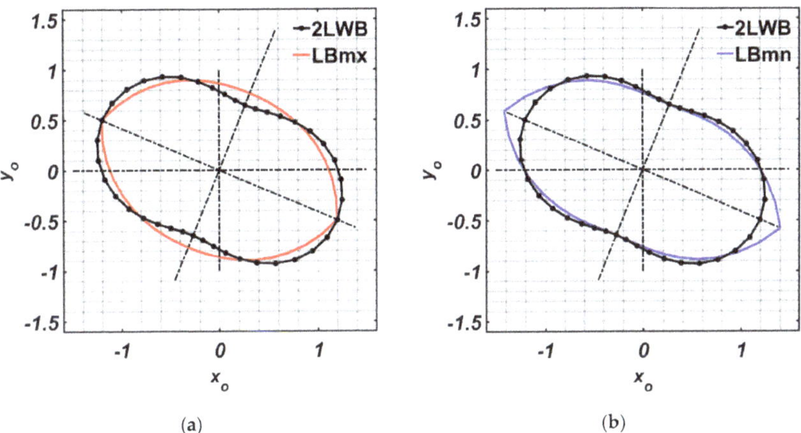

(a)   (b)

Figure 2. Comparison between the 2LWB and the LB clearances (the differences have been expressly highlighted by assigning $B = 0.3$). (**a**) conditions of equal mean values and maxima (LBmx); (**b**) conditions of equal mean values and minima (LBmn).

2.2. Rotor-Bearing Model

After suitable manipulation, the equations of motion of the rotor-bearing system are eventually written in a dimensionless way

$$\begin{cases} x'' + \frac{\mathcal{E}}{s}x' = U\cos\tau + \frac{\sigma^*_{m,D}}{2\Theta s}f_{i,x} \\ y'' + \frac{\mathcal{E}}{s}y' = U\sin\omega\tau - \frac{1}{s^2} + \frac{\sigma^*_{m,D}}{2\Theta s}f_{i,y} \\ x''_R = \frac{1}{\chi\kappa}\frac{\sigma^*_{m,D}}{2\Theta s}\left(f_{o,x}\frac{2\rho_\mu\Lambda^3 v\Theta_R}{\kappa^2} - f_{i,x}\right) \\ y''_R = \frac{1}{\chi\kappa}\frac{\sigma^*_{m,D}}{2\Theta s}\left(f_{o,y}\frac{2\rho_\mu\Lambda^3 v\Theta_R}{\kappa^2} - f_{i,y}\right) - \frac{1}{\kappa s^2} \end{cases} \tag{8}$$

The significance of the symbols is given in the relative section. The inner film force components are evaluated, according to the isoviscous, laminar, π-film model, under the infinitely short-bearing approximation, (πSBA), through the analytical procedure reported in [23], suitably adapted to the inner bearing, where both the ring and the journal rotate with different speeds (Appendix A). The ratio $\hat{\omega} = \omega_R/\omega$ between these angular speeds has been expressed in approximate way [24], so as to be independent of the rotor speed

$$\hat{\omega} = \frac{\omega_R}{\omega} = \frac{1}{1 + \frac{v^3}{k}\rho_\mu} \tag{9}$$

Expressions of the outer film force components, limiting to the CB layout, could be obtained similarly to the inner ones. A semi-analytical, more general procedure is instead adopted, which holds in the whole different geometries, i.e., CB, 2LWB, and LB. According to the SBA and the selected bearing geometry, the oil pressure is analytically obtained in the whole points of a grid with M circumferential rows × N axially oriented columns, suitably fixed to represent the outer film in discretized way.

After discarding the negative values, in accord to the Gumbel cavitation model, a numerical integration over the same grid yields the oil force components (Appendix A).

2.3. Linear Stability Analysis

A linear stability analysis is carried out through the linearized equations written about stationary equilibrium points. With this aim, the bearing modulus $\sigma^*_{m,D}$ and the Sommerfeld number ratio Σ

$$\sigma^*_{m,D} = \mu \frac{\Omega L D}{W} \frac{R^2}{C^2} \frac{L^2}{D^2}, \quad \Sigma = \frac{(1+\chi)k^2}{\rho_\mu \hat{\omega} \Lambda^3 v} \tag{10}$$

are assumed to represent a distinctive operating set of the system parameters (i.e., main geometrical parameters, load, and dynamic viscosity of the lubricant). In particular, $\sigma^*_{m,D}$ is related to the modified Sommerfeld numbers $\sigma_{m,D}$ and $\sigma_{mo,D}$, which respectively characterize the current operating conditions within the inner and the outer film, by means of the relationships

$$s = \frac{\sigma_{m,D}}{\sigma^*_{m,D}}, \quad \sigma_{m,D} = \Sigma \sigma_{mo,D}. \tag{11}$$

Two intervals, respectively I_σ of $\sigma^*_{m,D}$ and I_s of s, are suitably assigned, so that at each $\sigma^*_{m,D} \in I_\sigma$, the rotor speed s is varied in I_s. This way, different sets $(s, \sigma_{m,D}, \sigma_{mo,D})$ are fixed in turn by means of Equation (11). The static equilibrium point $P_E \equiv (x, y, 0, 0, x_o, y_o, 0, 0)_E$ regarding each set is hence obtained by solving the system

$$\begin{cases} f_{i,x} = 0 \\ f_{i,y} = \frac{1}{\sigma_{m,D}} \\ f_{o,x} = 0 \\ f_{o,y} = \Sigma f_{i,y} \end{cases} \tag{12}$$

Above equation system is obtained from Equation (8) by setting derivatives and unbalance equal to zero. It must be observed that the inner force $f_{i,y}$ can be evaluated here directly through the expression of $\sigma_{m,D}$ obtained by adaptation of the plain bearing case [24] and reported in Appendix A. Once P_E is obtained, the following linearized equations of motion hold in its neighborhood

$$s^2 \begin{bmatrix} I_2 & 0 \\ 0 & \chi I_2 \end{bmatrix} \begin{Bmatrix} \xi'' \\ \eta'' \\ \kappa \xi'' R \\ \kappa \eta'' R \end{Bmatrix} + \Theta \begin{bmatrix} Q & -Q \\ -Q & Q + pq \end{bmatrix} \begin{Bmatrix} \xi' \\ \eta' \\ \kappa \xi' R \\ \kappa \eta' R \end{Bmatrix} + \begin{bmatrix} P & -P \\ -P & P + \rho \Theta_R p \end{bmatrix} \begin{Bmatrix} \xi \\ \eta \\ \kappa \xi R \\ \kappa \eta R \end{Bmatrix} = \begin{Bmatrix} 0 \\ 0 \\ 0 \\ 0 \end{Bmatrix}. \tag{13}$$

The submatrices p, P, q and Q are written through the stiffness and damping coefficients, as explained in Appendix B. Further manipulation of Equation (13) (Appendix B) yields the eigenvalues problem

$$D\hat{Z} = \lambda^* \hat{Z}. \tag{14}$$

Provided that the stationary equilibrium point P_E is hyperbolic, the instability threshold is represented by conditions where some λ^*, representing a real eigenvalue or a pair of conjugate imaginary eigenvalues, turn out to have null real part.

2.4. Choice of the Parameter Values for the Analysis

Numerical simulation has been carried out for different operating conditions, fixed in turn by choosing the parameter values with suitable criterion, in order to limit the case study work. The values $\pm(3/4, 1/2, 1/4) \cdot \pi$ and 0 were assumed for the orientation angle φ, together with a wave amplitude $B = 0.2$. The last quantity, employed within Equations (6) and (7) for the LBmx and LBmn profiles, yields an ellipticity ratio p_E respectively equal to 0.262 and 0.408. The LB-clearance ratio

$\Psi_S = \overline{C}_H / \overline{C}_V = 1/(1 - p_E)$, with $\overline{C}_V = \overline{C}_H - \overline{d}$, represents a parameter that is frequently used too in literature and assumes values 1.36 and 1.69, respectively in the above two cases. The parameters ρ_μ, Λ, κ, v, and χ were given values as follows: $\rho_\mu = 1$; $\Lambda = 1$; $\kappa = 1, 2$; $v = 1.32, 1.5$; $\chi = 0.025, 0.043$, 0.075. Some of these values have been adopted in [5,6]. When the outer bearing is a 2LWB or an LB, v and κ are respectively evaluated with reference to the outer radius of the ring, which is maintained comparing the different geometries, and to the clearance C_o. A grid with M = 21 rows × N = 221 columns has been used to discretize the outer film.

3. Results

3.1. Steady-State Behavior: A Preliminary Comparison

Figure 3a,b illustrate the journal eccentricity ratio ε vs. the eccentricity ratio ε_0 of the ring, respectively in two cases relative to CB and LB mount. The parameter values $v = 1.2$, $\chi = 0.001$, $p_E = 0.5$, and $\kappa = 0.7$ and 1.3 have been adopted in order to carry out a comparison to the data from [16], also shown in figures. The CB data have been obtained here as well as in [16] as a limit case of the LB geometry. To the author's knowledge and according to the short literature review given above in the introductory section, the bearing model dealt with in [16] is so far the only one presenting some similarity with the FRB adopted in the present study. Regarding the plots put in comparison in Figure 3a,b, it is worth pointing out the differences between the present LB model and that adopted in [16], where a finite different method-scheme and a finite bearing model with L/D = 1 have been used to characterize the steady-state performances of an FRB with outer LB. In the light of these remarks, the comparison may be assumed acceptable with exception of the LB layout and $\kappa = 0.7$.

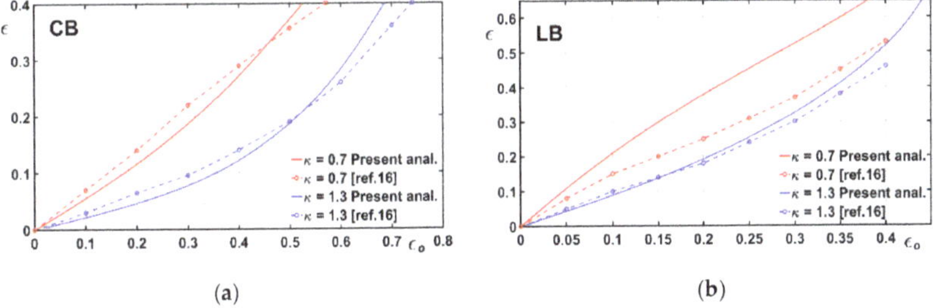

Figure 3. Journal eccentricity ratio ε vs. the ring eccentricity ratio ε_0: comparison to results from [16]. (a) CB geometry. (b) LB geometry with $p_E = 0.5$.

3.2. Linear Stability Analysis: The CB Case

Some check about the reliability of the semi-analytical procedure addressed to the linear stability analysis, as described in the above section, has been carried out with reference to the operation of a rotor on FRBs with outer CB. In this case, the stability responses obtained in both semi-analytical and analytical way (see for example [5,6]) have been put in comparison. Figure 4a shows the results (the analytical curves are reported in black) for three different κ values and the remaining parameters fixed as in [6]. The plots in figure show some differences in the outcomes of both procedures, particularly for the higher branches obtained with $\kappa = 1.5$ and 2. In this regard, the following remarks are worth to be considered. The analytical curves have been here obtained expressing the inner and outer fluid film forces as well as the inner and outer stiffness and damping coefficients in full analytical way, according to [5]. On the other hand, the stability borders obtained in semi-analytical way are calculated through a procedure that implies: the integration of the pressure distribution on a finite mesh; the account of cavitation through cut off of the negative values; the calculation of stiffness and damping coefficients by means of numerical derivatives. In both cases, the occurrence of pure imaginary eigenvalues, which is crucial for the determination of the instability threshold, has been carried out numerically. According to these aspects, the differences in the outcomes, as they appear in Figure 4a, receive some justification.

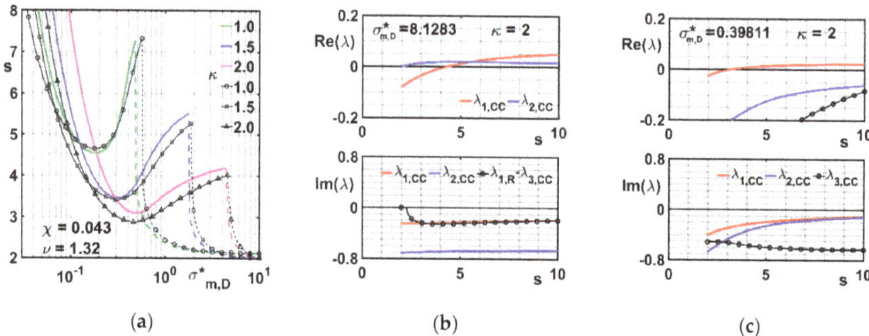

Figure 4. Results for the CB layout. (**a**): Stability thresholds: numeric (in color) and analytic (in black). (**b,c**): partial portraits of the eigenvalues behavior vs. s for $\kappa = 2$ in the LSR, at $\sigma^*_{m,D} = 8.13$ (**b**) and the HSR, at $\sigma^*_{m,D} = 0.398$ (**c**).

In Figure 4a, it may be observed that each κ-case is represented by a curve with two distinct branches, so as to organize the whole area under exam in two regions, a left higher stability region (HSR) and a right lower stability region (LSR). The branch in the HSR, at lower values of the bearing modulus $\sigma^*_{m,D}$, represents stability loss very likely due to supercritical Hopf bifurcations, whereas the branch in the LSR, characterized by light-load conditions, would very likely represent instability originated by subcritical bifurcations [6]. These behaviors are due to the play of the eight eigenvalues at varying the rotor speed s. In this regard, it is worth remarking that: (1) the set of eigenvalues is generally made by real eigenvalues (REs) $\lambda^*_{p,R}$, $p = 1, P$, and complex conjugate eigenvalues (CCEs) $\lambda^*_{q,CC}$, $q = 1,$ Q; (2) the number of eigenvalues in each subset can change from s to s, on condition, in the present system, that $P + 2Q = 8$. As an example, Figure 4b,c, which refer to the curve in Figure 4a numerically obtained for $\kappa = 2$, illustrate the conditions occurring at the onset of instability when $\sigma^*_{m,D} = 8.13$ and $\sigma^*_{m,D} = 0.398$, respectively. Both figures show the behavior of the real parts $\mathrm{Re}(\lambda)$ of the eigenvalues (upper diagrams) and that of the imaginary parts $\mathrm{Im}(\lambda)$ in the interval $s = 2 \div 10$ (for better clarity, only the non-positive $\mathrm{Im}(\lambda)$ have been plotted). Furthermore, owing to the different magnitudes of the eigenvalues parts, only partial data, i.e., those relevant to the Hopf bifurcations under analysis, have been reported. That being said, from Figure 4b it can be inferred that the loss of stability at $\sigma^*_{m,D} = 8.13$, occurs at s around 2, when the pair $\lambda^*_{2,CC}$ crosses the imaginary axis from the negative to the positive half-plane. It can be seen that, at the same speed, there are a second pair $\lambda^*_{1,CC}$ of CCEs and a RE $\lambda^*_{1,R}$, the latter one changing soon after, when s is about 2.2, into a further pair of CCEs. At the onset of instability, the absolute value of the imaginary part of the eigenvalues that determine the bifurcation, i.e., $\lambda^*_{2,CC}$, is higher than that of $\lambda^*_{1,CC}$. Differently in the HSR, when $\sigma^*_{m,D} = 0.398$, Figure 4c shows that the loss of stability, about $s = 3$, is due to the eigenvalue $\lambda^*_{1,CC}$, which possesses the lowest absolute value of the imaginary part with respect to the further CCEs (two of which, i.e., $\lambda^*_{2,CC}$ and $\lambda^*_{3,CC}$, are depicted in the upper and bottom diagrams of the figure). The above conditions, regarding the magnitude of the imaginary parts of the critical eigenvalues, have been observed in [6] to be frequently associated to sub-critical (Figure 4b) and super-critical (Figure 4c) bifurcations. Nevertheless, the same conditions cannot represent a criterion to decide whether the bifurcation is of the former or the latter type, nor in the present analysis are adopted the appropriate tools, like continuation or center manifold analysis, in order to solve the issue.

The change from LSR scenario to the HSR one manifests with a jump, as it can be inferred from the plots in Figure 4a and can be explained as follows, still making reference to Figure 4b,c, without providing further data for the sake of conciseness. According to the above remarks, the Hopf bifurcations in the LSR (Figure 4b) and in the HSR (Figure 4c) are due to different critical eigenvalues, i.e., $\lambda^*_{2,CC}$ and $\lambda^*_{1,CC}$, respectively. As the value of the bearing modulus $\sigma^*_{m,D}$ is decreased from 10 to 2 (compare Figure 4b,c), the $\mathrm{Re}\!\left(\lambda^*_{1,CC}\right)$ remains over the zero axis (i.e., the pair $\lambda^*_{1,CC}$ is continuing to cross the imaginary axis). Differently $\mathrm{Re}\!\left(\lambda^*_{2,CC}\right)$ diminishes gradually in value, till to abandon the zero axis and to position entirely beyond it. Owing to its slope, when $\mathrm{Re}\!\left(\lambda^*_{2,CC}\right)$ separates from the zero axis and ceases to be critical (around $\sigma^*_{m,D} = 4.5$), there is a jump from the critical s value that is due to $\lambda^*_{2,CC}$, say $s^*_{2,E}\!\left(\lambda^*_{2,CC}\right)$ this *end* value, to the critical s value that is due to $\lambda^*_{1,CC}$, say $s^*_{1,I}\!\left(\lambda^*_{1,CC}\right)$ this *initial* value, placed upward in the interval. The difference $s^*_{1,I}\!\left(\lambda^*_{1,CC}\right)\text{-}s^*_{2,E}\!\left(\lambda^*_{2,CC}\right)$ justifies the discontinuity in the observed threshold curve for $\kappa = 2$ in Figure 4a. Similar considerations hold for the remaining curves.

3.3. Linear Stability Analysis: The 2LWB and LB Cases

When the outer bearing is changed from the CB to the 2LWB or LB shapes, the dependence of the stability curves on the bearing orientation φ and the level of deviation from the circular shape (respectively represented in both types by the amplitude B and the ellipticity parameter p_E) turns out to be apparent. Figure 5a,b report the stability charts obtained adopting the 2LWB with $B = 0.2$, $\varphi = -(1,2,3)\!\cdot\!\pi/4$, $\upsilon = 1.32$ and $\chi = 0.043$; the clearance ratio κ was given the values of 1 and 2, respectively in both figures. The curves in each figure show similarities to the plots of Figure 4a and are mostly characterized by the presence of two distinct branches, corresponding to HSR and LSR. In each figure, the threshold curve obtained with the CB layout, operating under the same respective parameter values, has been added in order to allow comparison to the lobe solutions, so that a general increase in stability, gained by the non-circular geometry, may be observed. In fact, the threshold branches corresponding to the 2LWB are higher than in the CB case, with the only exception of the $\varphi = -3\pi/4$ orientation. This effect is particularly remarkable in the HSR and even more when the clearance ratio κ is equal to 1, whereas the dotted-line branches in the LSR turn out to be very close each to the other. Also worth of notice is the broadening of the HSR, at the expense of the LSR, when κ is raised from 1 to 2. Yet, comparing in each figure the curves corresponding to different angles, it can be inferred that the apparent improvement of the stable behavior in the HSR is higher the less the slope with respect to the horizontal layout assigned through the φ angle. Nevertheless, the orientation angle $\varphi = -3\pi/4$ yields a some contrasting stability response: (1) poor when $\kappa = 1$, i.e., worse than that of the CB mount, in a relatively broad interval of $\sigma^*_{m,D}$ and, moreover, without a distinct separation between HSR and LSR; (2) generally better than the CB response when $\kappa = 2$, with a threshold even higher than in the whole comparing cases, in a restricted interval of the bearing modulus, placed just downstream the jump up to the higher branch.

Figure 5c–f illustrate the stability charts obtained replacing the 2LWB with an LBmx bearing (Figure 5c,d) or an LBmn one (Figure 5e,f). In order to ease the comparison between geometries, the whole set of Figure 5a–f is organized by rows, i.e., the figures on a same row are obtained with the same parameter values. In the whole cases, it can be seen that the LB curves (Figure 5c–f) are more gathered, with partial overlapping, and closer to the CB branches than in the 2LWB mount (compare Figure 5a,b). From comparison, it can also be inferred a general, moderate decrease of the stability thresholds gained by the lemon profiles, as well as the clear similarities to the respective 2LWB results (compare figures on a same row), mainly consisting in similar partitions of the diagram area in HSR and LSR. Some attention may be paid to examining the performances of the different angular orientations, with particular reference to the HSR. The positioning of the LB curves, almost generally above the CB ones, is still to indicate better stability with respect to the circular geometry. Nevertheless, the relative gathering of the branches denotes a weak influence of the φ angle in this set of values, even though some differences in the performance quality can be observed comparing the results pertaining to a same given φ value in the LBmx and LBmn geometries (compare, for instance, the branches with $\varphi = -3\pi/4$ in Figure 5c,d respectively to those in Figure 5e,f).

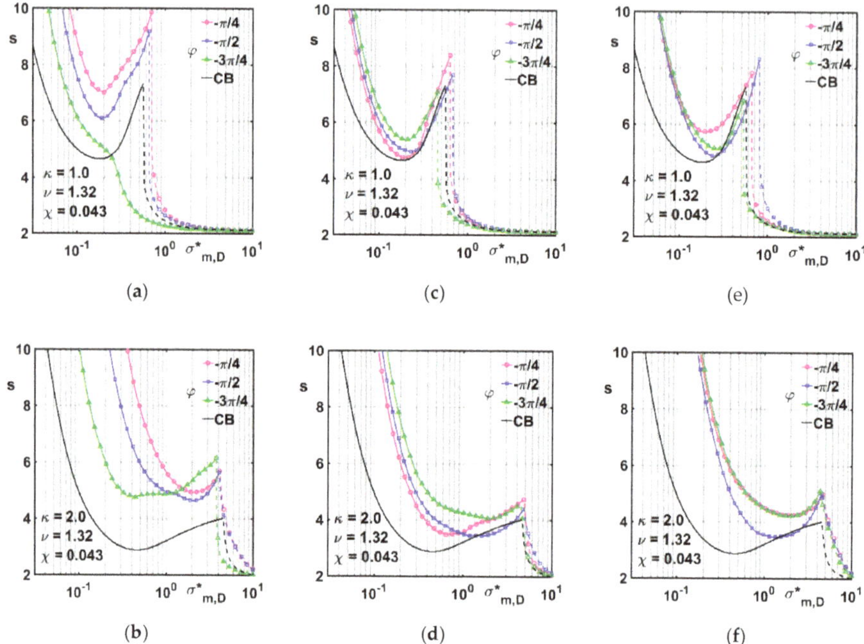

Figure 5. Stability charts for CB and non-circular geometries, $\varphi = -(1,2,3)\cdot\pi/4$, $v = 1.32$, $\chi = 0.043$, $\kappa = 1$ (**a,c,e**), $\kappa = 2$ (**b,d,f**). (**a,b**): 2LWB; (**c,d**): LBmx; (**e,f**): LBmn.

A separate analysis is addressed just to the mentioned 2LWB case with $\varphi = -3\pi/4$ and $\kappa = 1$, in order to throw light onto the characteristic continuity, i.e., without the typical jump from HSR to LSR or vice-versa, exhibited by the corresponding curve in Figure 5a. Figure 6a,b make it possible to assess that $\lambda^*_{2,CC}$ represents the critical eigenvalue both in the LSR and the HSR (compare Figure 5a). Owing to coalescences between the different eigenvalues, observed in the $\mathrm{Re}(\lambda)$ and $\mathrm{Im}(\lambda)$ diagrams at varying $\sigma^*_{m,D}$ in its interval, $\mathrm{Re}\!\left(\lambda^*_{2,CC}\right)$ maintains its position over the zero axis, so that the critical value of s^* (i.e., the speed at which the pair $\lambda^*_{2,CC}$ crosses the imaginary axis) varies continuously without jumps and intervention of different eigenvalues. It is also worth noticing that $\mathrm{Im}\!\left(\lambda^*_{2,CC}\right)$ changes its relative magnitude with respect to the other eigenvalues when passing from LSR to the HSR, as it can be inferred comparing the bottom diagrams in Figure 5a,b.

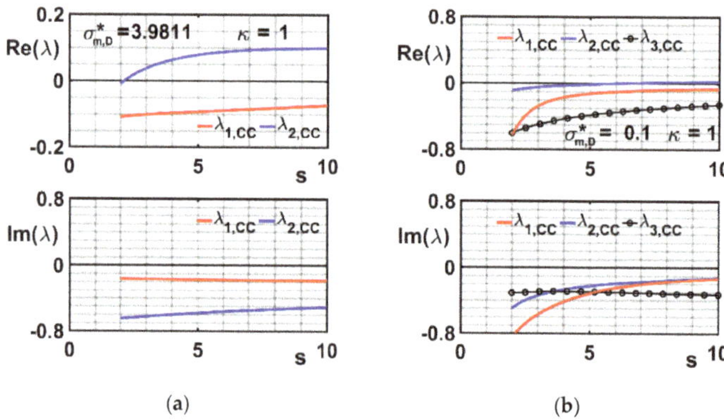

Figure 6. Partial portraits of the eigenvalues behavior vs. s for $\kappa = 2$, at $\sigma^*_{m,D} = 3.98$ (**a**) and $\sigma^*_{m,D} = 0.1$ (**b**).

The effects of an increase in the radius ratio v from 1.32 to 1.5 can be observed from Figure 7a–f. As in the previous set of figures, in each row it is possible to compare the 2LWB data (left) to the LBmx (center) and LBmn ones (right), obtained under the same parameter values. Observation and comparison of the plots in these figures make it possible to conclude that the above modification in the v value keeps unchanged the characteristics already observed in Figure 5a–f, while determining an increased stability, as confirmed by the raising of the whole branches. On the other hand, it was observed that a moderate change in the value of the mass ratio yielded no sensible modification in the charts depicted in Figure 5a–f. Related results to support this remark were obtained assuming $\chi = 0.025$ and 0.075 and are not presented here for the sake of brevity.

The stability charts obtained with the same parameter values of Figure 5a–f, but adopting a further set of angular positions, i.e., $\varphi = (0, 1, 2, 3, 4) \cdot \pi/4$, are reported in Figures 8a–c and 9a–f. The values $\kappa = 1$ and $\kappa = 2$ were respectively adopted in the former and the latter set of figures. A general decay of stability, partly lessened at higher loads, is shared by the different curves, which show substantial similarities in the three, compared geometries. It is apparent that, whatever the noncircular bearing shape, when $\kappa = 1$ the threshold curves for $\varphi = \pi/2$ and $\varphi = 3\pi/4$ are positioned quite below the CB curve. Differently, some higher branches of the curves calculated with $\varphi = 0$ and $\varphi = \pi/4$ overstep the CB stability border when $\sigma^*_{m,D}$ is about less than 0.2. Similar behavior characterize the data for $\varphi = \pi$ with restriction to the LBmx and LBmn profiles (Figure 8b,c). A further remark can be made about the shape of the curves, which are deprived of distinct separation between the HSR and LSR, with the only exception of the 2LWB and LBmx branches at $\varphi = 0$.

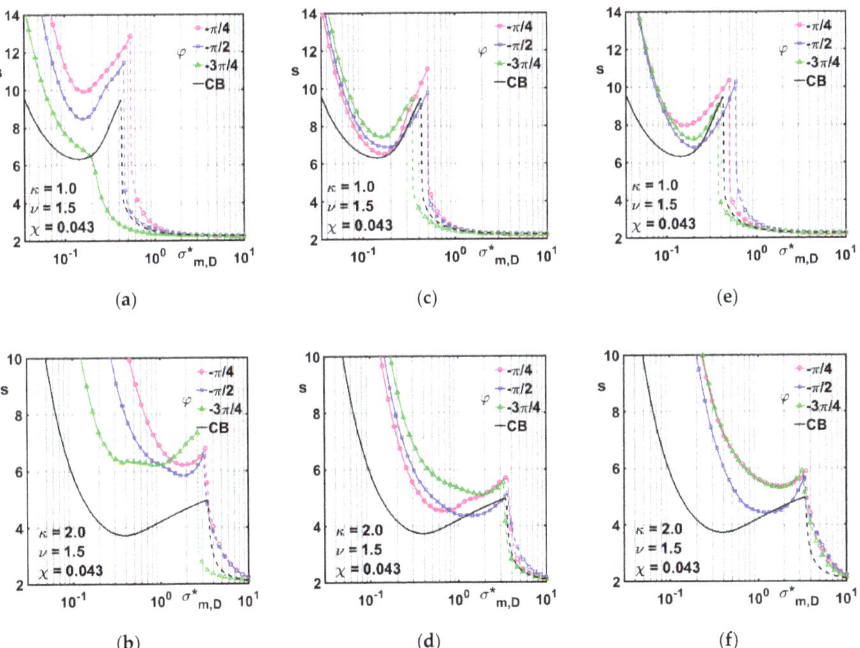

Figure 7. Stability charts for CB and non-circular geometries, $\varphi = -(1, 2, 3) \cdot \pi/4$, $v = 1.5$, $\chi = 0.043$, $\kappa = 1$ (**a,c,e**), $\kappa = 2$ (**b,d,f**). (**a,b**): 2LWB; (**c,d**): LBmx; (**e,f**): LBmn.

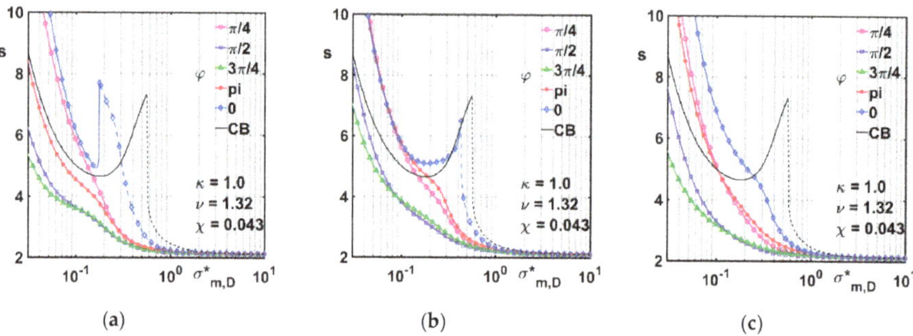

Figure 8. Stability charts for CB and non-circular geometries, $\varphi = (0,1,2,3,4)\cdot\pi/4$, $\upsilon = 1.32$, $\chi = 0.043$ and $\kappa = 1$. (**a**): 2LWB; (**b**): LBmx; (**c**): LBmn.

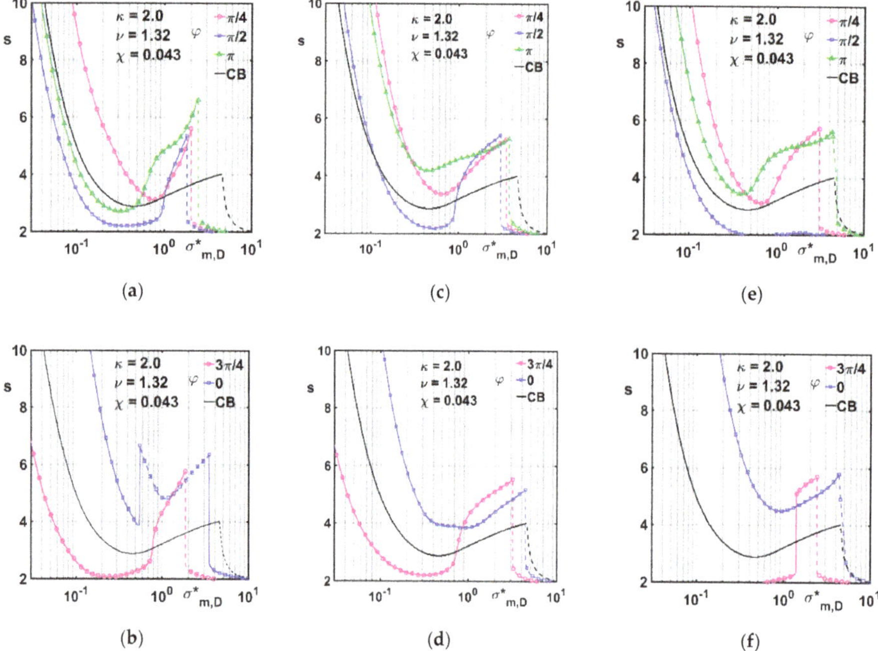

Figure 9. Stability charts for CB and non-circular geometries, $\varphi = (0,1,2,3,4)\cdot\pi/4$, $\upsilon = 1.32$, $\chi = 0.043$ and $\kappa = 2$. (**a**,**b**): 2LWB; (**c**,**d**): LBmx; (**e**,**f**): LBmn.

When $\kappa = 2$ (Figure 9a–f) a higher spread, with respect to the results of Figure 8a–c, is observed in the stability performances. Angular positions with $\varphi = \pi/2$ and $3\pi/4$ yield a remarkable lowering of the limit curves. This circumstance is particularly evident for the LBmn profile. Conversely, the best performance pertains to the $\varphi = 0$ orientation: this horizontal layout of the outer bearing, makes the stability borders in Figure 9b,d,f comparable to the higher branches depicted in Figure 5b,d,f, respectively. Even in this case, this outcome appears more evident when the LBmn shape is adopted. A further comment is to be addressed to the discontinuities exhibited in Figure 9b,f, which could represent, at first glance, numerical artifacts. Nevertheless, an inspection into the behavior of the critical eigenvalues clarifies the point. Figure 10a–e show the partial portraits of the eigenvalues obtained in the interval of rotor speed s, taken at different, close values of the bearing modulus. In particular, Figure 10a–c, relative to a sequence of $\sigma^*_{m,D}$ values around 3.5, show how the critical eigenvalue is initially represented by $\lambda_{3,CC}^*$ (Figure 10a,b, top) and successively (Figure 10c, top) by $\lambda_{2,CC}^*$. The passage from the one eigenvalue to the other takes place with a jump in the critical value s^* of s, which justifies the right discontinuity in the $\varphi = 0$ curve of Figure 9b. Further downward, when $\sigma_{m,D}^*$ is around 0.54 another jump takes place, as confirmed by the left discontinuity in the $\varphi = 0$ curve of Figure 9b. Differently from the previous case, the jump in the critical value s^* from 6.7 c.a. to 4.2 c.a. when $\sigma_{m,D}^* \simeq 0.54$ is due to the only $\lambda_{2,CC}^*$ eigenvalue, as confirmed by the behavior summarized in the top diagrams of Figure 10d,e.

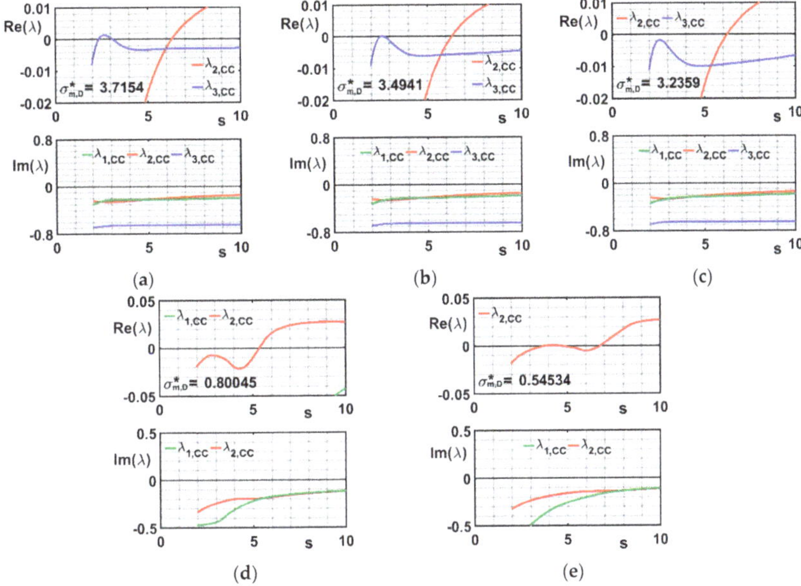

Figure 10. Behavior of the eigenvalues (partial portraits) related to the $\varphi = 0$ curve in Figure 9b, taken at successive values of $\sigma_{m,D}^*$. (**a**–**c**): sequence involved in the right discontinuity about $\sigma_{m,D}^* = 3.5$; (**d**,**e**): sequence involved in the left discontinuity about $\sigma_{m,D}^* = 0.54$.

3.4. Non-Linear Analysis with Brute-Force Method

The above analysis of stability suffers from the limits due to the several simplifying hypotheses that have been assumed, among which the linearization of the fluid film forces is a distinctive feature. The brute-force approach, based on the numerical integration of the system's ODE, represents a first, classic tool at disposal to get further insight into the dynamics under analysis [23], especially in the post-bifurcation scenario, before recurring to possible different methods, like continuation analysis or center manifolds analysis [6,18,19]. However, it is worth remarking that the brute-force method can cast light on the investigated dynamics, though restricting to the stable solutions and on condition to evaluate suitably the sensibility to initial conditions, so as to detect the presence of possibly coexisting solutions. Nevertheless, regarding the investigation carried out in the above section, this approach can effectively play a supporting role, mainly consisting in a survey of the bifurcation dynamics predicted through the linear stability analysis. Particularly, the stability loss can be verified by observing the positions occupied by the journal and ring centers within their respective clearances, as the speed is gradually increased in the low values range. In fact, when the bifurcation manifests, their equilibrium point-status is replaced by a motion condition as an effect of the onset of a limit cycle. Both the speed at which the transition sets in and the way the orbits' size increase can be suitably appraised through the numerical simulation.

On these premises, two examples have been chosen from the case-studies presented in the above section, in order to show the use of the brute-force procedure in the present context. In each example, for a given set of κ, υ, χ, and $\sigma^*_{m,D}$ values, the rotor speed is assigned in steps within a suitable interval. At each step, the numerical integration of Equation (8), written in the absence of unbalance, is carried out by means of an ode15s MATLAB routine, up to attaining a steady condition after that the initial transient is damped out. This procedure is repeated for different FRB layouts, whose responses are eventually compared in terms of journal and ring center orbits, minimum film thickness, and *SI* index. The last quantity is specifically related to the orbits. Besides contrasting the orbits obtained from a layout to the other at some chosen s values, the said index, defined as

$$SI = \sqrt{\Delta x \cdot \Delta y} \tag{15}$$

where

$$\Delta x = x_U - x_L, \ \Delta y = y_U - y_L, \ x_U = \max\{x\}, \ x_L = \min\{x\}, \ y_U = \max\{y\}, \ y_L = \min\{y\} \tag{16}$$

and $\{x\}$ and $\{y\}$ are the x and y coordinates of the orbit points, makes it possible to evaluate approximatively the orbit dimensions along the examined interval of speed.

The two selected examples are characterized by the following sets of parameters:

Set (1) $\sigma^*_{m,D} = 0.1$, $\kappa = 2$, $\upsilon = 1.32$, $\chi = 0.043$, layouts: CB, 2WLB with $\varphi = -\pi/4, 3\pi/4$ rad (reference to Figures 5b and 9b);

Set (2) $\sigma^*_{m,D} = 0.22$, $\kappa = 1$, $\upsilon = 1.32$, $\chi = 0.043$, layouts: CB, 2WLB with $\varphi = 0, \pi/4$ rad (reference to Figure 8a).

In the examples, intervals: $1.18 \div 15.4$ ($10 \div 130$ krpm) or $1.18 \div 21.3$ ($10 \div 190$ krpm) of the s rotor speed, in steps $\Delta s = 0.355$ (3 krpm), have been assumed.

Figure 11a,b depict the behavior of the *SI* index respectively for the above Set1 and Set2 conditions. Each curve in the diagrams starts at a speed value that is generally not coinciding with the exact critical one, but turns out to be placed just downstream of it, owing to the step resolution and the fact that the index assumes null values in the presence of equilibrium points. The *SI* behavior shown in the plots makes it possible to infer that:

- The orbits described by the ring center in the outer bearing are larger than those described by the journal center. In particular, the orbits of the ring center for the CB are generally larger than in the other layouts. An opposite outcome appears when comparing the orbits of the journal bearing, which are, for the circular geometry, smaller than in the 2LWB examples.
- Under the Set 1 conditions, the bifurcation manifests at rotor speeds that are quite different each to the other (Figure 11a). The 2WLB, $\varphi = -\pi/4$ layout presents a critical speed about $s = 15$, remarkably higher than those that occur about $s = 6$ and $s = 4$, respectively for the CB and the 2WLB $- \varphi = 3\pi/4$ cases. This outcome is quite congruent with the predictions that can be deduced from Figures 5b and 9b for these three layouts.
- Adopting the Set 2 (Figure 11b), the critical speeds obtained for the different FRBs are closer than in the Set 1 conditions. The lowest one, at about $s = 4.5$, is obtained when the outer bearing, i.e., the FRB housing, has a lobe profile with a $\varphi = \pi/4$ slope. In the other two cases, represented by the CB and 2WLB $- \varphi = 0$ layouts, the transition to the limit cycle manifests at about $s = 5$. Comparison to Figure 8a makes it possible to assess a sufficient congruence between the data in these conditions too, despite the jump exhibited in the same figure by the threshold curve relative to the lobe bearing with $\varphi = 0$ slope.

An insight into the results presented in Figure 11a,b can be achieved through inspection of the trajectories described by the journal and ring centers at different speeds and depicted in Figure 12a,c,e, for conditions of Set 1, and in Figure 12b,d,f for the Set 2 ones. A first indication about the stability performances that pertain to the different layouts, is obtained by noticing the long-lasting permanence of the equilibrium point at speed increase, which manifests in some cases with respect to the other ones. This remark complies particularly with the behavior of the 2WLB $- \varphi = -\pi/4$ reported in Figure 12e.

The wide orbits of the centers, detected for $s = 16$, contrast with the minute orbits obtained when $s = 15$. Differently, the other two geometries analyzed under the same Set 1 conditions (Figure 12a,c) turn out to exhibit big orbits well before, when $s = 7$.

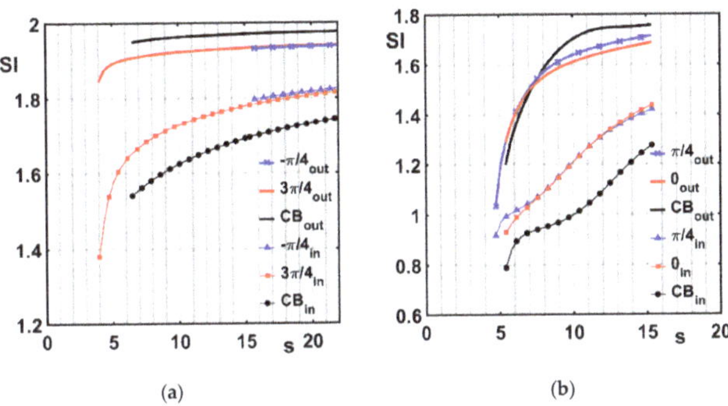

Figure 11. Behavior of the *SI* index relative to the orbits of the journal and ring centers (respectively: 'in' and 'out' subscripts): examples for the CB and 2LWB layouts. (**a**) Set 1 conditions and layouts; (**b**) Set 2 conditions and layouts.

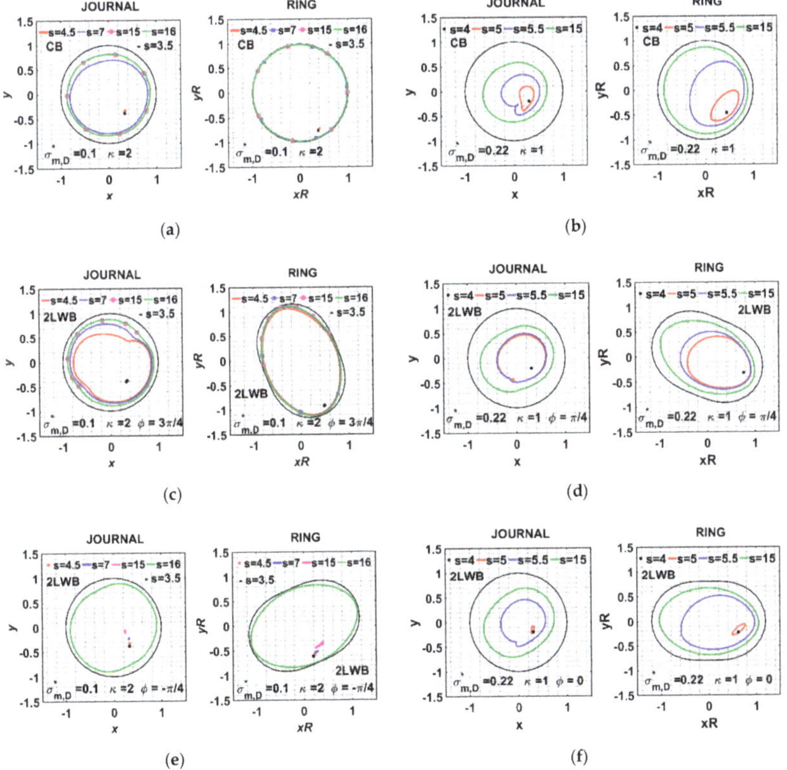

Figure 12. Orbits of the journal and ring centers for different FRB layouts and rotor speeds. (**a,c,e**) Conditions and layouts as in Figure 11 (Set 1), for $s = 3.5, 4.5, 7, 15$, and 16. (**b,d,f**) Conditions and layouts as in Figure 11b (Set 2), for $s = 4, 5, 5.5$, and 15.

The data of Figure 12a,c,e also confirm that the trajectories of the ring center in the CB layout are on average closer to the housing wall with respect to the compared lobe geometries, as inferred from observation of the *SI* curves in Figure 11a. The opposite result, relative to the magnitude of the journal bearing orbits, which appears to be relatively lesser in the CB case than in the compared geometries, is also verified. Figure 12b,d,f depict the trajectories obtained when conditions are those of Set 2. The critical speed occurs for s values between 4 and 5 in the whole three bearing types. Yet, the transition to the precession motion appears slightly anticipated by the $2\text{WLB} - \varphi = \pi/4$ layout with respect to the CB and the $2\text{WLB} - \varphi = 0$ ones. Furthermore, comparing the data obtained at $s = 4$ and 5 in Figure 12d,f to those relative to $s = 3.5$ and 4.5 in Figure 12c and $s = 15$ and 16 in Figure 12e, it can be said that the transition in the Set 2 cases manifests very likely some more gradually than in the Set 1 examples. Further observation of the data in Figure 12a–f makes it possible to notice the presence of higher harmonics that affect, at some degree, the journal orbits downstream of the stability loss, particularly under the Set 2 conditions.

The brief portrait given in the present section is completed here below with an observation of the minimum thickness behavior, as reported in Figure 13a,b. The whole curves in the plots present an initial branch that raises with the rotor angular velocity, from the initial $s = 2$ value up to the critical speed. This behavior is justified by the gradual centering of the steady equilibrium points within the bearing clearances, which evolves as far as the speed is increased. The difference in the loading conditions from Set 1 and Set 2 explains the different range of values pertaining to the curves reported in the lower part of Figure 13a,b and relative to the inner bearing. It is worth observing that the minimum thickness assumes practically the same values, whatever the FRB layout in the examples. The curves shows quite clearly the jump that is due to the stability loss, with indications that agree with those inferred from Figures 11 and 12. The significant jumps that affect the minimum film thickness in the inner bearing under Set 1 are also worth of remark. This outcome appears to be reasonably explained in terms of the said equilibrium point-centering and load condition effects. The example represented by the $2\text{WLB} - \varphi = -\pi/4$ case in Figure 13a is representative in this regard.

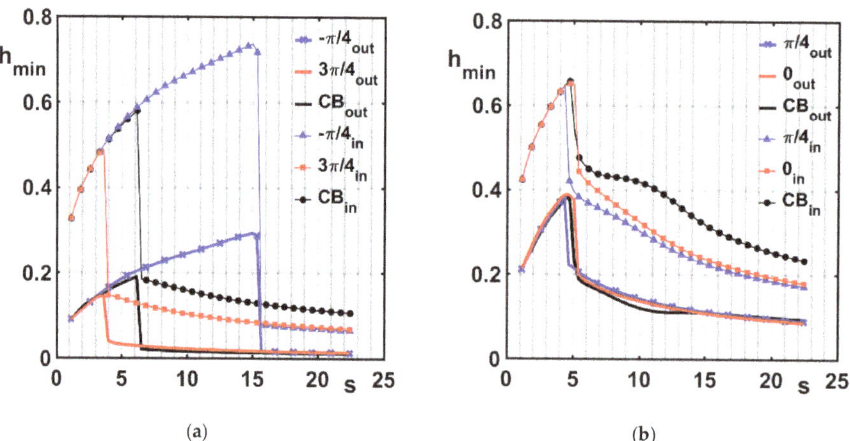

(a)   (b)

Figure 13. Minimum film thickness within the inner and outer bearings. (**a**) Set 1 conditions and layouts. (**b**) Set 2 conditions and layouts.

4. Discussion and Conclusions

The paper presents a theoretical investigation that shows how the instability threshold of a perfectly balanced, horizontally placed, rigid rotor on FRBs is dependent on the shape of the FRB housing. With this aim, two geometries, respectively characterized by a two-lobe wave shape and a lemon one, have been considered in place of the circular profile commonly adopted for the outer plain bearing. The angular orientation and the preload level are distinctive parameters of these two unsymmetrical profiles. Shaping the bore of journal bearings by means of multi-lobe, dammed or stepped profiles is a well-known, low cost practice to gain a better rotor stability with respect to the basic circular geometry. Past research carried out by the author on this topic and the relative scarceness of studies about the specific use of noncircular geometries within FRBs motivated the present study, carried out with a linear approach. In this sense, and through the extension to different bearing layouts, it particularly follows in the wake of some previous works [5,6] focused on the behavior of the common FRB.

The results make it possible to assess the incidence of the noncircular shape and, particularly, of the angular orientation of the profile, on the stability response of the rotor system, with positive or negative effects that vary from case to case. In particular, the adoption of FRBs with outer 2LWB, makes it possible to obtain better stability performances with respect to the CB layout when negative values of the φ angle are employed. The highest instability thresholds are obtained with the negative angle that presents the lesser absolute value among the adopted ones, i.e., $\varphi = -\pi/4$. Above this value, i.e., when φ is null or positive, the performance gradually worsens, with a sensible lowering of the curves occurring beyond $\varphi = \pi/4$. These remarks make it possible to infer that, regarding the analyzed type of stability loss, the response of the 2LWB layout is better than the CB one, on condition that the bearing is not too much inclined above the horizontal, with a peak in the counter-clockwise range of angular displacements. The worst responses are instead obtained with the $\varphi = \pi/2$ and $\varphi = 3\pi/4$ clockwise positions. These circumstances are likely in relation to the assumed vertical direction of the external load and the counter-clockwise, positive angular speed of the journal. The influence due to modifications of the radius and clearance ratios, in terms of extension and position of the HSR and LSR, are in line with those observed in the CB case, whereas varying the mass ratio in the range 0.025–0.075 yields no sensible changes.

When the LBmx and LBmn shapes are adopted in the outer FRB bearing, the stability charts appear rather similar to those obtained in the corresponding 2LWB cases. The influence of negative φ values is less pronounced than in the 2LWB layout, despite a general improvement of stability may be detected, with curves that appear more gathered and close to the CB ones adopted for reference. Nevertheless, the results of these specific inclinations show some differences with respect to those described about the 2LWB layout, particularly in the case of the LBmx profile. A higher degree of similarity may instead be assessed in the stability charts obtained for the 2LWB and the two LB layouts, from case to case, when positive φ values are assigned.

The inspection of the eigenvalues behavior has proved to be an effective tool in order to justify some specific features of the threshold curves. The recourse to brute-force integration, in a few examples related to the case-studies considered in the previous analysis, has given further insight into the examined dynamics. This part of the investigation has been carried out through observation of the journal and ring orbits and evaluation of the minimum film thickness, with further use of a suitable parameter correlated to the orbit dimensions.

The several, simplifying hypotheses in the analysis and, above all, the restriction to the Hopf bifurcation, which takes place from the steady state equilibrium of the perfectly balanced rigid rotor, mark the limits of the investigation. Nevertheless, the obtained results highlight significant effects of the non-circular geometry, as it has been assumed here within the FRB layout, which could reasonably encourage further research on the topic.

Funding: This research received no external funding.

Conflicts of Interest: The author declares no conflict of interest.

Nomenclature

Symbols	Description
B	dimensionless wave amplitude of the 2LWB
C	clearance of the inner bearing [m]
$\overline{C}_H$	parameter of the LB bore [m]
C_o	clearance of the outer bearing (CB bore, reference for the 2LWB and LB cases) [m]
$\overline{Cl}_{o,WB}(\delta,\varphi)$	clearance of the outer bearing (2LWB bore) [m]
$\overline{Cl}_{o,LB\_1,2}(\delta,\theta)$	clearance of the outer bearing (LB bore) [m]
$\overline{d}$	parameter of the LB bore [m]
D	diameter of the inner bearing [m]
$\left\{\begin{array}{c} f_{i,x} \\ f_{i,y} \end{array}\right\}$	$= \dfrac{2\Theta}{\sigma_{m,D}W}\left\{\begin{array}{c} F_{i,x} \\ F_{i,y} \end{array}\right\}$: dimensionless force components of the inner fluid film
$\left\{\begin{array}{c} f_{o,x} \\ f_{o,y} \end{array}\right\}$	$= \dfrac{1}{\sigma_{mo,D}(W+W_R)}\left\{\begin{array}{c} F_{o,x} \\ F_{o,y} \end{array}\right\}$: dimensionless force components of the outer fluid film
$F_{i,x}, F_{i,y}$	force components of the inner fluid film [N]
$F_{o,x}, F_{o,y}$	force components of the outer fluid film [N]
g	gravity acceleration [m/s^2]
h	dimensionless fluid film thickness of the inner bearing
L	axial length of the inner bearing [m]
m, m_R	semi-mass of the rotor, mass of the ring [kg]
p	oil film pressure [N]
$p*$	$= \mu\overline{\omega}\dfrac{R^2}{C^2}\dfrac{L^2}{D^2}$: reference pressure in the internal oil film [N]
p_E	$= \overline{d}/\overline{C}_H = d/C_H$: lemon-bearing ellipticity ratio
p_o^*	$= \mu_o\omega_R\dfrac{R_o^2}{C_o^2}\dfrac{L_o^2}{D_o^2}$: reference pressures in the outer oil film [N]
R	radius of the inner bearing [m]
s	$= \omega/\Omega$: dimensionless rotor speed
t	time [s]
W, W_R	half-weight of the rotor, weight of the ring [N]
x, y	dimensionless coordinates of the journal (Oxy: right-hand reference with horizontal x-axis)
$\overline{x}, \overline{y}$	coordinates of the journal center [m]
$\hat{x}, \hat{y}$	$= x - \kappa x_R,\ y - \kappa y_R$: dimensionless relative positions between journal and ring
z	dimensionless axial coordinate of the bearing
α	angular position of the film border (Appendix A)
δ	circumferential coordinate
γ	dimensionless pressure of the oil film
ζ	$= z/L$: dimensionless axial coordinate
θ	$= \varphi/2$: orientation angle of the LB profile (Figure 1b)
Θ	$= \omega/\overline{\omega} = (1+\hat{\omega})^{-1}$
κ	$= C_o/C$: clearance ratio
λ^*	eigenvalue (Appendix B)
Λ	$= L_o/L$: axial length ratio
μ	absolute viscosity of the lubricant in the inner bearing [Pa s]
ρ_μ	$= \mu_o/\mu$: viscosity ratio
$\sigma_{m,D}$	$= \dfrac{\mu\omega LD}{W}\dfrac{R^2L^2}{C^2D^2}$: Sommerfeld number in the inner oil film
$\sigma_{mo,D}$	$= \dfrac{\mu_o\omega_R L_o D}{W+W_R}\dfrac{R_o^2}{C_o^2}\dfrac{L_o^2}{D_o^2}$: Sommerfeld number in the outer oil film
$\sigma_{m,D}^*$	$= \dfrac{\mu\Omega LD}{W}\dfrac{R^2L^2}{C^2D^2}$: bearing parameter
Σ	$= \sigma_{m,D}/\sigma_{mo,D} = \dfrac{(1+\chi)\kappa^2}{\rho_\mu\hat{\omega}\Lambda^3 \upsilon}$: Sommerfeld number ratio
τ	$= \omega t$: dimensionless time
υ	$= R_o/R$: radius ratio
φ	angular orientation of the noncircular outer bearing (Figure 1a)
χ	$= m_R/m = W_R/W$: mass ratio
Ψ_S	$= \overline{C}_H/\overline{C}_V = 1/(1-p_E)$: lemon-bearing clearance ratio

ω	rotor angular speed [rad s^{-1}]
$\overline{\omega}$	$= \omega + \omega_R$ [rad s^{-1}]
$\hat{\omega}$	$= \omega_R/\omega$: angular speed ratio
Ω	$= \sqrt{g/C}$: reference angular frequency [rad s^{-1}]
$_o$	subscript for the quantities relative to the outer bearing
$_R$	subscript for the quantities relative to the ring
$'$	dimensionless time derivative

Appendix A

Appendix A.1 Inner Bearing

Dimensionless Reynolds Equation with SBA:

$$\frac{\partial^2 \gamma}{\partial \zeta^2} = 24 \frac{\left(\hat{x} - 2\Theta\hat{y}'\right)\sin\delta - \left(\hat{y} + 2\Theta\hat{x}'\right)}{(1 - \hat{x}\cos\delta - \hat{y}\sin\delta)^3}, \tag{A1}$$

Pressure within the oil film:

$$\gamma = 3\frac{\left(\hat{x} - 2\Theta\hat{y}'\right)\sin\delta - \left(\hat{y} + 2\Theta\hat{x}'\right)\cos\delta}{(1 - \hat{x}\cos\delta - \hat{y}\sin\delta)^3}\left(4\zeta^2 - 1\right), \; p = p^*\gamma, \; p^* = \mu\overline{\omega}(R^2/C^2)(L^2/D^2) \tag{A2}$$

Dimensionless expression of the force components of the inner oil film:

$$\left\{ \begin{array}{c} f_{i,x} \\ f_{i,y} \end{array} \right\} = \frac{2\Theta}{\sigma_{m,D}W}\left\{ \begin{array}{c} F_{i,x} \\ F_{i,y} \end{array} \right\} = E\left(3V\left\{ \begin{array}{c} x \\ y \end{array} \right\} - G\left\{ \begin{array}{c} \sin\gamma \\ -\cos\gamma \end{array} \right\} - 2S\left\{ \begin{array}{c} \cos\gamma \\ \sin\gamma \end{array} \right\}\right)$$

$$\alpha = \arctan\left((\hat{y} + 2\Theta\hat{x}')/(\hat{x} - 2\Theta\hat{y}')\right) - (\pi/2)\sin\left((\hat{y} + 2\Theta\hat{x}')/(\hat{x} - 2\Theta\hat{y}')\right) - (\pi/2)\sin(\hat{y} + 2\Theta\hat{x}')$$

$$E1 = \left((\hat{x} - 2\Theta\hat{y}')^2 + (\hat{y} + 2\Theta\hat{x}')^2\right)^{0.5}, E2 = 1 - \hat{x}^2 - \hat{y}^2, E3 = \hat{y}\cos\alpha - \hat{x}\sin\alpha, E = -E1/E2,$$

$$S1 = \hat{x}\cos\alpha - \hat{y}\sin\alpha, S2 = 1 - S1^2, G = (2/E2^{0.5})\left(\arctan(E3/E2^{0.5}) + \pi/2\right), V1 = 2 + GE3,$$

$$V = V1/E2. \tag{A3}$$

Sommerfeld number for the inner oil film:

$$\sigma_{m,D} = \frac{1}{1 + \hat{\omega}}\frac{2\left(1 - \varepsilon^2\right)^2}{\varepsilon\,\sqrt{\pi^2(1 - \varepsilon^2) + 16\varepsilon^2}} = \Theta\sigma_{m,D\_simp} \tag{A4}$$

σ_{m,D_simp} is the number relative to the simple plain bearing.

Appendix A.2 Outer Bearing

Dimensionless Reynolds Equation with SBA:

$$\frac{\partial^2 \gamma_o}{\partial \zeta^2} = 24\frac{1}{h_o^3}\left(\frac{\partial h_o}{\partial \delta} + \frac{2}{\hat{\omega}}\frac{\partial h_o}{\partial \tau}\right), \; \gamma_o = p_o/p_o^*, \; p_o^* = \mu_o\omega R\frac{R_o^2}{C_o^2}\frac{L_o^2}{D_o^2} \tag{A5}$$

Dimensionless oil film pressures:

$$\text{CB}: \gamma_o = 3\frac{\left(x_o - 2\hat{\omega}^{-1}y'_o\right)\sin\delta - \left(y_o + 2\hat{\omega}^{-1}x'_o\right)\cos\delta}{(1 - x_o\cos\delta - y_o\sin\delta)^3}\left(4\zeta^2 - 1\right) \tag{A6}$$

$$\text{2LWB}: \gamma_o = 3\frac{\left(x_o - 2\hat{\omega}^{-1}y'_o\right)\sin\delta - \left(y_o + 2\hat{\omega}^{-1}x'_o\right)\cos\delta - 2B\sin(2\delta + \varphi)}{[1 - x_o\cos\delta - y_o\sin\delta + B\cos(2\delta + \varphi)]^3}\left(4\zeta^2 - 1\right) \tag{A7}$$

$$\text{LBmx}: \gamma_o = 3\frac{\left(x_o - 2\hat{\omega}^{-1}y'_o\right)\sin\delta - \left(y_o + 2\hat{\omega}^{-1}x'_o\right)\cos\delta + (-1)^k B\frac{\pi}{2}\cos(\delta + \theta)}{\left\{1 - x_o\cos\delta - y_o\sin\delta + B\left[1 + (-1)^k\frac{\pi}{2}\sin(\delta + \theta)\right]\right\}^3}\left(4\zeta^2 - 1\right) \tag{A8}$$

$$\text{LBmn}: \quad \gamma_o = 3\frac{\left(x_o - 2\hat{\omega}^{-1}y'_{\,o}\right)\sin\delta - \left(y_o + 2\hat{\omega}^{-1}x'_{\,o}\right)\cos\delta + (-1)^k p_E \frac{1-B}{1-p_E}\cos\delta}{\left\{\frac{1-B}{1-p_E}\left[1 + (-1)^k p_E \sin(\delta+\theta)\right] - x_o\cos\delta - y_o\sin\delta\right\}^3}\left(4\zeta^2 - 1\right) \tag{A9}$$

The following equation expresses the components in dimensionless way

$$\left\{ \begin{array}{c} f_{o,x} \\ f_{o,y} \end{array} \right\} = \frac{1}{\sigma_{mo,D}(W + W_R)}\left\{ \begin{array}{c} F_{o,x} \\ F_{o,y} \end{array} \right\} = \Psi_{SH}\left\{ \begin{array}{c} f_{SHS,x} \\ f_{SHS,y} \end{array} \right\}, \tag{A10}$$

where

$$\Psi_{SH} = -\frac{3\pi}{4(N-1)(M-1)}, \quad \left\{ \begin{array}{c} f_{SHS,x} \\ f_{SHS,y} \end{array} \right\} = \sum_{i=1}^{M-1}\sum_{j=1}^{N-1}\left[\left(\hat{p}_{o(i,j)} + \hat{p}_{o(i+1,j)} + \hat{p}_{o(i,j+1)} + \hat{p}_{o(i+1,j+1)}\right)\left\{ \begin{array}{c} \cos\delta_{(2j+1)/2} \\ \sin\delta_{(2j+1)/2} \end{array} \right\}\right] \tag{A11}$$

Appendix B

Main quantities adopted in Equation (13):

$$\underline{\underline{I_2}} = \left[\begin{array}{cc} 1 & 0 \\ 0 & 1 \end{array} \right], \; \underline{\underline{Q}} = \left[\begin{array}{cc} d_{\xi\xi} & d_{\xi\eta} \\ d_{\eta\xi} & d_{\eta\eta} \end{array} \right], \; \underline{q} = \left[\begin{array}{cc} d_{o,\xi\xi} & d_{o,\xi\eta} \\ d_{o,\eta\xi} & d_{o,\eta\eta} \end{array} \right], \; \underline{\underline{P}} = \left[\begin{array}{cc} k_{\xi\xi} & k_{\xi\eta} \\ k_{\eta\xi} & k_{\eta\eta} \end{array} \right], \; \underline{p} = \left[\begin{array}{cc} k_{o,\xi\xi} & k_{o,\xi\eta} \\ k_{o,\eta\xi} & k_{o,\eta\eta} \end{array} \right]$$

$$\rho = \frac{1+\chi}{\kappa\Theta_R}, \; \Theta_R = \omega_R/\overline{\omega}.$$

The matrices $\underline{\underline{Q}}$ and $\underline{\underline{P}}$ can be deduced from ([5], Appendix A), with the remarks

$d_{\xi\xi} = A_1, d_{\xi\eta} = -A_2, d_{\eta\xi} = -A_3, d_{\eta\eta} = A_4, k_{\xi\xi} = B_1, k_{\xi\eta} = k_{\eta\xi} = -B_2, k_{\eta\eta} = B_4$.

The elements of $\underline{q}$ and $\underline{p}$ are numerically obtained by means of finite difference approximations of the partial derivatives of the outer film force around the equilibrium point.

With use of the following quantities

$$\underline{\underline{M_m}} = s^2\left[\begin{array}{c:c} \underline{\underline{I_2}} & 0 \\ \hdashline 0 & \chi\underline{\underline{I_2}} \end{array} \right], \; \underline{\underline{M_D}} = \Theta\left[\begin{array}{c:c} \underline{\underline{Q}} & -\underline{\underline{Q}} \\ \hdashline -\underline{\underline{Q}} & \underline{\underline{Q}} + \rho\underline{q} \end{array} \right], \; \underline{\underline{M_k}} = \left[\begin{array}{c:c} \underline{\underline{P}} & -\underline{\underline{P}} \\ \hdashline -\underline{\underline{P}} & \underline{\underline{P}} + \rho\Theta_R\underline{p} \end{array} \right], \; \underline{z} = \left\{ \begin{array}{c} \xi \\ \eta \\ \kappa\xi_R \\ \kappa\eta_R \end{array} \right\} \tag{A12}$$

Equation (13) can be equivalently written

$$\underline{\underline{M_m}}\underline{z}'' + \underline{\underline{M_D}}\underline{z}' + \underline{\underline{M_k}}\underline{z} = 0 \tag{A13}$$

and furthermore

$$\underline{\underline{U}}\underline{Z}' + \underline{\underline{V}}\underline{Z} = \underline{0} \tag{A14}$$

where

$$\underline{\underline{U}} = \left[\begin{array}{c:c} \underline{\underline{M_d}} & \underline{\underline{M_m}} \\ \hdashline \underline{\underline{M_m}} & 0 \end{array} \right], \; \underline{\underline{V}} = \left[\begin{array}{c:c} \underline{\underline{M_k}} & 0 \\ \hdashline 0 & -\underline{\underline{M_m}} \end{array} \right], \; \underline{Z} = \left\{ \begin{array}{c} \underline{z} \\ \underline{z}' \end{array} \right\}. \tag{A15}$$

Let the solution of Equation (A13) be given as $\underline{Z}(\tau) = \underline{\hat{Z}}e^{\lambda\tau}$, so as to write

$$\left[\lambda\underline{\underline{U}} + \underline{\underline{V}}\right]\underline{\hat{Z}} = 0. \tag{A16}$$

After premultiplication by $\underline{\underline{V}}^{-1}$ it is obtained

$$\left[\lambda\underline{\underline{D}} + \underline{\underline{I}}\right]\underline{\hat{Z}} = 0, \text{ with}: \; \underline{\underline{D}} = \underline{\underline{V}}^{-1}\underline{\underline{U}} \tag{A17}$$

and eventually

$$\underline{\underline{D}}\underline{\hat{Z}} = \lambda^*\underline{\hat{Z}}, \tag{A18}$$

which represents the eigenvalue problem with $\lambda^* = -\lambda^{-1}$.

References

1. Schweizer, B. Total instability of turbocharger rotors—Physical explanation of the dynamic failure of rotors with full-floating ring bearings. *J. Sound Vib.* **2009**, *328*, 156–190. [CrossRef]
2. Tian, L. Investigation into Nonlinear Dynamics of Rotor-Floating Ring Bearing Systems in Automotive Turbochargers. Ph.D. Thesis, University of Sussex, Brighton, UK, 2012.
3. Shaw, M.C.; Nussdorfer, T.J. *An Analysis of the Full-Floating Journal Bearing*; NACA Report 866; National Advisory Committee for Aeronautics: Washington, DC, USA, 1947.
4. Orcutt, F.K.; Ng, C.W. Steady-state and dynamic properties of the floating ring journal bearing. *ASME J. Lubr. Tech.* **1968**, *90*, 243–253. [CrossRef]
5. Tanaka, M.; Hori, Y. Stability characteristics of floating bush bearings. *ASME J. Lubr. Tech.* **1972**, *94*, 248–256. [CrossRef]
6. Boyaci, A.; Hetzler, H.; Seemann, W.; Proppe, C.; Wauer, J. Analytical bifurcation analysis of a rotor supported by floating ring bearings. *Nonlinear Dyn.* **2009**, *57*, 497–507. [CrossRef]
7. Hatakenaka, K.; Tanaka, M.; Suzuki, K. A theoretical analysis of floating bush journal bearing with axial oil film rupture being considered. *ASME J. Tribol.* **2002**, *124*, 494–505. [CrossRef]
8. Chasalevris, A. Finite length floating ring bearings: Operational characteristics using analytical methods. *Tribol. Int.* **2016**, *94*, 571–590. [CrossRef]
9. Gunter, E.J.; Chen, W.J. Dynamic Analysis of a Turbocharger in Floating Bushing Bearings. In Proceedings of the ISCORMA-3, 3rd International Symposium on Stability Control of Rotating Machinery, Cleveland, OH, USA, 19–23 September 2005.
10. Inagaki, M.; Kawamoto, A.; Abekura, T.; Suzuki, A.; Rübel, J.; Starke, J. Coupling Analysis of Dynamics and Oil Film Lubrication on Rotor—Floating Bush Bearing System. *J. Syst. Des. Dyn.* **2011**, *5*, 461–473. [CrossRef]
11. Tian, L.; Wang, W.J.; Peng, Z.J. Nonlinear effects of unbalance in the rotor-floating ring bearing system of turbochargers. *Mech. Syst. Signal. Process.* **2013**, *34*, 298–320. [CrossRef]
12. Schweizer, B. Dynamics and stability of turbocharger rotors. *Arch. Appl. Mech.* **2010**, *80*, 1017–1043. [CrossRef]
13. Bonello, P. Transient modal analysis of the non-linear dynamics of a turbocharger on floating ring bearings. *Proc. Inst. Mech. Eng. Part J J. Eng. Tribol.* **2009**, *223*, 79–93. [CrossRef]
14. San Andrés, L.; Rivadeneira, J.C.; Chinta, M.; Gjika, K.; LaRue, G. Nonlinear Rotordynamics of Automotive Turbochargers: Prediction and Comparison to Test Data. *ASME J. Eng. Gas Turbines Power* **2007**, *129*, 488–493. [CrossRef]
15. Eling, R.; van Ostayen, R.; Rixen, D. Multilobe Floating Ring Bearings for Automotive Turbochargers. In Proceedings of the 9th IFToMM International Conference on Rotor Dynamics, Milan, Italy, 22–25 September 2014; Mechanisms and Machine Science 21. Pennacchi, P., Ed.; Springer: Cham, Switzerland, 2015; pp. 821–833.
16. Soni, S.; Vakharia, D.P. A steady-state performance analysis of a non-circular cylindrical floating ring journal bearing. *Proc. Inst. Mech. Eng. Part J J. Eng. Tribol.* **2017**, *231*, 41–56. [CrossRef]
17. Bernhauser, L.; Heinisch, M.; Schörgenhumer, M.; Nader, M. The Effect of Non-Circular Bearing Shapes in Hydrodynamic Journal Bearings on the Vibration Behavior of Turbocharger Structures. *Lubricants* **2017**, *5*, 6. [CrossRef]
18. Adiletta, G.; Mancusi, E.; Strano, S. Nonlinear behaviour analysis of a rotor on two-lobe wave journal bearings. *Tribol. Int.* **2011**, *44*, 42–54. [CrossRef]
19. Adiletta, G. An insight into the dynamics of a rigid rotor on two-lobe wave squeeze film damper. *Tribol. Int.* **2017**, *116*, 69–83. [CrossRef]
20. Pinkus, O. Analysis of elliptical bearings. *Trans. ASME* **1956**, *78*, 965–973.
21. Ott, H.H.; Buchter, J.-E.; Frey, U. Wellenlage, Reibung und Öldurchsatz von Radial-Gleitlagern mit zwei zusammengeschobenen Kreisschalen (Zitronenspiellager). *Schweiz. Bauztg.* **1969**, *87*, 207–212.
22. Myers, C.J. Bifurcation theory applied to oil whirl in plain cylindrical journal bearings. *ASME J. Appl. Mech.* **1984**, *51*, 244–250. [CrossRef]

23. Capone, G. Descrizione analitica del campo di forze fluidodinamico nei cuscinetti cilindrici lubrificati. *Energ. Elettr.* **1991**, *3*, 105–110.
24. Lang, O.R.; Steinhilper, W. *Gleitlager: Berechnung und Konstruktion von Gleitlagern mit Konstanter und Zeitlich Veränderlicher Belastung*, 1st ed.; Springer: Berlin/Heidelberg, Germany, 1978; pp. 304–307.

Publisher's Note: MDPI stays neutral with regard to jurisdictional claims in published maps and institutional affiliations.

lubricants

Article

Run-Up Simulation of a Semi-Floating Ring Supported Turbocharger Rotor Considering Thrust Bearing and Mass-Conserving Cavitation

Christian Ziese *, Cornelius Irmscher, Steffen Nitzschke, Christian Daniel and Elmar Woschke

Institute of Mechanics, Otto von Guericke University Magdeburg, 39106 Magdeburg, Germany;
cornelius.irmscher@ovgu.de (C.I.); steffen.nitzschke@ovgu.de (S.N.); christian.daniel@ovgu.de (C.D.);
elmar.woschke@ovgu.de (E.W.)
* Correspondence: christian.ziese@ovgu.de; Tel.: +49-391-67-52885

Abstract: The vibration behaviour of turbocharger rotors is influenced by the acting loads as well as by the type and arrangement of the hydrodynamic bearings and their operating condition. Due to the highly non-linear bearing behaviour, lubricant film-induced excitations can occur, which lead to sub-synchronous rotor vibrations. A significant impact on the oscillation behaviour is attributed to the pressure distribution in the hydrodynamic bearings, which is influenced by the thermo-hydrodynamic conditions and the occurrence of outgassing processes. This contribution investigates the vibration behaviour of a floating ring supported turbocharger rotor. For detailed modelling of the bearings, the Reynolds equation with mass-conserving cavitation, the three-dimensional energy equation and the heat conduction equation are solved. To examine the impact of outgassing processes and thrust bearing on the occurrence of sub-synchronous rotor vibrations separately, a variation of the bearing model is made. This includes run-up simulations considering or neglecting thrust bearings and two-phase flow in the lubrication gap. It is shown that, for a reliable prediction of sub-synchronous vibrations, both the modelling of outgassing processes in hydrodynamic bearings and the consideration of thrust bearing are necessary.

Keywords: run-up simulation; semi-floating ring bearing; thrust bearing; two-phase flow cavitation; oil-whirl and oil-whip

Citation: Ziese, C.; Irmscher, C.; Nitzschke, S.; Daniel, C.; Woschke, E. Run-Up Simulation of a Semi-Floating Ring Supported Turbocharger Rotor Considering Thrust Bearing and Mass-Conserving Cavitation. *Lubricants* **2021**, *9*, 44. https://doi.org/10.3390/lubricants 9040044

Received: 27 January 2021
Accepted: 12 April 2021
Published: 16 April 2021

Publisher's Note: MDPI stays neutral with regard to jurisdictional claims in published maps and institutional affiliations.

1. Introduction

Turbochargers have become indispensable in combustion engines due to economic, ecological and efficiency reasons. The engine's charge exchange work is no longer sufficient to satisfy the required demands, so additional charging is necessary. The rotor's principle design consists of a compressor and turbine wheel taken up by a common shaft. Due to the flow of exhaust gases to the turbine wheel, the kinetic energy of the gases is used to set the rotor in a rotational motion. Simultaneously on the compressor side, the fresh air is compressed and supplied to the combustion chamber. Consequently, the main aim of turbocharging is to increase the power and combustion efficiency of the engine by increasing the mean pressure while reducing fuel consumption if possible.

An essential factor for compliance with the requirements is the support of the rotor. This is preferably achieved by using hydrodynamic bearings. Compared to roller bearings, hydrodynamic bearings have a simpler and more cost-efficient design as well as more favourable thermo-hydrodynamic operating conditions. A disadvantage is their highly non-linear behaviour, which can even lead to additional rotor excitations in terms of sub-synchronous vibration [1–11]. Woschke et al. [1,2] examined the excitation mechanism in context with the current natural frequencies and damping of the rotor. By solving the eigenvalue problem at the current operating point, including the non-linear bearing properties, a Campbell diagram with the corresponding damping is plotted and the rotor excitation via the half-whirl frequency is evaluated over the entire run-up. It is shown that an oil-whip can occur if a weakly damped natural frequency is excited by an oil-whirl. The current rotor natural frequencies depend on the mass and stiffness distributions of the rotor as well as on the stiffness and damping properties of the individual bearings [2]. With the occurrence of oil-whips, increased vibration amplitudes, wear and even failure of the rotor can occur. For this reason, a reliable prediction of the critical sub-synchronous vibrations is necessary.

Concerning the stiffness and damping properties of hydrodynamic bearings, the kinematics of shaft and housing, the temperature distribution and outgassing processes have significant influences. Temperature changes lead to thermal gap changes. In addition, the lubricant properties such as the viscosity are non-linearly dependent on it. Due to thermal boundary conditions and the weak thermal conductivity of common lubricating oils, the viscosity can vary in all three spatial directions, which must be taken into account owing to the aforementioned non-linear temperature dependence. For this reason, Dowson derived the generalised Reynolds equation [12], which allows the consideration of a viscosity that varies over the gap height. In the following years, thermo-hydrodynamic phenomena were thoroughly investigated, which is comprehensively documented in the reviews [13–16]. In more recent publications, a transient consideration of the three-dimensional thermal phenomena is increasingly carried out (see, e.g., [17–19]). This is a requirement for the investigation of vibration phenomena.

Outgassing processes describe the presence of gas in the lubricating film with the consequence of developing a multi-phase flow [20]. In hydrodynamic bearings, vapour and gas cavitation as well as aeration can occur in particular. Vapour cavitation appears if the hydrodynamic pressure falls below the vapour pressure so that the lubricant changes its aggregate state. In contrast, gas cavitation is a diffusion-induced process in which the gas dissolved in the lubricant undergoes a phase transition if the pressure falls below the saturation pressure. Besides vapour and gas cavitation, air from the environment can also penetrate the lubrication gap via the bearing edges (aeration) [21]. With the occurrence of two-phase flow, a partially filled lubricating gap is established, which leads to a softer bearing behaviour and consequently favours the appearance of sub-synchronous rotor vibrations. For modelling outgassing processes, the two-phase model is applied in this contribution. It is a mass-conserving cavitation model, which makes it possible to consider outgassing processes depending on the hydrodynamic pressure and the lubricant film temperature [3,22–25]. For completeness, the cavitation theories according to Jakobsson Floberg and Olsson (JFO-model) or its efficient numerical implementation by Elrod [11,26–29] and the model of bubble dynamics [30] may also be mentioned.

In addition to the stiffness and damping behaviour of the individual bearings, the arrangement and type of bearings also influence the current rotor natural frequency. The radial displacement of the rotor is mainly supported by journal bearings. Consequently, this bearing type contributes to the stiffness and damping in horizontal and vertical rotor direction, while the tilting stiffness is limited. In terms of tilting stiffness and damping, thrust bearings contribute essentially. The rotor tilting results in an asymmetrical pressure distribution at the thrust bearing so that a torque is created, which counteracts the rotor tilting. Thus, the thrust bearing provides a further part to the stiffness and damping, which can influence the current natural frequencies of the rotor. Investigations of the influence of thrust bearings on a full-floating ring supported turbocharger rotor are carried out in [31–36]. A variation of the axial load and thrust bearing design parameters is made in [32]. The simulations show that with increasing axial load, the start frequency of the oil-whip can be shifted to higher speed ranges. This is because smaller lubrication gaps can occur at the thrust bearing so that the hydrodynamic pressure and consequently the stiffness and damping of the bearing increase. In comparison, the number of bearing segments seems to have only a minor influence.

This contribution deals with the influence of outgassing processes as well as the impact of the thrust bearing on the occurrence of sub-synchronous vibrations for a semi-floating ring supported turbocharger rotor. Due to the design, the rotor's centre of gravity is close to the turbine bearing, which results in a general tendency to high rotor tilting during operation. Under these circumstances, it is essential to consider the thrust bearing in order to predict the sub-synchronous vibrations and the rotor amplitudes reliably. To investigate the individual effects on the rotor's response behaviour, various models of the hydrodynamic bearings are chosen and their influences are compared with the results of the run-up measurement. The level of detail includes the Half-Sommerfeld solution without thrust bearing up to the two-phase model with thrust bearing. For detailed modelling of the thermo-hydrodynamic states of the journal bearings, the Reynolds equation with mass-conserving cavitation according to the two-phase model and the three-dimensional energy equation for the temperature distribution in the lubricating film as well as the heat conduction equation for the shaft, bushing and housing are solved. Regarding the thrust bearing, the Reynolds equation with centrifugal flow and mass-conservation cavitation is solved.

2. Theoretical Fundamentals

The examined rotor consists of an elastic shaft, the compressor and turbine wheel, as well as the sealing disk, auxiliary bearing and thrust ring (cf. Figure 1). To model the elastic deformations of the shaft, the Finite-Element-Method (FEM) according to the Timoshenko beam theory [37] is applied. The FE-nodes are positioned at relevant shaft shoulders as well as at the bearings in order to determine the lubrication gap. Furthermore, the components assembled on the shaft can be defined as rigid bodies and their inertia properties are assigned to the corresponding FE-nodes. The total mass of the rotor is m_{Rot} = 20 kg and its centre of gravity is located close to the turbine bearing. To take the unbalance condition of the rotor into account, the compressor and turbine wheel have an unbalance distribution (u_C, u_T) with an angular offset. In addition to the rotor, the compressor and turbine side floating rings exist as separate rigid bodies within a multi-body environment. The interaction between the individual components is realised via the resulting forces and torques of the hydrodynamic films. These are marked blue in the figure. As a result, there is an interaction between housing and bushing (outer lubrication gap) and between bushing and shaft (inner lubrication gap) at the journal bearing. The forces and torques occurring at the thrust bearings act between the bearing housing (inertial system) and the thrust ring or auxiliary bearing.

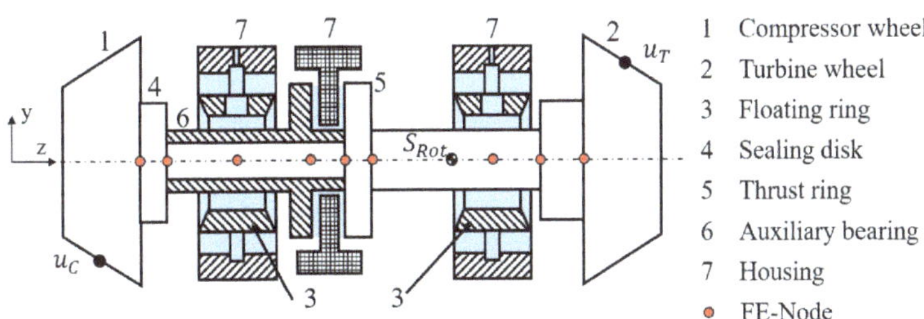

Figure 1. Schematic representation of the rotor model.

For the radial support of the rotor, floating ring bearings with outer squeeze-film damping are used. Due to the prevention of the rotational movement of the bushing around the longitudinal rotor axis, a hydrodynamic pressure can only be generated at the outer gap if there is a squeezing of the lubricant. To ensure a sufficient lubricant supply, the outer film has a circumferential groove, so the oil is distributed evenly over the entire circumference. Furthermore, the oil can flow through the connecting channels of the bushing to the inner lubrication gap and also build up a hydrodynamic pressure. Accordingly, the lubricating films are coupled to each other, whereby the lubricant can flow either to the inner or to the outer gap, depending on the operating bearing condition. It should be mentioned that the inner lubrication gap has a multi-lobe geometry consisting of three segments and both the inner and outer gaps are unsealed. A schematic illustration of the bearing with considered fluid flows is shown in Figure 2.

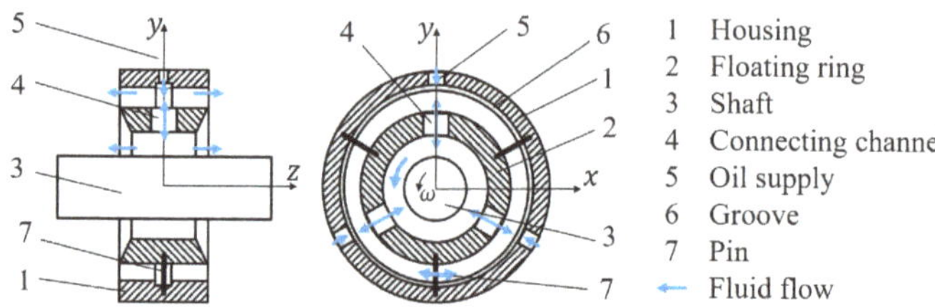

Figure 2. Schematic representation of the floating ring bearing with considered fluid flows.

For the axial support of the rotor, thrust bearings are used. With regard to the design, each segment has a wedge, flat and groove area, whereby the oil is supplied through holes in each groove area. To avoid the oil flowing off at the outer edge, the thrust bearings are additionally sealed. A sealing effect is achieved by the fact that the outer bearing edge has no contour. Due to the turbocharger design, the axial lubrication films on the compressor and turbine sides can be considered separately. A schematic design of the thrust bearing can be seen in Figure 3.

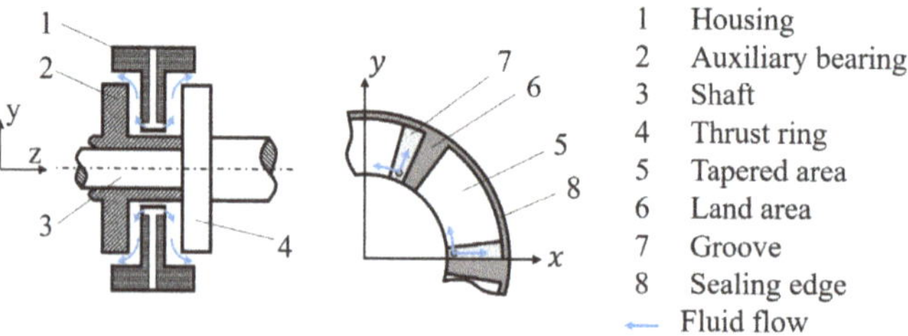

Figure 3. Schematic representation of the thrust bearing with considered fluid flows.

The following sections present the theoretical fundamentals for determining the pressure distribution in hydrodynamic bearings by solving the Reynolds equation taking mass-conserving cavitation into account. Besides the influence of cavitation, thermodynamic processes in the lubricating film and at supported elements as well as their coupling to each other are explained more in detail.

2.1. Time Integration

With respect to the simulation of transient processes, time integration and the implementation of hydrodynamics is described first. Within the multi-body system (MBS), the differential equation of motion

$$\underline{\underline{M}}(y) \cdot \ddot{y} + \underline{h}_\omega(\dot{y}) + \underline{h}_{el}(y, \dot{y}) = \underline{h}_e(t, y, \dot{y}) \tag{1}$$

is solved, where $\underline{\underline{M}}$ is equal to the inertia matrix and y represents the state vector of the bushings and the degrees of freedom of the elastic rotor. In addition, the gyroscopic effects, centrifugal and Coriolis forces $\underline{h}_\omega$ as well as the stiffness and damping properties of the shaft $\underline{h}_{el}$ are taken into account. External forces such as the unbalance or resulting forces of the hydrodynamics are summarised in $\underline{h}_e$. The workflow of time integration is shown in Figure 4.

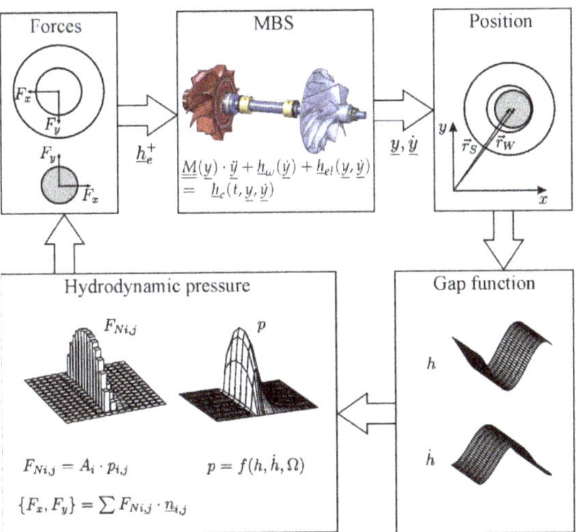

Figure 4. Workflow of time integration, according to Woschke [1].

Starting from the rotor position at the current time, the lubrication gap h und its time derivation $\dot{h}$ can be determined via the kinematic states of the shaft and housing ($\vec{r}_S$, $\vec{r}_H$). With knowledge of the gap, the hydrodynamic pressure distribution is calculated by solving the Reynolds equation. Subsequently, an integration of the pressure over the bearing surface provides the resulting bearing forces acting on the shaft and housing (F_x, F_y). With regard to the thrust bearing, both thrust forces and torques are considered due to the asymmetrical pressure distribution. With knowledge of the forces and torques acting on the rotor, the right-hand side of Equation (1) and finally the acceleration vector can be calculated. By executing a time integration procedure, the state vector $(\underline{y}, \underline{\dot{y}})$ at the next time step is known.

The described procedure of time integration corresponds to an online solution of the Reynolds equation. The pressure distribution is calculated at each time step, which provides the advantage of an exact determination of bearing forces depending on the current kinematic state of the supporting elements. Consequently, it is not necessary to generate stiffness and damping tables and interpolate the table data to calculate the acting forces.

2.2. Reynolds Equation for Journal and Thrust Bearings

To determine the pressure distribution in journal bearings, a simplified representation of the Navier–Stokes equation is used [1,2,26]. It is assumed that the lubricant is incompressible and has a laminar flow behaviour. Furthermore, the oil properties can be variable due to the lubricant film temperature. The implementation of the bearing-specific assumptions leads to the generalised Reynolds equation [12,24] for journal bearings

$$0 = \underbrace{\frac{\partial}{\partial x}\left(G\frac{\partial p}{\partial x}\right) + \frac{\partial}{\partial y}\left(G\frac{\partial p}{\partial y}\right)}_{\text{Poiseuille-flow}} + \underbrace{\frac{\partial}{\partial x}(F_{CS}u_S + F_{CH}u_H)}_{\text{Couette-flow}} + \underbrace{\frac{\partial}{\partial t}(F_\rho)}_{\text{Squeeze-film flow}} \tag{2}$$

with

$$F_0 = \int_{z=0}^{h} \frac{1}{\eta}\,dz\,, \quad F_1 = \int_{z=0}^{h} \frac{z}{\eta}\,dz\,, \quad I_0 = \int_{z=0}^{z} \frac{1}{\eta}\,dz\,, \quad I_1 = \int_{z=0}^{z} \frac{z}{\eta}\,dz \tag{3}$$

and

$$G = \int_{z=0}^{h} \rho \left[I_1 - \frac{F_1}{F_0} I_0 \right] dz \,, \tag{4}$$

$$F_{CS} = \int_{z=0}^{h} \rho \frac{I_0}{F_0} dz \,, \tag{5}$$

$$F_{CH} = \int_{z=0}^{h} \rho \left(1 - \frac{I_0}{F_0} \right) dz \,, \tag{6}$$

$$F_\rho = \int_{z=0}^{h} \rho \, dz \,. \tag{7}$$

In Equations (2)–(7), p is the hydrodynamic pressure, h is the lubricant gap height and ρ, η are the density and viscosity of the lubricant. Due to the assumption of a significantly smaller lubrication gap height compared to other bearing dimensions, cartesian coordinates with x and y can be used for the bearing circumference and width direction. Furthermore, the circumferential velocity on the surface of the shaft u_S and housing u_H can be used as boundary conditions for the fluid flow. The time derivative is needed to take transient effects into account.

A simplification of Equation (2) is achieved by assuming constant lubricant properties over the gap height. As a consequence, Equations (3)–(7) can be calculated, leading to the Reynolds equation

$$0 = \underbrace{- \frac{\partial}{\partial x} \left(\frac{\rho h^3}{12\eta} \frac{\partial p}{\partial x} \right) - \frac{\partial}{\partial y} \left(\frac{\rho h^3}{12\eta} \frac{\partial p}{\partial y} \right)}_{\text{Poiseuille-flow}} + \underbrace{\frac{\partial}{\partial x} \left(\rho h \frac{u_S + u_H}{2} \right)}_{\text{Couette-flow}} + \underbrace{\frac{\partial}{\partial t}(\rho h)}_{\text{Squeeze-film flow}} \,. \tag{8}$$

At this point, it should be mentioned that the lubricant properties need to be averaged over the gap height to use Equation (8). Using a film-averaged temperature to determine viscosity would lead to a reduced pressure distribution and softer bearing behaviour. For the numerical implementation, the three-dimensionally distributed lubricant properties are determined first as a function of oil temperature and subsequently averaged over the gap height. The oil properties are averaged after the non-linear properties have been taken into account.

Within this contribution, Equation (8) is chosen to determine the pressure distribution in journal bearings because the computing times are lower compared to the generalised Reynolds equation. This is due to the integrals within the generalised Reynolds equation, which have to be calculated for each time step of the time integration. To show the validity of the simplified Reynolds equation, the spectrograms for the run-up simulations with the generalised Reynolds equation are in the Appendix A.

To determine the pressure distribution in thrust bearings, the generalised Reynolds equation with centrifugal force can also be used, but, for the same reasons as for the journal bearing, the lubricant properties can be averaged over the gap height, resulting in the Reynolds equation

$$0 = \underbrace{- \frac{\partial}{r\partial r} \left(r \frac{\rho h^3}{12\eta} \frac{\partial p}{\partial r} \right) - \frac{\partial}{r\partial\varphi} \left(\frac{\rho h^3}{12\eta} \frac{\partial p}{r\partial\varphi} \right)}_{\text{Poiseuille-flow}} + \underbrace{\frac{\partial}{r\partial\varphi} \left(\rho h \frac{u_{\varphi S} + u_{\varphi H}}{2} \right)}_{\text{Couette-flow}} + \underbrace{\frac{\partial}{\partial t}(\rho h)}_{\text{Squeeze-film flow}} \tag{9}$$

$$+ \underbrace{\frac{\partial}{r\partial r} \left(\frac{\rho^2 h^3}{\eta} \left(\frac{u_{\varphi S}^2}{40} + \frac{u_{\varphi S} u_{\varphi H}}{30} + \frac{u_{\varphi H}^2}{40} \right) \right)}_{\text{Centrifugal flow}} \,.$$

Due to the bearing design, centrifugal flows have to be taken into account because they provide an additional flow component in radial direction. In general, centrifugal flows can be induced via the hydrodynamic pressure gradient and via the circumferential velocities of the fluid. In this contribution, it is assumed that the latter influence has a more significant effect. The lubrication gap is in the fifth and seventh power in the case of the pressure-induced centrifugal flow, while the lubrication gap is in the third power when circumferential velocity is considered [38–40].

By solving the Reynolds equation, the hydrodynamic pressure distribution is known. However, cavitation phenomena occur in the area of the opening lubrication gap, which lead to the generation of a multiphase flow. With regard to the modelling of outgassing processes, the focus is on diffusion-induced gas generation with the implementation of the two-phase model. Therein, the gas phases dissolved in the lubricant and those present separately in the lubrication gap are considered. Detailed explanations of the two-phase model can be found in [23,24,41], so a summary is provided within this contribution.

2.3. Two-Phase Flow Cavitation

For further examinations, a control volume within a partially filled lubrication gap is considered (cf. Figure 5). The phase transition takes place at the surface of the bubble.

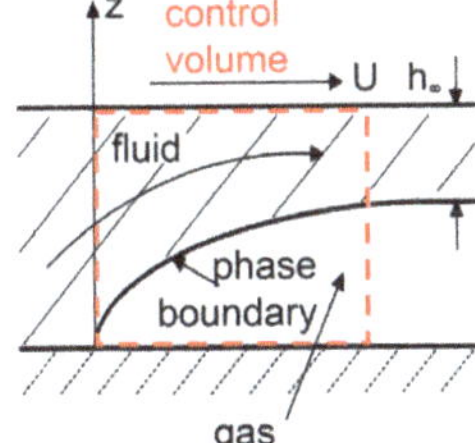

Figure 5. Partially filled lubrication gap with control volume, according to Dowson [20].

Based on Figure 5, it becomes clear that the lubricant properties are significant for the pressure build-up, while the properties of the gas phase can be neglected. Consequently, within the Reynolds equation (Equations (8) and (9)), effective lubricant properties

$$\eta_{eff} = \eta_{liq}F + \eta_g(1-F) \approx \eta_{liq}F \tag{10}$$

$$\rho_{eff} = \rho_{liq}F + \rho_g(1-F) \approx \rho_{liq}F$$

are used. The lubricant fraction F represents the ratio of the oil volume (hatched area) in relation to the size of the control volume. For further calculations, the bubble content r is introduced, which describes the ratio of the bubble volume to the oil volume as follows

$$F = \frac{V_{oil}}{V_{CV}} = \frac{r}{1+r} \tag{11}$$

$$r = \frac{V_B}{V_{oil}} \quad . \tag{12}$$

The basis for implementing the two-phase model is the determination of total gas mass m_B, which is composed of the masses dissolved and undissolved in the lubricant

$$m_B = m_{B\,dis} + m_{B\,undis} = const \quad . \tag{13}$$

For determining the undissolved gas masses, the application of the ideal gas law (cf. Equation (14)) is sufficient, whereas, for the dissolved gas masses, the Bunsen solubility (Henry–Dalton law) has to be taken into account as well (cf. Equations (15) and (16)). The Henry–Dalton law describes the solubility of gases in lubricants and provides the possibility of idealised modelling of diffusion-induced phase transformations. The gas volume $V_{B\,dis}$ dissolved in the lubricant is linearly dependent on the existing oil volume

V_{oil} and the ratio of hydrodynamic pressure p to a reference pressure p_0. For the lubricants ISO-VG 32-220, the Bunsen coefficient is $\alpha_B = 0.08 - 0.09$ [23].

$$m_{B\,undis} = V_B \frac{p}{R\,T} \tag{14}$$

$$V_{B\,dis} = \alpha_B V_{oil} \frac{p}{p_0} \tag{15}$$

$$m_{B\,dis} = V_{B\,dis} \frac{p}{RT} = \alpha_B V_{oil} \frac{p^2}{p_0 R\,T} \tag{16}$$

Inserting Equations (14) and (16) into Equation (13) with inclusion of the bubble content by Equation (12) gives the total gas masses

$$m_B = \left(r + \alpha_B \frac{p}{p_0} \right) \frac{V_{oil}\,p}{R\,T} = const. \tag{17}$$

The assumption of a constant total gas mass provides the possibility to compare the current operating bearing condition with a reference one [24,41]. With regard to the reference condition, the pressure p_0, lubricant film temperature T_0 and reference bubble content r_0 are known, so the total mass of gas can be calculated via Equation (17). With knowledge of the current operating bearing state (hydrodynamic pressure p and oil temperature T at the current step of time integration), the current bubble content and lubricant fraction can be determined

$$m_{B\,0} = m_B \tag{18}$$

$$r = r_0 \frac{p_0 T}{p T_0} - \alpha_B \frac{p T_0 - p_0 T}{p T_0} \tag{19}$$

$$F = \frac{p}{(r_0 + \alpha_B) p_0 \dfrac{T}{T_0} + (1 - \alpha_B) p} \tag{20}$$

With the occurrence of gas cavitation, a partially filled lubrication gap is established, which depends on the hydrodynamic pressure and lubricant film temperature. Restrictions of the cavitation model result from the simplified modelling of phase transitions between dissolved and undissolved gas masses. Here, the equilibrium between the gas phases is instantaneous, which means that the inertia of the bubble growth is neglected. To take the bubble inertia into account, the solution of the Rayleigh–Plesset equation is recommended. This is a differential equation derived from the Navier–Stokes equation, which determines the size of the bubbles as a function of the pressure, surface tensions and inertial effects. In terms of an efficient implementation of outgassing processes, the evaluation of the two-phase model is sufficient for the investigation of hydrodynamically supported rotors. Furthermore, it should be mentioned that surface damage caused by cavitation is not considered. The focus is on the generation of a two-phase flow.

For the numerical implementation of the two-phase model, the introduction of a pressure-related lubricant fraction

$$F_D = \frac{F}{p} = \frac{1}{(r_0 + \alpha_B) p_0 \dfrac{T}{T_0} + (1 - \alpha_B) p} \tag{21}$$

is recommended. This procedure provides the opportunity to apply stabilisation methods in the cavitation domain (First Order Upwind) and methods for efficient solution of the Reynolds equation with mass-conserving cavitation (Newton–Raphson method).

In summary, the Reynolds equation with mass-conserving cavitation according to the two-phase model for the journal bearing is

$$0 = -\frac{\partial}{\partial x}\left(\frac{\rho_{liq}\,h^3}{12\,\eta_{liq}}\frac{\partial p}{\partial x}\right) - \frac{\partial}{\partial y}\left(\frac{\rho_{liq}\,h^3}{12\,\eta_{liq}}\frac{\partial p}{\partial y}\right) + \frac{\partial}{\partial x}\left(\rho_{liq}h\frac{u_S + u_{II}}{2}F_D\,p\right) + \frac{\partial}{\partial t}\left(\rho_{liq}h\,F_D\,p\right) \tag{22}$$

and for the thrust bearing taking centrifugal flow into account

$$0 = -\frac{\partial}{r\partial r}\left(r\frac{\rho_{liq}h^3}{12\eta_{liq}}\frac{\partial p}{\partial r}\right) - \frac{\partial}{r\partial\varphi}\left(\frac{\rho_{liq}h^3}{12\eta_{liq}}\frac{\partial p}{r\partial\varphi}\right) + \frac{\partial}{r\partial\varphi}\left(\rho_{liq}h\frac{u_{\varphi S} + u_{\varphi H}}{2}F_D\,p\right) + \frac{\partial}{\partial t}\left(\rho_{liq}h F_D\,p\right)$$

$$+\frac{\partial}{r\partial r}\left(\frac{\rho_{liq}^2 h^3}{\eta_{liq}}\left(\frac{u_{\varphi S}^2}{40} + \frac{u_{\varphi S}\,u_{\varphi H}}{30} + \frac{u_{\varphi H}^2}{40}\right)F_D\,p\right) \quad . \tag{23}$$

2.4. Temperature Model

In addition to the Reynolds equation, the energy and heat conduction equations are also needed to describe the bearing condition with regard to the temperature distribution in the lubricant film and on the shaft and housing, respectively. The oil temperature is calculated by

$$\underbrace{\frac{\partial T}{\partial t}}_{\text{instat. t.}} + \underbrace{u\frac{\partial T}{\partial x} + v\frac{\partial T}{\partial y} + w\frac{\partial T}{\partial z}}_{\text{heat convection}} = \underbrace{\frac{\lambda}{\rho\,c_p}\left[\frac{\partial}{\partial x}\left(\frac{\partial T}{\partial x}\right) + \frac{\partial}{\partial y}\left(\frac{\partial T}{\partial y}\right) + \frac{\partial}{\partial z}\left(\frac{\partial T}{\partial z}\right)\right]}_{\text{heat conduction}} + \underbrace{\frac{\eta}{\rho\,c_p}\left[\left(\frac{\partial u}{\partial z}\right)^2 + \left(\frac{\partial v}{\partial z}\right)^2\right]}_{\text{lubricant film dissipation}}, \tag{24}$$

wherein T is the oil temperature; (u, v, w) denotes the fluid velocity in bearing circumferential, width and height direction; and η, ρ, λ, c_p are lubricant properties in terms of viscosity, density, thermal conductivity and specific heat capacity. The dissipative part of Equation (24) is due to the lubricant shearing resulting from the change of the fluid flow over the gap height $(\frac{\partial u}{\partial z}, \frac{\partial v}{\partial z})$. The gradients of the velocities are already known from the derivation of the Reynolds equation

$$\frac{\partial u}{\partial z} = \frac{1}{\eta}\left[\frac{\partial p}{\partial x}\left(z - \frac{F_1}{F_0}\right) + \frac{u_S - u_H}{F_0}\right] \quad , \tag{25}$$

$$\frac{\partial v}{\partial z} = \frac{1}{\eta}\frac{\partial p}{\partial y}\left(z - \frac{F_1}{F_0}\right) \quad . \tag{26}$$

With regard to heat convection, the velocities are calculated by

$$u(z) = \frac{\partial p}{\partial x}\left[I_1 - \frac{F_1}{F_0}\cdot I_0\right] + \frac{I_0}{F_0}(u_S - u_H) + u_H \quad , \tag{27}$$

$$v(z) = \frac{\partial p}{\partial y}\left[I_1 - \frac{F_1}{F_0}\cdot I_0\right] . \tag{28}$$

It should be mentioned that effective lubricant properties, according to Equation (10), have to be used. Otherwise, there will be an overestimation of dissipation and heat generation [42]. Furthermore, the velocity component w can be neglected since it is much smaller than the other ones and laminar flow is assumed.

To determine the three-dimensional temperature distribution at the shaft and housing, the heat conduction equation is evaluated

$$\frac{\partial T_{S,H}}{\partial t} = \frac{\lambda}{\rho c}\left[\frac{\partial}{r\partial\varphi}\left(\frac{\partial T_{S,H}}{r\partial\varphi}\right) + \frac{\partial}{r\partial r}\left(r\frac{\partial T_{S,H}}{\partial r}\right) + \frac{\partial}{\partial z}\left(\frac{\partial T_{S,H}}{\partial z}\right)\right] \tag{29}$$

in cylindrical coordinates, where $T_{S,H}$ is the temperature of shaft and housing, respectively; r, φ, z are the coordinates in radial, circumferential and bearing width direction; and ρ, c and λ represent the material properties of the bearing parts.

2.5. Coupling of Thermo- and Hydrodynamics

Considering Equations (22), (24) and (29), it becomes clear that a coupled system has to be solved to describe the thermo-hydrodynamic processes in journal bearings. The interaction consists of, on the one hand, the thermodynamics and hydrodynamics within the lubricating film and, on the other hand, of the heat transfer processes between the fluid and the shaft, bushing or housing. The interrelationships and holistic interactions are shown in Figure 6.

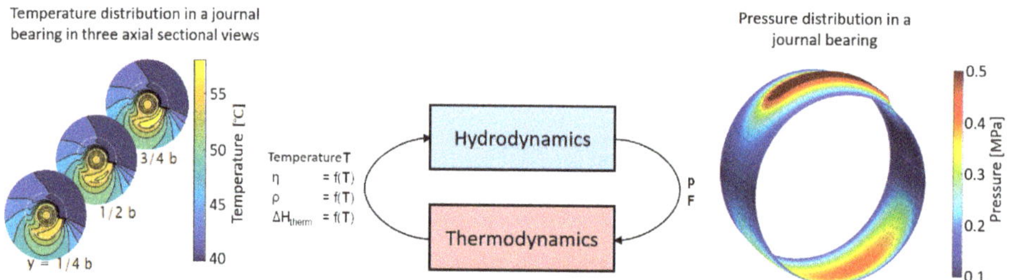

Figure 6. Coupling of thermodynamics and hydrodynamics.

With regard to the pressure distribution in journal bearings, Equation (22) is evaluated, wherein the hydrodynamic pressure depends on the temperature-dependent lubricant properties as well as the lubrication gap including thermal gap expansion and the oil temperature. Concerning the occurrence of outgassing processes, it should be noted that the lubricant fraction depends on both the hydrodynamic pressure and the lubricant film temperature (cf. Equation (20)). Thus, heating of the oil can lead to increasing bubble generation, so the lubricant fraction decreases and consequently outgassing processes are favoured. The occurrence of cavitation has a significant influence on the resulting forces and torques of the bearing and thus on the bearing stiffness and damping. With the solution of the Reynolds equation and two-phase model, the pressure and lubricant distributions are known so that the thermodynamic states can be determined as the next step.

With knowledge of the pressure and lubricant fraction, the fluid velocities in the lubrication gap and their change over the gap height can be calculated, so the dissipative part and the components of the heat convection of Equation (24) are determined. Subsequently, the temperature distribution in the lubricating film as well as at the shaft and housing is evaluated, in which a thermally fully coupled system is considered. Heat transfer between fluid and solid domain is based on a conjugate heat transfer model, in which heat conduction processes at the surface of the supported components are assumed. Heat transfer coefficients, which are difficult to determine, are not needed within the temperature calculations. With the solution of temperature distribution, the viscosity, density and thermal gap changes can be updated, which are again input variables for the hydrodynamics.

With regard to the shaft motion measurements, the turbocharger run-up was carried out slowly, so the assumption of a quasi-steady-state temperature distribution is valid. Furthermore, it is shown in [42] that solving the energy and heat conduction equation is not necessary at every time step, since thermal processes take place significantly more slowly than rotordynamic and hydrodynamic processes. Furthermore, solving these equations at defined output steps provides the advantage of saving computational time. The temperature is kept constant between the output steps.

2.6. Validation of Hydrodynamics and Thermodynamics

2.6.1. Journal and Floating Ring Bearing

Validation of the hydrodynamic model taking mass-conserving cavitation into account has already been done for a rigid shaft supported by journal bearings and for a Jeffcott rotor supported by full-floating ring bearings [41]. On the one hand, the shaft motion under cyclic load is evaluated and compared with the results of Ausas [28]. On the other hand, the run-up behaviour of a Jeffcott rotor with floating ring bearings is investigated, whereby the rotor's response under the assumption of Half-Sommerfeld solution and the two-phase model is compared with the results of Eling [19]. The influence of cavitation is shown by the occurrence of sub-synchronous oscillations over a larger speed range. In summary, the run-up simulation with mass-conserving cavitation shows better agreement with the shaft motion measurement.

The validation of the thermodynamics is carried out in [42]. There, a Jeffcott rotor with journal bearings is examined with regard to the influence of temperature assumptions on the generation of sub-synchronous oscillations. Run-up simulations are performed assuming isothermal operating conditions, a thermal lumped mass model and the complete solution of the three-dimensional energy and heat conduction equations. In summary, the rotor response based on the solution of energy and heat conduction equations shows the best agreement with the results of Eling [19]. Furthermore, it is shown that the solution of the energy equation does not have to be calculated in every time step, so the solution at defined time steps is also sufficient.

2.6.2. Thrust Bearing

In this section, the results of the validation for hydrodynamics at thrust bearing are shown. For this purpose, a comparison is made with the results of Hao [38]. The corresponding experimental investigations were carried out by Zhang [43]. The considered thrust bearing consists of 24 segments, where each of them has a flat and grooved area (cf. Figure 7). Furthermore, ambient pressure is assumed at bearing edges. Under these conditions, the lubricant fraction at inner and outer radius is equal to $F = 1$, which has the consequence that the oil is supplied to the lubrication gap via the bearing edges. The bearing design and the corresponding operating boundary conditions are summarised in Tables 1 and 2.

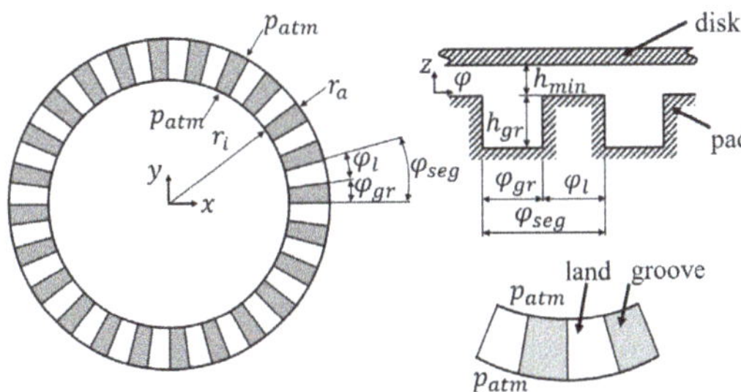

Figure 7. Bearing design and boundary conditions, according to Hao [38].

Table 1. Summary of lubrication gap, according to Hao [38].

Property	Name	Value
Number of segments		24
Inner radius	r_i	24.0 mm
Outer radius	r_a	32.0 mm
Angle of groove	φ_{gr}	8.4 deg
Angle of land	φ_l	6.6 deg
Groove depth	h_{gr}	10.2 μm

Table 2. Summary of operating condition, according to Hao [38].

Property	Name	Value
Viscosity	η_{liq}	0.083 Pas
Density	ρ_{liq}	812 kg/m^3
Temperature	T_{liq}	22.5 °C
Bunsen-Solubility	α_B	0.08
Bubble content	r_0	0.0–0.05
Angular velocity	n	333.98 rpm
Min. lubrication gap	h_{min}	15.4 µm

To determine the pressure distribution, the Reynolds equation with consideration of centrifugal inertia, according to Equation (9), is evaluated for stationary operating conditions and iso-viscous temperature distributions. In [38], the implementation of mass-conserving cavitation is analogous to the principle of dissolved and undissolved gas masses explained within this contribution. However, a temperature- and pressure-dependent Bunsen coefficient was assumed, which is approximated via an exponential approach. In contrast, within this contribution the Bunsen coefficient α_B is constant. Furthermore, it should be mentioned that a variation of the reference bubble content is carried out in order to illustrate the influence of already existing undissolved gas masses. The results of pressure and lubricant distribution at the bearing centre plane are shown in Figure 8.

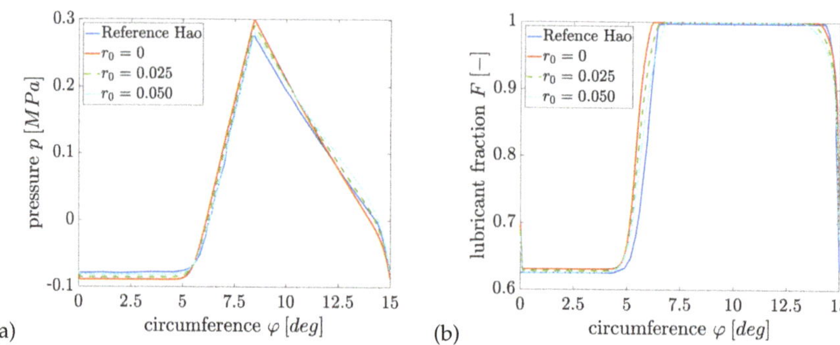

Figure 8. Gauge pressure (**a**) and lubricant fraction (**b**) at bearing centre plane r = 28 mm with h_{min} = 15.4 µm and n = 333.98 rpm.

In summary, the results show a good agreement with the determined pressure and lubricant distributions of [38]. Neglecting already undissolved air masses $r_0 = 0$, a maximum gauge pressure of $p_{max} = 0.3$ MPa is obtained between $\varphi = 6.2°$ and $13.9°$. At cavitation domain, a minimum lubricant fraction of $F_{min} = 0.63$ is observed. With increasing reference bubble content, the fraction of undissolved gas masses increases, so that the maximum pressure decreases and the pressure area covers a smaller range overall. For a bubble content of $r_0 = 0.025$, the maximum pressure is $p_{max} = 0.28$ MPa. The pressure area is located between $\varphi = 6.7°$ and $13.2°$ and a minimum lubricant fraction of $F_{min} = 0.62$ is determined.

The pressure and lubricant distribution over the entire bearing segment is shown in Figure 9. The cavitation domain occurs mainly at groove area. Since a constant pressure is assumed at bearing edges, a pressure gradient is created at inner and outer radius so that the lubricant can be transported into the gap and thus a pressure is built up at flat area.

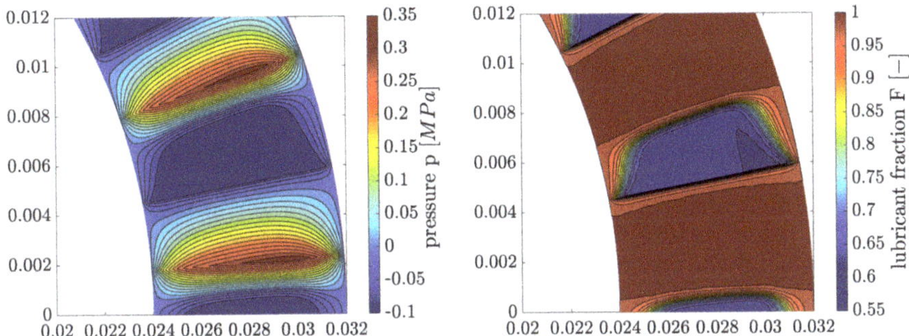

Figure 9. Gauge pressure (**left**) and lubricant fraction (**right**) at h_{min} = 15.3 µm and n = 333.98 rpm.

3. Results of Run up Simulations

To evaluate the rotor response behaviour, run-up simulations are compared with shaft motion measurements in the time and frequency domain. Furthermore, the influence of thrust bearings on the rotor vibrations is also examined. For this purpose, the normalised eccentricity and motion orbitals at compressor- and turbine-side bearings as well as the rotor tilting angle are discussed more in detail.

With regard to the hydrodynamic boundary conditions, constant supply pressure is assumed within the inlet holes. Due to the circumferential groove at the outer lubrication gap of the floating ring bearing, there is a constant supply pressure over the entire groove. Furthermore, the oil can flow out freely over the bearing edge since there are no lateral seals. In addition, thermal boundary conditions have to be taken into account. At the outer lubrication gap, the oil supply temperature is set within the circumferential groove. Since the angular positions of the connecting channels are the same as the positions of feeding holes and the lubricant is evenly distributed over the circumference within the groove area, an effective mixing temperature can be assumed within the supply area of the inner lubrication gap. Here, a perfect mixture of warm and fresh oil is supposed.

Besides the thermo-hydrodynamic boundary conditions, the run-up process is also considered. Within the shaft motion measurement, the run-up is relatively slow, so that the assumption of thermally steady-state operating condition is valid. In addition, thrust forces occur due to the flow of gases to the impellers. Within the simulation, the lateral forces are specified as look-up tables based on the measurement conditions. In addition, the total thrust bearing clearance is known. Within the simulation, the rotor starts from a minimum speed and accelerates constantly until the maximum speed is reached.

To evaluate the influence of cavitation and thrust bearings more precisely, simulations were carried out with various levels of detail of the bearings. As the lowest model accuracy (V1), a fully filled lubrication gap regardless of operating bearing condition (Half-Sommerfeld cavitation) is assumed and thrust bearings are neglected. An increase in the level of detail is achieved by considering outgassing processes, but thrust bearings are still neglected (V2). Consequently, a comparison of the rotor response behaviour between the model levels V1 and V2 shows the influence of cavitation. With regard to the modelling level (V3), Half-Sommerfeld cavitation is assumed again, but thrust bearings are included. As a result, a comparison between V1 and V3 shows the influence of thrust bearings on vibration behaviour. The highest model accuracy (V4) includes both mass-conserving cavitation and the consideration of thrust bearings.

3.1. Shaft Motion in Frequency Domain

First, the results of the shaft motion measurement are discussed in frequency domain (cf. Figure 10). The figure shows the rotor response frequency as a function of the angular velocity. From f_{Rot} = 0.48 of the maximum speed, sub-synchronous oscillation with a rotor response frequency f = 0.16 are detected. Besides the oil-whip, the rotor behaviour is also determined by the unbalance-induced vibrations. A synchronous resonance occurs at $f_{Rot} \approx 0.24$, whereby the amplitudes decrease after passing the resonance.

Figure 11 shows the results of the run-up simulations at various levels of detail of the bearings. Assuming a fully filled lubrication gap independent of the operating condition of the bearing and neglecting the tilting stiffness of the thrust bearings (V1), no sub-synchronous vibrations can be detected. On the one hand, a fully filled lubrication gap can lead to an overestimation of damping properties at squeeze-film damper and, on the other hand, it can lead to increased stiffness within the inner lubrication gap. Furthermore, neglecting the thrust bearing has an effect on the vibration mode of the rotor and thus also on the bearing stiffness and damping, which can be seen from the shaft orbits and the eccentricity in the bearing planes. Taking outgassing processes (V2) into account, sub-synchronous oscillation can already be predicted from a rotor speed of $f_{Rot} = 0.55$ and, thus, a better agreement with the measurements can be achieved. The occurrence of a two-phase flow leads to a reduction of bearing stiffness and damping, which favours lubricant film-induced excitations. The run-up simulations V1 and V2 were carried out without thrust bearings, with the consequence of neglecting the tilting stiffness and damping. To evaluate the influence of the tilting stiffness on the rotor response behaviour, the thrust bearing is added and Half-Sommerfeld cavitation at floating ring and thrust bearings is assumed (V3). The rotor response shows sub-synchronous oscillations from a speed of $f_{Rot} = 0.66$, whereby the oil-whip occurs later and with lower amplitude compared to the measurement. One reason for the occurrence of sub-synchronous vibrations can be that with the consideration of tilting stiffness, a rotor natural frequency is shifted closer to the half-whirl frequency of the oil at floating ring bearing. Consequently, an oil-whip can occur if a weakly-damped natural frequency of the rotor is excited. The highest level of detail is given by mass-conserving cavitation both in floating ring and thrust bearings (V4). Thus, the effects of the two-phase flow are superimposed on those of the thrust bearing, so that, in addition to the start frequency of the oil-whip $f_{Rot} = 0.50$, the rotor amplitude is also well predicted compared to the measurement. A summary of the rotor response behaviour is given in Table 3.

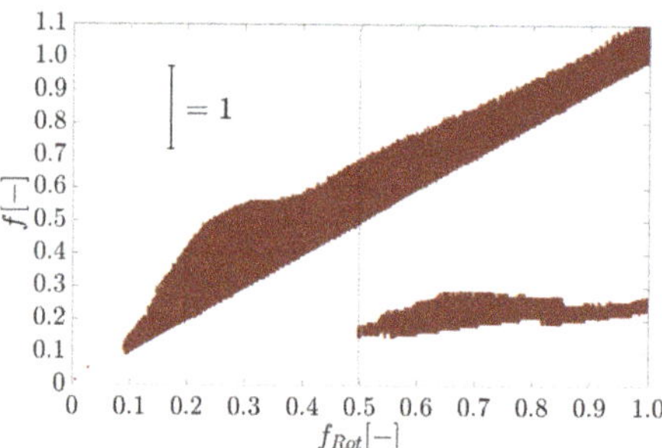

Figure 10. Spectrogram of the shaft motion measurement captured at the sealing disc.

Table 3. Summary of rotor response behaviour measured at sealing disk.

Description	Synchronous Resonance	Start of Oil-Whip	Response Frequency
Measurement	0.24	0.48	0.16
Half-Sommerfeld without thrust bearing (V1)	0.23	–	–
2PM without thrust bearing (V2)	0.23	0.55	0.15
Half-Sommerfeld with thrust bearing (V3)	0.23	0.66	0.18
2PM with thrust bearing (V4)	0.23	0.50	0.15

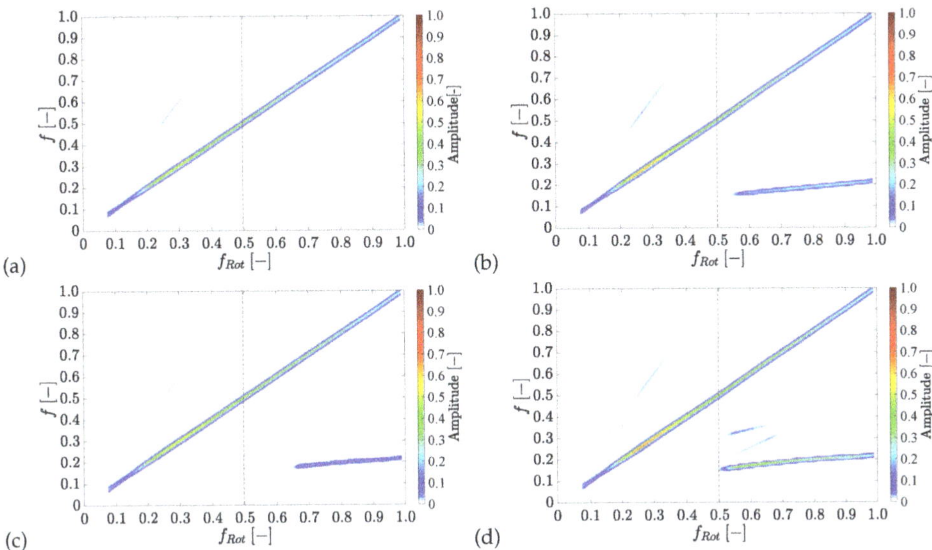

Figure 11. Spectrogram of the rotor response behaviour under the assumption of: (**a**) Half-Sommerfeld cavitation without thrust bearing (V1); (**b**) two-phase model without thrust bearing (V2); (**c**) Half-Sommerfeld cavitation with thrust bearing (V3); and (**d**) two-phase model with thrust bearing (V4).

3.2. Shaft Motion in Time Domain

Furthermore, the time signal of the horizontal rotor movement is evaluated (cf. Figure 12). The signal represents the oscillation of the rotor around its stationary position and is calculated by taking the half of the difference between upper and lower envelope curves of the time signal. Measured sub-synchronous oscillations can be observed from a speed of $f_{Rot} > 0.48$, whereby these are characterised by a stronger oscillation of the time signal. A comparison of the run-up simulations shows that a better agreement with the measurement can be achieved with increasing level of detail of the bearings. The run-up simulation with mass-conserving cavitation and without thrust bearing (V2) already showed a good congruence with respect to the start frequency of the oil-whip in the frequency domain, but the rotor oscillations are predicted too low. Finally, the simulation with consideration of thrust bearing and outgassing processes (V4) shows the best agreement with the measurement both for the start frequency of the oil-whip and with regard to the rotor oscillation magnitudes. Here, the oil-whip shows a clear increase of the vibration amplitude from a speed of $f_{Rot} > 0.50$.

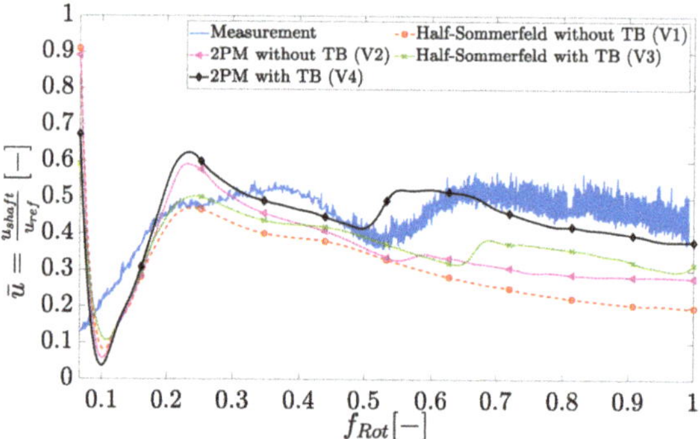

Figure 12. Horizontal shaft motion evaluated at sealing disc.

Discrepancies between measurement and simulations can be explained by the fact that a complete mixture between warm and cold oil is assumed within the supply area of the inner lubrication gap. A more detailed implementation of the mixing processes is based on a boundary layer model, in which a part of the warm oil can be transported over the supply area. Furthermore, it should be noted that the temperatures at thrust bearing have been specified via a look-up table. Spatial distributions of the temperatures are not taken into account. For this purpose, the three-dimensional energy equation for the lubricating film and the heat conduction equation for the shaft and housing must be solved in analogy to the thermodynamics of floating ring bearings.

3.3. Normalised Eccentricity and Orbits of Shaft and Floating Ring

To investigate the influence of the thrust bearing on the rotor response, the normalised eccentricities and the motion orbit of the shaft and bushings are examined (see Figures 13 and 14). For a better evaluation, simulations with mass-conserving cavitation and without thrust bearing (V2) and with thrust bearing (V4) are chosen.

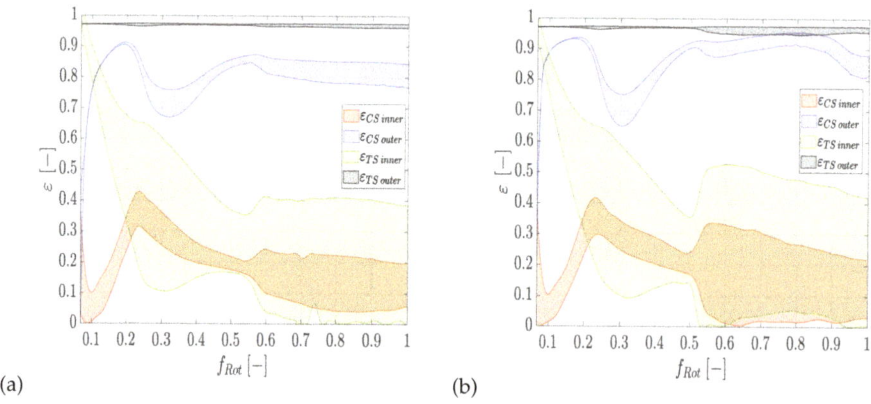

(a) (b)

Figure 13. Normalised eccentricity of the shaft and bushing considering mass-conserving cavitation: without thrust bearing (V2) (**a**); and with thrust bearing (V4) (**b**). Abbreviations: *CS*, compressor side gap; *TS*, turbine side gap.

Based on the normalised eccentricity of the bushing at turbine bearing (black area in Figure 13), it can be seen that there is a contact at outer lubrication gap over the entire run-up. This is due to the rotor's design since its centre of gravity is close to the turbine bearing (see Figure 1). In addition, the outer gap is designed as squeeze-film damper since the rotational movement of the bushing is prevented. The hydrodynamic pressure can only be built up if there is a sufficient squeezing of the lubricant. Otherwise, the bushing rests. From the motion orbit of the turbine-sided bushing (grey curve in Figure 14c,d), it is also clear to see that the bushing gets immediately into contact and increased oscillation can only be observed with the occurrence of sub-synchronous oscillations at $f_{Rot} > 0.5$ or during unbalance resonance.

The influence of thrust bearings can be seen, on the one hand, with regard to the normalised eccentricity of the compressor-sided bushing (blue area in Figure 13) and, on the other hand, on the motion orbit (grey curve in Figure 14a,b). Without thrust bearings, the normalised eccentricities of the bushing can be found between $\varepsilon \approx 0.78 - 0.86$ and the contact process occurs in the upper half $\phi = 60 - 90°$. Consequently, the rotor exhibits a high degree of tilting. Considering the tilting stiffness, the rotor tilting is limited, which results in, on the one hand, higher eccentricities between $\varepsilon \approx 0.88$ and 0.95 and, on the other hand, the motion orbit of the bushing occurs at the lower region $\phi = 240-270°$. Thus, the influence of the thrust bearing is clearly illustrated by the motion orbit of the compressor-side bushing. At this point, it should be mentioned that, with increasing eccentricity of the bushing, the anisotropy of the squeeze-film damper becomes more significant, which also has an influence on the rotor natural frequencies [10].

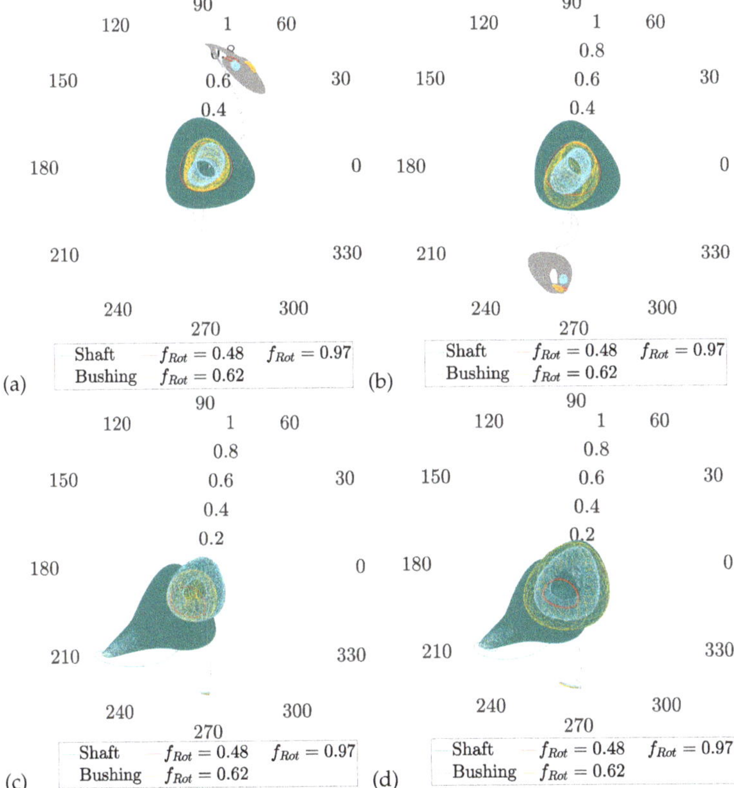

Figure 14. Normalised shaft and bushing orbit considering mass-conserving cavitation: (**a**) without thrust bearing at compressor bearing (V2); (**b**) with thrust bearing at compressor bearing (V4); (**c**) without thrust bearing at turbine bearing (V2); and (**d**) with thrust bearing at turbine bearing (V4).

3.4. Thrust Bearing

In this section, the forces and torques acting at thrust bearing are discussed (see Figures 15 and 16). With regard to the bearing forces, both thrust bearings are in equilibrium at lower speed range $f_{Rot} < 0.1$, since no thrust load is acting yet. Due to the transport of the lubricant into the narrowing gap, a hydrodynamic pressure can be built up at thrust bearing, which depends on the initial bearing clearance and thus the lubrication gap as well as on the angular velocity of the rotor. The resulting thrust forces are only equal if the thrust bearing is unloaded. Furthermore, the pre-loading of the bearing is due to the fact that the total axial bearing clearance remains constant. The axial equilibrium position of the shaft is within the bearing clearance. The axial force increases with rising rotor speed, which results in a reduction of the load on the compressor-sided and an additional load on the turbine-sided thrust bearing. The lubricant film-induced rotor excitations can also be observed in increasing oscillations of the minimum lubrication gap from $f_{Rot} > 0.52$.

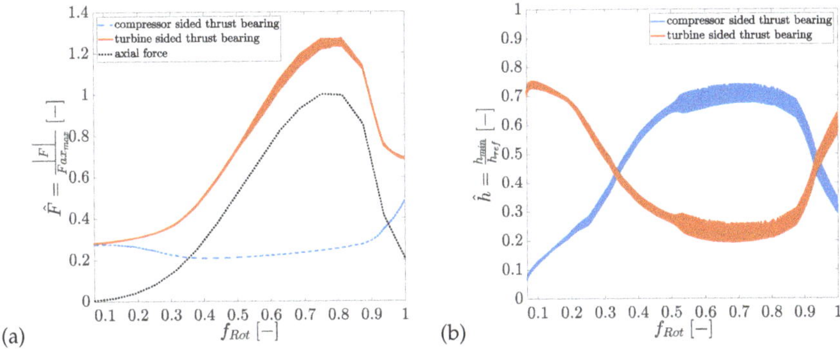

Figure 15. Thrust forces (**a**) and minimum lubrication gap (**b**) at thrust bearing (Bearing model level V4).

As mentioned above when evaluating the normalised eccentricity, the thrust bearing limits the rotor's tilting. For this reason, the bearing torque and the corresponding rotor tilting angle are evaluated (see Figure 16). If the thrust bearing is neglected, the rotor can tilt up to $\hat{\varphi} \approx 0.58$–$0.70$, whereas with consideration of the tilting stiffness, the rotor tilts by $\hat{\varphi} \approx 0.0$–$0.2$. The tilting angle refers to the maximum possible tilting angle of the rotor. Furthermore, the bearing torque shows that with the occurrence of sub-synchronous oscillations, increased tilting torques can occur, which contribute to the overall stiffness of the bearing.

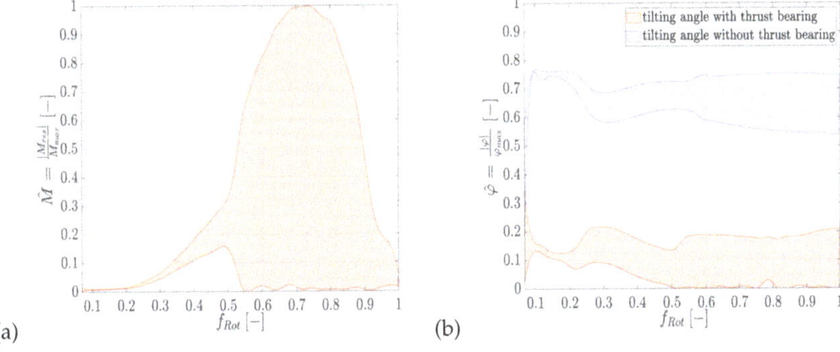

Figure 16. Thrust bearing torque (**a**) and tilting angle of the rotor (**b**) considering mass-conserving cavitation (V2 and V4).

4. Conclusions

In this contribution, the response behaviour of a semi-floating ring supported turbocharger rotor is investigated under the influence of both cavitation and thrust bearing support concerning the occurrence and amplitudes of sub-synchronous vibrations. The Reynolds equation with mass-conserving cavitation according to the two-phase model is used to determine the pressure and lubricant distribution in floating ring and thrust bearings. In addition to the hydrodynamic state, the thermodynamics at floating ring bearings are also considered. Here, the three-dimensional energy equation for the lubricant film and the heat conduction equation for the supported elements are evaluated as a fully coupled system.

To investigate the effects of cavitation and thrust bearing, various levels of detail of the bearing system are chosen and their impact on the rotor response is evaluated in time and frequency domain. As the simplest modelling, the thrust bearing is neglected and a fully filled lubrication gap (Half-Sommerfeld cavitation) is assumed (V1). Under these circumstances, no sub-synchronous vibrations could be predicted since an always fully filled gap overestimates the stiffness and damping of the floating ring bearings. A significantly better agreement with the measurements can be achieved by taking outgassing processes into account (V2). Compared to V1, the occurrence of cavitation leads to a softer bearing behaviour with lower damping and stiffness, which favours the occurrence of oil-whip phenomena. The results with mass-conserving cavitation show better agreement with regard to the starting frequency of the oil-whip, but the vibration amplitudes are still too low compared to the measurements. Subsequently, the influence of the thrust bearing is discussed more in detail. For this purpose, run-up simulations with thrust bearing and Half-Sommerfeld solution (V3) were carried out. Here, sub-synchronous vibrations are also observed at higher speed range. The occurrence of oil-whip can be explained by the fact that the additional tilting stiffness can shift the rotor natural frequencies closer to the oil whirling frequency of the floating ring bearing. The best agreement with the measurements is achieved when both thrust bearing and outgassing processes are taken into account. Here, the effects of a softer bearing behaviour and shifting of the rotor natural frequencies are superimposed.

Furthermore, the rotor's vibration mode was investigated using the normalised eccentricity and motion orbit of the bushing and shaft. For the considered rotor, the influence of thrust bearing can be seen by the compressor sided bushing motion. If the tilting stiffness and damping are neglected, the bushing is located at the upper range of the bearing housing so that the rotor has a significant tilting position. The consideration of tilting stiffness counteracts the rotor tilting so that the motion orbit of the bushing is located in lower region. In contrast, the bushing at turbine bearing does not show any significant influence from the thrust bearing since the centre of gravity of the rotor is close to the turbine bearing and thus higher radial loads are generally present. Consequently, the influence of the thrust bearing is more significant the more decentred the centre of gravity of the rotor is, and thus a possible tilting of the rotor during turbocharger operation is favoured.

In addition to the floating ring bearing state, the minimum lubrication gap, the tilting angle of the rotor and the resulting forces and torques at thrust bearing are evaluated. The already mentioned conclusions based on the bushing movements can also be obtained when the rotor tilting angle is evaluated.

Author Contributions: Conceptualization, C.Z., E.W.; methodology, C.Z.; software, C.Z. and C.I.; validation, C.Z. and C.I.; formal analysis, C.Z.; investigation, C.Z. and C.I.; resources, S.N., C.D. and E.W.; data curation, C.Z.; writing—original draft preparation, C.Z.; writing—review and editing, C.Z., C.I., S.N., C.D., E.W.; visualization, C.Z.; supervision, S.N., C.D. and E.W.; project administration, S.N., C.D. and E.W.; funding acquisition, E.W. All authors have read and agreed to the published version of the manuscript.

Funding: The research project (FVV project no. 1258) was performed by the Junior Professorship Fluid Structure Interaction in Multibody Systems (FSK) at the Institute of Mechanics of the Otto von Guericke University Magdeburg under the direction of Jun.-Prof. Dr.-Ing. Elmar Woschke and by the Chair of Technical Dynamics (LTD) at the Institute of Mechanics of the Otto von Guericke University Magdeburg under the direction of Prof. Dr.-Ing. habil. Jens Strackeljan. Based on a decision taken by the German Bundestag, it was supported by the Federal Ministry for Economic Affairs and Energy (BMWi) and the AIF (German Federation of Industrial Research Associations eV) within the framework of the industrial collective research (IGF) programme (IGF No. 18760 BR). The project was conducted by an expert group led by Dipl.-Ing. Thomas Klimpel (ABB Turbo Systems AG). The authors gratefully acknowledge the support received from the funding organisations, from the FVV (Research Association for Combustion Engines eV) and from all those involved in the project.

Institutional Review Board Statement: Not applicable.

Informed Consent Statement: Not applicable.

Data Availability Statement: Not applicable.

Conflicts of Interest: The authors declare no conflict of interest.

Nomenclature

c_p	heat capacity
f	ratio of the rotor response frequency and maximum rotational frequency
f_{Rot}	ratio of the rotor speed and maximum speed
F	lubricant fraction
F_C	coefficient of the Couette flow
F_D	pressure-related lubricant fraction
$F_{x/y}$	bearing forces
F_ρ	coefficient of the squeeze-film flow
F_0, F_1	integrals for the Reynolds equation
$\hat{F}$	normalised thrust force
G	coefficient of Poiseuille flow
h	lubricant gap height
$\underline{h_e}$	external forces
$\underline{h_{el}}$	elastic properties of the shaft (FEM)
$\underline{h_\omega}$	gyroscopic and inertia forces
$\hat{h}$	normalised axial gap
I_0, I_1	integrals for the Reynolds equation
m	mass
$\hat{M}$	normalised bearing torque
$\underline{\underline{M}}$	inertia matrix
p	hydrodynamic pressure
r	bubble content
r, φ, z	lubrication gap coordinates
$\vec{r}$	position vector
R	universal gas constant
t	time
T	lubricant temperature
$u_{C,T}$	unbalance
u_{ref}	reference shaft displacement
u_{shaft}	horizontal shaft motion
$\bar{u}$	normalised horizontal shaft motion
u, v, w	velocity components of the lubricant flow
V	volume
$\underline{y}$	state vector
x, y, z	lubrication gap coordinates
α_B	Bunsen coefficient
ε	normalised eccentricity
η	viscosity of the lubricant
λ	thermal conductivity
ρ	density
$\hat{\varphi}$	normalised tilting angle
atm	atmosphere/ambient
B	bubbles
C	compressor
CV	control volume
dis	dissolved
eff	effective properties
g	gaseous phase

H	housing
min	minimum
liq	liquid phase
S	shaft
T	turbine
undis	undissolved
0	reference state

Appendix A

The run-up simulations shown in Section 3.1 are based on the Reynolds equation (Equations (8) and (9)), where lubricant properties are averaged over the gap height. To consider the non-linear oil properties more accurately, the run-up simulations were repeated with generalised Reynolds equation (see Figure A1). In summary, the conclusions can also be adopted here. With increasing level of detail, better agreement with the measurements can be achieved. Consequently, the run-up simulation with two-phase model and thrust bearing shows the best agreement with the measurements. Compared to the simplified Reynolds equation, the overall simulation time was longer by a factor of 10. The rotor response behaviour is summarised in Table A1.

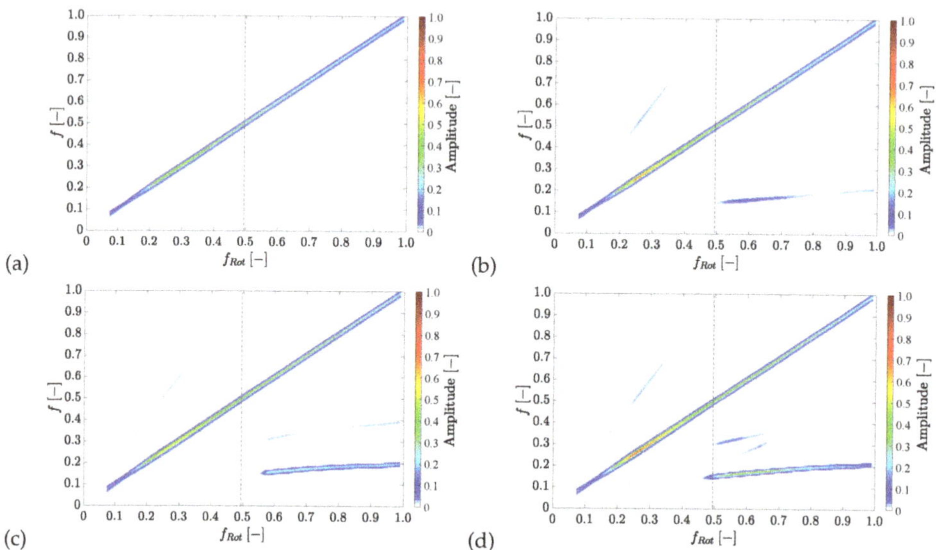

Figure A1. Spectrogram of the rotor response behaviour under the assumption of: (**a**) Half-Sommerfeld cavitation without thrust bearing (V1); (**b**) two-phase model without thrust bearing (V2); (**c**) Half-Sommerfeld cavitation with thrust bearing (V3); and (**d**) two-phase model with thrust bearing (V4). Run-up simulations were carried out with the generalised Reynolds equation.

Table A1. Summary of rotor response behaviour measured at sealing disk (Generalised Reynolds equation).

Description	Synchronous Resonance	Start of Oil-Whip	Response Frequency
Measurement	0.24	0.48	0.16
Half-Sommerfeld without thrust bearing (V1)	0.24	–	–
2PM without thrust bearing (V2)	0.23	0.51	0.14
Half-Sommerfeld with thrust bearing (V3)	0.24	0.56	0.15
2PM with thrust bearing (V4)	0.23	0.47	0.14

Discrepancies between the run-up simulations with simplified and generalised Reynolds equation result mainly from the integration of the non-linear oil properties via the clearance height. The generalised Reynolds equation takes these into account more precisely, which has an effect on the pressure build-up, the resulting bearing forces and finally the response behaviour of the rotor. Nevertheless, the run-up simulations with generalised and simplified Reynolds equation show similar results. One reason for this could be that the run-up was carried out with high oil inlet temperatures (as in the measurement). Thus, the oil viscosity is in a range where temperature changes cause only minor changes in viscosity. More detailed investigations of the temperature influence on the pressure distribution are described in [17,44–46].

References

1. Woschke, E.; Daniel, C.; Nitzschke, S. Excitation mechanisms of non-linear rotor systems with floating ring bearings-simulation and validation. *Int. J. Mech. Sci.* **2017**. [CrossRef]
2. Daniel, C.; Woschke, E.; Nitzschke, S.; Göbel, S.; Strackeljan, J. Determinismus der subharmonischen Schwingungen in gleitgelagerten Turbomaschinen. In Proceedings of the 12. Magdeburger Maschinenbau-Tage, Magdeburg, Germany, 30 September–1 October 2015.
3. Nguyen-Schäfer, H. *Rotordynamics of Automotive Turbochargers, Chapter 7.5*; Choi, S.-B., Duan, H., Fu, Y., Sun, J.-Q., Eds.; Springer: Cham, Switzerland, 2015. [CrossRef]
4. Nguyen-Schäfer, H. *Aero and Vibroacoustics of Automotive Turbochargers*; Springer: Berlin, Germany, 2013; Chapter 6.3. [CrossRef]
5. Schweizer, B. Total instability of turbocharger rotors-Physical explanation of the dynamic failure of rotors with full-floating ring bearigs. *J. Sound Vib.* **2008**. [CrossRef]
6. Schweizer, B. Oil whirl, oil whip and whirl/whip synchronization occurring in rotor systems with full-floating ring bearings. *Nonlinear Dyn.* **2009**. [CrossRef]
7. Muszynska, A. Whirl and Whip—Rotor/Bearing Stability Problems. *J. Sound Vib.* **1986**, *110*, 443–462. [CrossRef]
8. Smolik, L.; Hajzman, M.; Byrtus, M. Investigation of bearing clearance effects in dynamics of turbochargers. *Int. J. Mech. Sci.* **2016**. [CrossRef]
9. San Andres, L.; Rivadeneira, J.C.; Gjika, K.; Groves, C.; LaRue, G. Rotordynamics of Small Turbochargers Supported on Floating Ring Bearings—Highlights in Bearing Analysis and Experimental Validation. *ASME J. Tribol.* **2007**. [CrossRef]
10. Fuchs, A.; Klimpel, T.; Schmied, J.; Rohne, K.H. Comparison of measured and calculated vibrations of a turbocharger. In Proceedings of the SIRM 2017, Graz, Austria, 15–17 February 2017.
11. Nowald, G.E. Numerical Investigation of Rotors in Floating Ring Bearings Using Co-Simulation. Ph.D. Thesis, Technical University Darmstadt, Darmstadt, Germany, 2018.
12. Dowson, D. A generalized Reynolds equation for fluid film lubrication. *Int. J. Mech. Sci.* **1962**, *4*. [CrossRef]
13. Khonsari, M.M. A review of thermal effects in hydrodynamic bearings part I: Slider and thrust bearings. *ASLE Trans.* **1987**, *30*. [CrossRef]
14. Khonsari, M.M. A review of thermal effects in hydrodynamic bearings part II: Journal bearings. *ASLE Trans.* **1987**, *30*. [CrossRef]
15. Pinkus, O. *Thermal aspects of fluid film tribology*; Amer Society of Mechanical: New York, NY, USA, 1990.
16. Fillon, M.; Frêne, J.; Boncompain, R. Historical aspects and present development on thermal effects in hydrodynamic bearings. In Proceedings of the 13th Leeds-Lyon Symposium on Tribology, Leeds, UK, 8–12 September 1987.
17. Paranjpe, R.; Han, T. A transient thermohydrodynamic analysis including mass conserving cavitation for dynamically loaded journal bearings. *J. Tribol.* **1995**, *117*. [CrossRef]
18. Li, Y.; Liang, F.; Zhou Y.; Ding, S.; Du, F; Zhou, M. Numerical and experimental investigation on thermohydrodynamic performance of turbocharger rotor-bearing system. *Appl. Therm. Eng.* **2017**, *121*. [CrossRef]
19. Eling, R. Torwards Robust Design Optimization of Automotive Turbocharger Rotor-Bearing Systems. Ph.D Thesis, Delft University of Technology, Delft, The Netherlands, 2018. [CrossRef]
20. Dowson, D.; Taylor, C.M. Cavitation in bearings. *Ann. Rev. Fluid Mech.* **1979**, *11*, 33–65. [CrossRef]
21. San Andres, L.; Diaz, S. Flow visualization and forces from a squeeze film damper operating with natural air entrainment. *ASME J. Tribol.* **2003**. [CrossRef]
22. Li X.; Song, Y.; Hao, H.; Gu, C. Cavitation Mechanism of Oil-Film Bearing and Development of a New Gaseous Cavitation Model Based on Air Solubility. *ASME J. Tribol.* **2012**, *134*. [CrossRef]
23. Fuchs, A. Schnelllaufende Radialgleitlagerungen im Instationären Betrieb. Ph.D. Thesis, Technical University Carolo Wilhelmina zu Braunschweig, Braunschweig, Germany, 2002.
24. Mermertas, Ü. Nichtlinearer Einfluss von Radialgleitlager auf die Dynamik Schnelllaufender Rotoren. Ph.D. Thesis, Technical University Clausthal, Clausthal, Germany, 2007.
25. Grando, F. P.; Priest, M.; Prata, A.T. A two-phase flow approach to cavitation modelling in journal bearings. *Tribol. Lett.* **2006**, *21*. [CrossRef]

26. Nitzschke, S. Instationäres Verhalten Schwimmbuchsengelagerter Rotoren unter Berücksichtigung Masseerhaltender Kavitation. Ph.D. Thesis, Otto von Guericke University, Magdeburg, Germany, 2016.
27. Nitzschke, S.; Woschke, E.; Daniel, C. Application of Regularised Cavitation Algorithm for Transient Analysis of Rotors Supported in Floating Ring Bearings. In Proceedings of the 10th International Conference on Rotor Dynamics—IFToMM, Rio de Janeiro, Brazil, 23–27 September 2018; Volume 4, pp. 371–387.
28. Ausas, R.; Jai, M.; Buscaglia C. A mass-conserving algorithm for dynamical lubrication problems with cavitation. *J. Tribol.* **2009**, *131*. [CrossRef]
29. Kumar, A.; Booker, J.F. A Finite Element Cavitation Algorithm. *ASME J. Tribol.* **1991**, *113*. [CrossRef]
30. Song, Y.; Gu, C. Development and Validation of a Three-Dimensional Computational Fluid Dynamics Analysis for Journal Bearings Considering Cavitation and Conjugate Heat Transfer. *AMSE J. Eng. Gas Turbines Power* **2015**, *137*. [CrossRef]
31. Chatzisavvas, I. Efficient Thermohydrodynamic Radial and Thrust Bearing Modeling for Transient Rotor Simulations. Ph.D. Thesis, TU Darmstadt, Darmstadt, Germany, 2018.
32. Chatzisavvas, I.; Boyaci, A.; Schweizer, B. Influence of hydrodynamic thrust bearings on the nonlinear oscillations of high-speed rotors. *J. Sound Vib.* **2016**, *380*. [CrossRef]
33. Chatzisavvas, I.; Koutsovasilis, P.; Schweizer, B. Influence of the oil temperature of thrust bearings on the vibratory behavior of small turbochargers. In Proceedings of the 11th International Conference on Engineering Vibration, Ljubljana, Slovenia, 7–10 September 2015. [CrossRef]
34. Chatzisavvas, I.; Koutsovasilis, P. On the influence of thrust bearing on the nonlinear rotor vibrations of turbochargers. In Proceedings of ASME Turbo Expo 2016: Turbomachinery Technical Conference and Exposition GT2016, Seoul, Korea, 13–17 June 2016. [CrossRef]
35. Li, S.; Tuzcu, S.; Klaus, M.; Rienäcker A.; Schwarze, H. Analyse der Einflüsse der hydrodynamischen Axiallagerung auf das rotordynamische Verhalten eines PKW-Abgasturboladers. In Proceedings of the SIRM 2015–11. Internationale Tagung Schwingungen in rotierenden Maschinen, Magdeburg, Germany, 23–25 February 2015.
36. Koutsovasilis, P. Automotive turbocharger rotordynamics: Interaction of thrust and radial bearings in shaft motion simulation. *J. Sound Vib.* **2019**, *455*. [CrossRef]
37. Greenhill, L.M.; Bickford, W.B.; Nelson, H.D. Conical Beam finite Element For rotordynamics Analysis. *J. Vib. Acoust.* **1985**, *107*, 421–430. [CrossRef]
38. Hao Z.; Gu, C. Numerical modeling for gaseous cavitation of oil film and non-equilibrium dissolution effects in thrust bearing. *Tribol. Int.* **2014**. [CrossRef]
39. Song Y.; Ren, X.; Gu, C.; Li, X. Experimental and numerical studies of cavitation effects in a tapered land thrust bearing. *J. Tribol.* **2015**. [CrossRef]
40. Dousti S.; Allaire, P. A thermohydrodynamic approach for signle film and double-film floating disk fixed thrust bearings verified with experiments. *Tribol. Int.* **2019**. [CrossRef]
41. Ziese, C.; Nitzschke, S.; Woschke, E. Run up simulation of a full-floating ring supported Jeffcott-rotor considering two-phase flow cavitation. *Arch. Appl. Mech.* **2020**. [CrossRef]
42. Irmscher, C.; Nitzschke, S.; Woschke, E. Transient thermo-hydrodynamic analysis of a laval rotor supported by journal bearings with respect to calculation times. In Proceedings of the SIRM 2019—13th International Conference on Dynamics of Rotating Machines, Copenhagen, Denmark, 13–15 February 2019.
43. Zhang, J.; Meng Y. Direct observation of cavitation phenomenon and hydrodynamic lubrication analysis of textured surfaces. *State Key Lab. Tribol.* **2011**. [CrossRef]
44. Shyu, S.H.; Li, F.; Jeng, Y.R.; Lee, W.R.; Hsieh, S.J. THD effects of static performance characteristic of infinitely wide turbulent journal bearing. *Soc. Tribol. Lubr. Eng.* **2010**. [CrossRef]
45. Shyu, S.H.; Jeng, Y.R.; Li, F. A legendre collocation method for thermohydrodynamic journal-bearing problems with Elrod's cavitation algorithm. *Tribol. Int.* **2007**. [CrossRef]
46. Boncompain, R.; Fillon, M.; Frene, J. Analysis of thermal effects in hydrodynamics bearings. *J. Tribol.* **1986**, *108*, 219–224. [CrossRef]

 *lubricants*

Article

Operating Behavior of Sliding Planet Gear Bearings for Wind Turbine Gearbox Applications—Part I: Basic Relations

Thomas Hagemann *, Huanhuan Ding, Esther Radtke and Hubert Schwarze

Institute of Tribology and Energy Conversion Machinery, Clausthal University of Technology, 38678 Clausthal-Zellerfeld, Germany; ding@itr.tu-clausthal.de (H.D.); radtke@itr.tu-clausthal.de (E.R.); schwarze@itr.tu-clausthal.de (H.S.)
* Correspondence: hagemann@itr.tu-clausthal.de; Tel.: +49-5323-722-469

Abstract: The application of sliding planet gear bearings in wind turbine gearboxes has become more common in recent years. Assuming practically applied helix angles, the gear mesh of the planet stage causes high force and moment loads for these bearings involving high local loads at the bearing edges. Specific operating behavior and suitable design measures to cope with these challenging conditions are studied in detail based on a thermo-hydrodynamic (THD) bearing model. Radial clearance and axial crowning are identified as important design parameters to reduce maximum pressures occurring at the bearing edges. Furthermore, results indicate that a distinct analysis of the gear mesh load distribution is required to characterize bearing operating behavior at part-load. Here, operating conditions as critical as the ones reached at nominal load might occur. Wear phenomena can improve the shape of the gap in the circumferential as well as in axial direction incorporating a significant reduction of local maximum pressures. The complexity of the combination of these aspects and the additionally expected impact of structure deformation gives an insight into the challenges in the design processes of sliding planet gear bearings for wind turbine gearbox applications.

Keywords: planet gear bearing; journal bearing; misalignment; mixed friction; wear; wind turbine gearbox

Citation: Hagemann, T.; Ding, H.; Radtke, E.; Schwarze, H. Operating Behavior of Sliding Planet Gear Bearings for Wind Turbine Gearbox Applications—Part I: Basic Relations. *Lubricants* **2021**, *9*, 97. https://doi.org/10.3390/lubricants9100097

Received: 30 July 2021
Accepted: 21 September 2021
Published: 1 October 2021

Publisher's Note: MDPI stays neutral with regard to jurisdictional claims in published maps and institutional affiliations.

1. Introduction

Generally, two different design concepts exist for wind turbine power trains. The first one foregoes a transmission between rotor and generator, and therefore, generator speed equals the one of the main rotor. Whereas this direct drive design provides high robustness of the power train, it involves high demands on generator design as the rotating frequency of the main rotor is in the magnitude of 1% or even less of the net frequency. The second design approach applies a gearbox between the rotor and generator shaft, which increases the speed on the generator side and reduces the previously mentioned discrepancy to approximately 50%. This enables a more compact generator design. Both concepts feature specific advantages and disadvantages that are comprehensively discussed by Ragheb and Ragheb [1]. Gearboxes failures contribute 10% to 20% of wind turbine downtimes [2,3]. Most commonly, wind turbines gearboxes feature one or two planetary and one spur gear stage due to lightweight and compact design requirements. Here, gear tooth as well as bearing damages represent the main issues leading to downtimes [4]. In recent years, the application of sliding bearings in wind turbine power trains has become more common, in particular, in the case of the planet gear bearing. The sliding planet gear bearing requires less radial space than the conventionally used rolling element bearings and enables flexible modification of gear rim thickness. Moreover, it is not susceptible to typical damage mechanisms that occur in rolling bearings and are caused by excitation from the gear mesh. However, several challenging phenomena in low speed and high load planet gear bearings exist that are partly specific for this tribologic system and that have not been studied in detail yet.

Figure 1 explains the basic design of a typical planet stage for wind turbine gearbox applications investigated in this study. The torque generated by the kinetic energy of the wind drives the main rotor that is connected by a hub to the planet carrier. The carrier rotates with the main rotor speed. Most commonly, four or five planets are attached to the carrier and distribute the driving torque to the same number of mesh contacts. The planets rotate in a non-rotating ring gear due to the carrier speed and drive a sun gear that is connected to the output shaft of the planet stage. Torque decreases from the rotor to the generator side while rotating frequencies increase according to the particular transmission ratios. Commonly, additional gear stages follow this first low-speed stage.

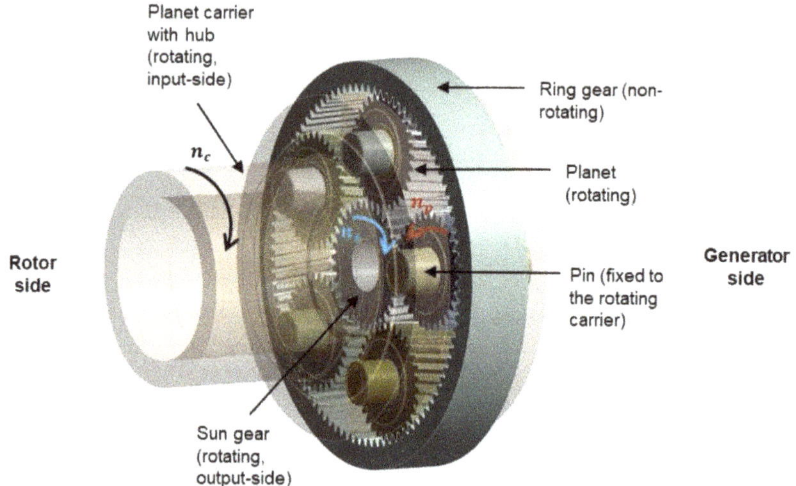

Figure 1. Basic design and kinematic properties of a planet stage for a wind turbine gearbox.

Figure 2 shows the red signed forces on the planet as well as the yellow signed components of the mesh forces. In the case of the straight gear in Figure 2a, the mesh forces feature a tangential F_t and radial F_r component. As the sum of the radial force components is zero, the bearing force F_{sc} is equal to the sum of the tangential forces. Both components of the mesh force contribute to an oval deformation of the planet. The magnitude of the oval shape of deformation strongly depends on the ratio between mesh forces and rim thickness and influences the lifetime of rolling element bearings [5,6]. In the case of the helical gear in Figure 2b, additional axial force components F_{ax} exist that involve a moment load about y-axis. This moment load M_{sc} has to be restored by the contact pressure in the bearing and is a function of the helix angle of the helical gear. An additional moment about the x-axis occurs if the load distribution on the tooth flanks becomes inhomogeneous in the lateral direction and causes eccentric points of application of resulting mesh forces. Figure 2c depicts an assembled helical planet gear with its contact partners in the front view of a planet stage.

Restoring the moment load requires a significant misalignment between planet and pin to generate the moment by fluid and asperity pressure in the bearing contact. This creates a load condition that distinguishes the planet gear bearing from most other applications, which are primarily loaded by radial forces. Bouyer and Fillon [7] and Sun and Changlin [8] investigate the impact of misalignment between shaft and stator for journal bearings and report on the well-known effect of edge loading. Whereas Bouyer and Fillon [7] experimentally apply different vertical torque levels parallel to load direction, Sun and Changlin [8] theoretically study the impact of the same misalignment angle applied at different angular journal positions and variable eccentricities.

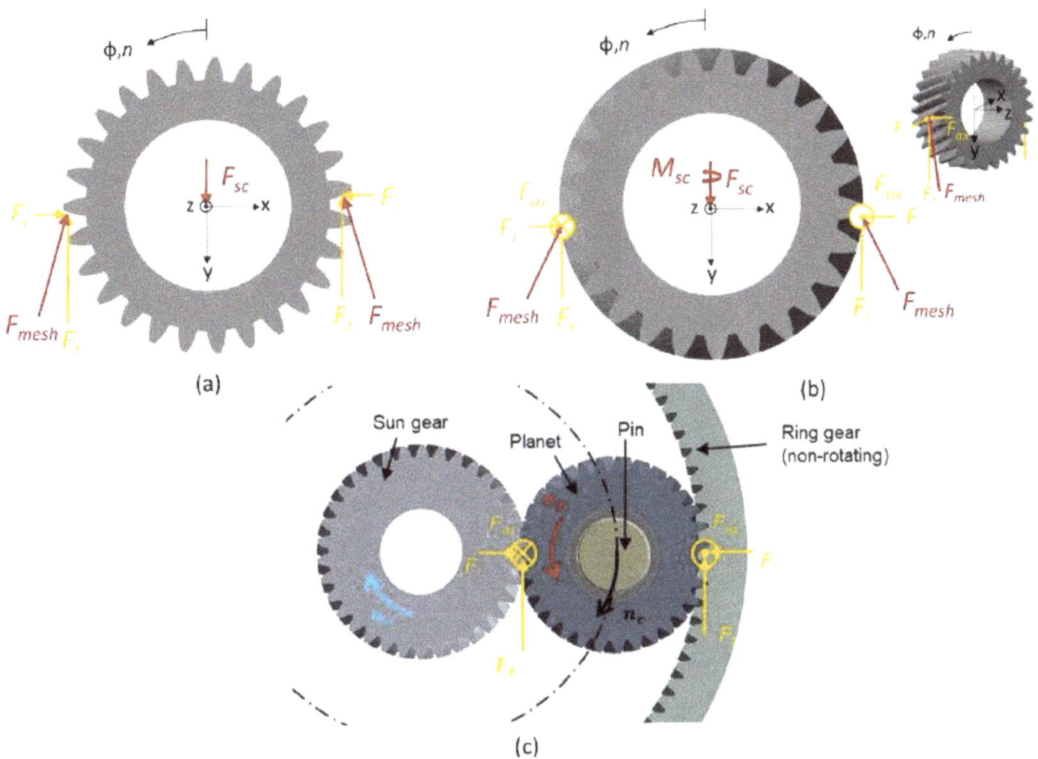

Figure 2. Mesh forces and bearing load for (**a**) straight gear, (**b**) helical gear, and (**c**) the helical gear assembled in the planet stage.

Depending on the concrete application, deformation can have positive as well as negative impact on the load-carrying capacity of sliding bearings. If elasto-hydrodynamic deformation is dominant, a rise of the film thickness compared to the hydrodynamic theory exists [9]. Conversely, Hagemann et al. [10] show a reduction of film thickness for a large five-pad tilting-pad journal bearing due to consideration of thermo-mechanical deformation in their analysis and validate the results with experimental data. This decrease in load capacity is traced to the thermally and mechanically induced axial bending of the tilting pads. Prölß [11] extends and adjusts the sliding bearing model presented by the authors in [10,12] to planet gear bearings. He determines a significant impact of deformation on predicted operating behavior due to the above-described load situation.

This investigation focuses on a basic understanding of the impact of different parameters on the operating behavior of sliding planet gear bearings as this type of bearing has received increasing importance, especially within recent years. While numerous publications on conventional journal bearings under low speed and high load conditions exist, e.g., [13–15], the specific properties of planet gear bearings have not been comprehensively described in the literature with the exception of [11] to the authors' knowledge. However, understanding their specific kinematics, load situations, and deformation behavior is essential for the development of a robust planet gear bearing design, in particular in the case of helical gears. Many investigations focus on nominal or maximum load conditions but part- and under-load can also contribute bearing damages as Garabedian et al. [16] show for rolling element planet gear bearings of a wind turbine gearbox. Therefore, aspects of the impact of part-load conditions on the performance of sliding planet gear bearings are also studied. All subsequent analyses in the first part of this study assume rigid macro geometries for pin and planet, while the second part [17] focuses on the impact of structure deformation on predicted results. Moreover, all results represent static equilibriums and neglect dynamic force components that are of minor interest for the investigated load cases.

2. Materials and Methods

2.1. Governing Equations and Bearing Model

The authors describe the governing equations of the bearing model including the consideration of oil supply effects in [10,12]. The generalized average Reynolds Equation (1) for laminar flow is extended by the flow factors according to Patir and Cheng [18] to consider the impact of surface roughness on hydrodynamic flow in the thin film region. Due to the exclusive static load conditions investigated in this study, the squeeze term is neglected in Equation (1).

$$\frac{\partial}{r_j \partial \phi}\left(\Phi_x^p \cdot F_2 \frac{\partial p_{hyd}}{r_j \partial \phi}\right) + \frac{\partial}{\partial z}\left(\Phi_z^p \cdot F_2 \frac{\partial p_{hyd}}{\partial z}\right)$$
$$= U_{planet}\frac{\partial}{r_j \partial \phi}\left(\Theta\left(h - \frac{F_1}{F_0}\right)\right) - \frac{U_{planet}}{2}\frac{\partial\left(\Theta R_q \Phi_x^s\right)}{r_j \partial \phi} \tag{1}$$

Herein, Φ_x^p and Φ_z^p are the pressure flow factors, and Φ_x^s is the shear flow factor. Equation (2) describes the local thickness of the lubricant gap h considering alignment angles between pin and pad φ_x and φ_z.

$$h(\phi, z) = C_R - e \cdot cos(\phi - \gamma) + z \cdot \left(-\varphi_x \cdot cos(\phi) + \varphi_y \cdot sin(\phi)\right) + \Delta h(\phi, z) \tag{2}$$

In contrast to the conventional journal bearing, the outer solid body rotates, and the pin is fixed. Therefore, the pin features a stationary temperature profile whereas temperature in circumferential direction homogenizes in the planet due to sufficiently high rotational speeds. Equation (3) includes the governing equations of the temperature model of the planet gear bearing for its three components. Figure 3 visualizes the differences between the discretization of the solution domains of journal and planet gear bearing for the thermal bearing model.

$$\text{Lubricant gap}: \quad c \cdot \rho \cdot \left(u \frac{1}{r_j}\frac{\partial T}{\partial \phi} + v\frac{\partial T}{\partial y} + w\frac{\partial T}{\partial z}\right) = \eta \cdot \left[\left(\frac{\partial u}{\partial y}\right)^2 + \left(\frac{\partial w}{\partial y}\right)^2\right] + \cdots$$

$$\frac{1}{r_j}\frac{\partial}{\partial \phi}\left(\lambda \frac{1}{r_j}\frac{\partial T}{\partial \phi}\right) + \frac{\partial}{\partial y}\left(\lambda \frac{\partial T}{\partial y}\right) + \frac{\partial}{\partial z}\left(\lambda \frac{\partial T}{\partial z}\right)$$

$$\text{Pin}: \quad \frac{1}{r}\frac{\partial}{\partial \phi}\left(\lambda_{pin}\frac{1}{r}\frac{\partial T}{\partial \phi}\right) + \frac{1}{r}\frac{\partial}{\partial r}\left(\lambda_{pin} r\frac{\partial T}{\partial r}\right) + \frac{\partial}{\partial z}\left(\lambda_{pin}\frac{\partial T}{\partial z}\right) = 0 \tag{3}$$

$$\text{Planet}: \quad \frac{1}{r}\frac{\partial}{\partial r}\left(\lambda_{planet} r\frac{\partial T}{\partial r}\right) + \frac{\partial}{\partial z}\left(\lambda_{planet}\frac{\partial T}{\partial z}\right) = 0$$

Herein u, v, and w are the flow velocities in the lubricant gap. In comparison to the journal bearing, the three-dimensional part of the solution domain is adjusted to the new requirements and the interface of the three to the two-dimensional section of the solution domain is shifted from the journal radius in Figure 4a to the inner radius of the planet in Figure 4b. At each internal fluid–solid interface, conjugate heat transfer boundary conditions are valid. Temperature-dependent lubricant viscosity is determined using the dynamic viscosity η_0 at the reference temperature T_0 by the equation of Falz [19].

$$\eta(T) = \eta_0 \cdot \left(\frac{T}{T_0}\right)^{-l} \tag{4}$$

A periodicity boundary condition in the circumferential direction is applied for the solution of Equation (1) and the three-dimensional parts of Equation (3) at the particular boundaries of the solution domain. Moreover, the thermal bearing model assumes convective heat transfer boundary conditions for the free solid body surfaces. For the solution of Reynolds and energy equation, a conservative finite difference scheme (Finite Volume Method (FVM)) is applied. The solution of the generalized average Reynolds

Equation (1) is based on the difference scheme proposed by Elrod [20]. The combined convection and diffusion problem of the energy equation is stabilized by the Hybrid scheme according to the description in [21]. References [10,12] describe further details on numerical implementation, including flow charts of the entire numerical procedure.

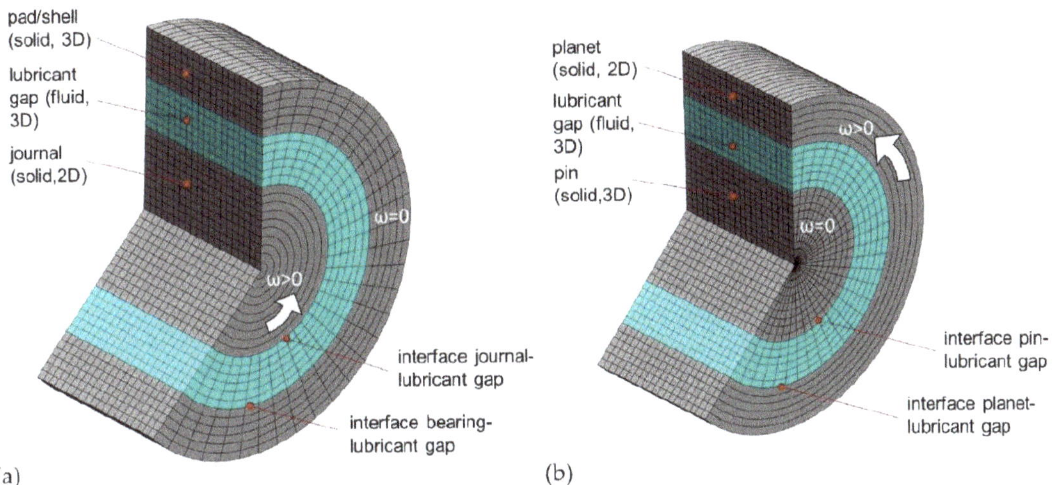

Figure 3. Numerical grid for temperature evaluation for the (**a**) journal bearing and (**b**) planet gear bearing.

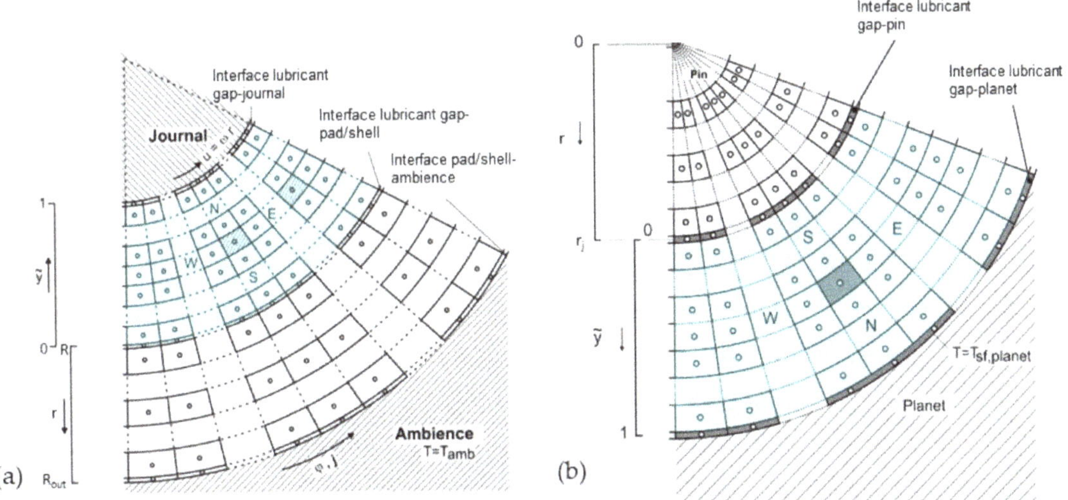

Figure 4. Cross section of the three-dimensional solution domain of the temperature analysis (**a**) journal bearing, and (**b**) planet gear bearing at a certain axial position.

The eccentricities e_x, e_y and angular alignment φ_x, φ_y between the centerlines of planet and pin are determined using stiffness matrix **C** defined by Equation (5) to minimize the error between mesh forces and bearing contact forces in order to find the static state of equilibrium. The coefficients of **C** represent the sum of hydrodynamic and solid contact stiffness coefficients in the mixed friction regime while solid contact stiffness is equal to zero in the hydrodynamic regime.

$$C = \begin{pmatrix} C_{xx} & C_{xy} & C_{x\varphi_x} & C_{x\varphi_y} \\ C_{yx} & C_{yy} & C_{y\varphi_x} & C_{y\varphi_y} \\ C_{\varphi_x x} & C_{\varphi_x y} & C_{\varphi_x \varphi_x} & C_{\varphi_x \varphi_y} \\ C_{\varphi_y x} & C_{\varphi_y y} & C_{\varphi_y \varphi_{yx}} & C_{\varphi_y \varphi_y} \end{pmatrix} = C_{hyd} + C_{solid} \tag{5}$$

Hydrodynamic stiffness coefficients are derived from a linear perturbation of generalized average Reynolds Equation (1). For this purpose, a linear impact of film thickness modification on fluid film pressure due to translational (X,Y) or angular (φ_X, φ_Y) rigid body displacements relative to the current position described by e_x, e_y, φ_x, and φ_y is assumed according to Equation (6).

$$h^*(\phi, z) = h(\phi, z) + X \sin\phi + Y \cos\phi - z \cdot \varphi_X \cdot \cos\phi + z \cdot \varphi_Y \sin\phi \tag{6}$$

Solid body stiffness directly follows from the local gradient of the relation between the mean distance of the contact partners and the asperity contact forces. A highly time efficient iterative scheme based on the idea of Waltermann [22] and extended to consider moment loads adjusts solid body positions and minimizes the differences between the outer forces and moments and the ones restored by the fluid film and the solid contact pressure.

The authors comprehensively validated the theoretical THD bearing model and its implementation with test data in previous investigations for fixed-pad and tilting-pad journal bearings, e.g., [10,12,23]. As no test data for planet bearings is available, this step must be omitted here. However, the correct implementation of the impact of misalignment, which is of particular interest in this study, was not part of previous validation and is verified with data from the literature.

2.2. Verification of the Numerical Procedure

Results of Sun and Changlin [8] are used to verify the numerical implementation of the THD model in the planet gear bearing code with a special focus on misalignment. Table 1 includes the data of the investigated journal bearing.

Table 1. Bearing parameters [8].

Parameter	Value
Geometrical properties	
Number of pads, -	1
Nominal diameter, mm	60
Bearing width, mm	66
Radial clearance, µm	30
Pad sliding surface preload, -	0.0
Static analysis parameters	
Rotational speed, rpm	3000
Lubricant properties	
Lubricant dynamic viscosity, mPas	9.0

Due to the isoviscous approach of the study, no impact of the modified kinematics in the planet gear bearing code has to be expected. Figure 5 shows excellent agreement between the two analyses for the local film pressure distribution in the case of the aligned bearing (a,b) as well as for the highest investigated level of misalignment (c,d). While maximum pressures in [8] are given with 33.06 MPa for the aligned case and 143.34 MPa for the misaligned case, they are predicted with 33.30 and 144.2 MPa in this study. Additionally, the resulting moments of the fluid forces due to different levels of misalignment in Figure 6 nearly match for both analyses and verify the basic implementation of the THD model in the planet gear bearing code.

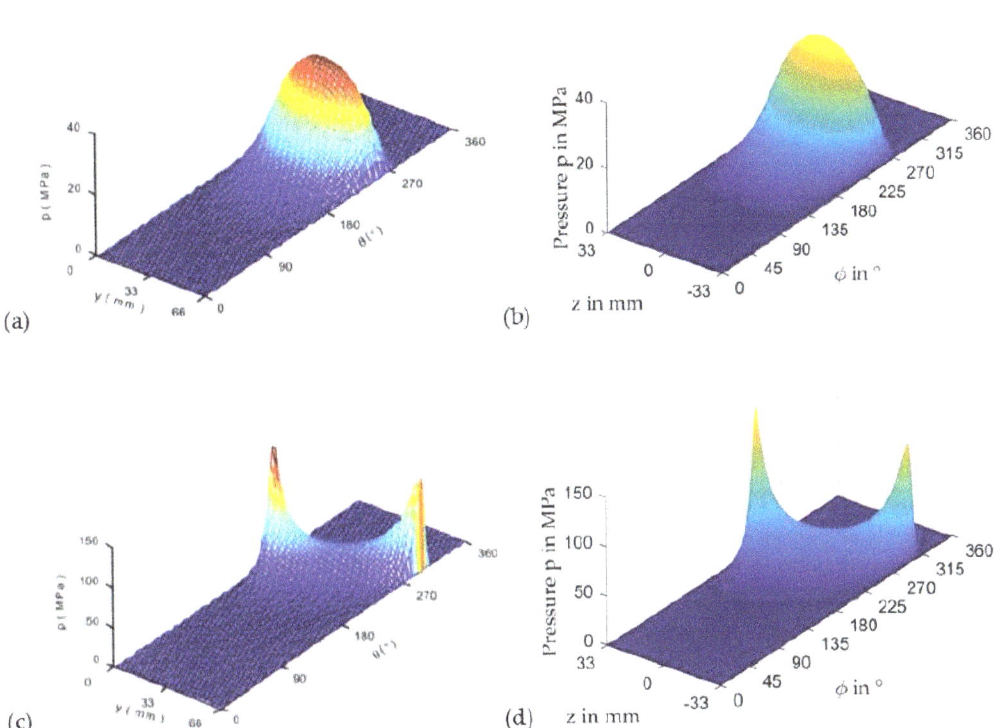

Figure 5. Pressure distributions determined by Sun and Changlin [8] (**a**,**c**) and the planet gear bearing code (**b**,**d**) for misalignment angles of $\gamma = 0.0°$ (**a**,**b**) and $\gamma = 0.03°$ (**c**,**d**).

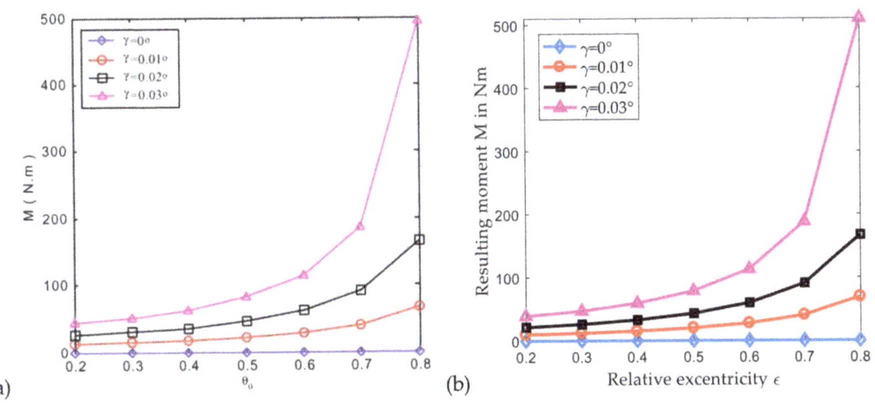

Figure 6. Resulting moment for different levels of misalignment angles determined by Sun and Changlin [8] (**a**) and the planet gear bearing code (**b**).

2.3. Flow Factors and Solid Contact Pressure

The detailed prediction of flow factors and solid body pressure is not an aim of this study. In fact, these parameters are introduced to apply a consistent bearing model with representative parameters. Flow factors are analyzed based on the average Reynolds equation according to [18]. Results are given in Figure 7a. An application of the theory of Greenwood and Williamson [24] provides a relation between the mean film thickness on the node of the numerical grid and solid contact pressure due to the interaction of the asperities that is shown in Figure 7b. The contact surface features a root mean square roughness value of $R_q = 2.83$ μm and matches the one earlier presented by the authors in [25]. A turning process generates the surface topography that features small grooves in the circumferential direction. Solid body contact pressure is determined based on the formula proposed by Hu et al. [26]. Consequently, it slightly deviates from the ones presented in [25], which are evaluated differently. The concrete course of the solid body load-carrying capacity has exemplary character in this investigation. Therefore, the contradiction that the theory of Greenwood and Williamson presupposes a normal distribution of the surface roughness that does not exist for the investigated contact surface is accepted here.

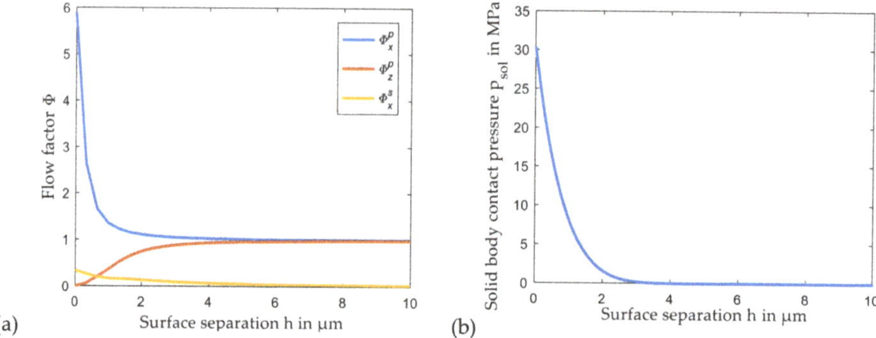

Figure 7. Pressure and shear flow factors (**a**) and solid body contact pressure (**b**).

2.4. Investigated Sliding Planet Gear Bearing

Table 2 includes the basic analysis parameters of the investigated planet gear bearing. The values are of the typical magnitude of planet gear bearings applied in 3 MW wind turbine gearboxes.

Table 2. Planet gear bearing parameters.

Parameter	Value
Geometrical properties	
Number of pads, -	1
Nominal diameter, mm	250
Pitch circle diameter, mm	499
Bearing width, mm	300
Angular span of lube oil pocket, degrees	20.5
Width of lube oil pocket, mm	260
Radial clearance, μm	138
Pad sliding surface preload, -	0.0
Static analysis parameters	
Nominal rotational speed, rpm	30
Nominal bearing load, kN	900
Lubricant supply temperature, °C	60
Lube oil supply pressure, MPa	0.2

Table 2. *Cont.*

Parameter	Value
Lubricant properties	
Lubricant	ISO VG 320
Lubricant density kg/m^3	865 @ 40 °C
Lubricant specific heat capacity kJ/(kg·K)	2.0 @ 20 °C
Lubricant thermal conductivity, W/(m·K)	0.13

2.5. Bearing Loads Due to Mesh Forces

According to the illustrations in Figure 2, the mesh forces generate the load of the planet gear bearing. The static radial bearing load F_{sc} is equal to the sum of the tangential mesh forces that result from the driving torque T_d and the contact between ring gear and planet gear and between planet gear and sun gear, respectively.

$$F_{sc} = 2 \cdot F_t = \frac{4 \cdot T_d}{d_1} \tag{7}$$

Using the normal pressure angle α_n and the helix angle of the helical gear β, Equation (8) defines the axial and radial components of the mesh forces.

$$F_r = F_t \cdot \frac{\tan \alpha_n}{\cos \beta}$$
$$F_{ax} = F_t \cdot \tan \beta \tag{8}$$

Neglecting special features of tooth flank geometry and the impact of deformation, axial as well as radial mesh forces of the two contacts compensate for each other and do not contribute to static bearing force. However, the eccentricity of axial forces causes a moment load M_{sc} that has to be restored by the bearing. This moment load follows according to Equation (9).

$$M_{sc} = F_{ax} \cdot d_1 \tag{9}$$

Consequently, moment load M_{sc} depends linearly on the driving torque and, additionally, it is a nearly linear function of the helix angle of the helical gear. Restoring this moment by the bearing contact requires an alignment of pin and planet relative to each other. The impact of this effect on planet gear bearing operating behavior is considered as a key or contributing effect throughout the entire subsequent investigations.

Figure 8 illustrates static bearing force and moment load as a function of the relative input torque T_r. Here, 100% T_r correspond to the nominal load situation with a bearing force load of 900 kN according to Table 2. While the force load in Figure 8a is independent of the helix angle, the moment load in Figure 8b strongly depends on it. The distribution of tooth forces to more than one contact due to a profile overlap is neglected in this study.

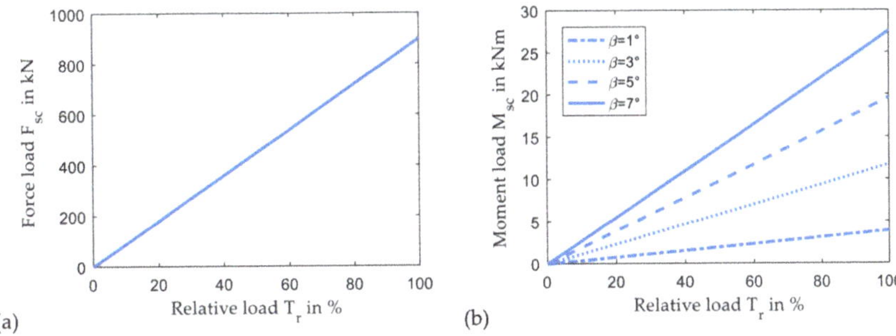

Figure 8. Force (**a**) and moment load (**b**) due to relative mesh load.

3. Results

Wind turbine gearboxes are commonly equipped with helical gears in the planet stage. The low level of dynamic excitation due to the continuous tooth mesh reduces the danger of fatigue damages. As the forces in the axial thrust washers are comparably low, the moments resulting from the gear mesh have to be restored in the radial part of the bearing. This specific load conditions cause operating conditions and design measures that are studied subsequently.

3.1. Impact of Helix Angle of the Helical Gear on Operating Conditions in the Lubricant Gap

To investigate the impact of helix angle of the helical gear on operating conditions in the lubricant gap, a variation of this parameter is conducted in a range between $\beta = 0°$ and $8°$, assuming a homogenous load distribution on each tooth flank. Consequently, no eccentricity of the tangential mesh forces exists, and according to the conventions in Figure 2b, the resulting moment restored by the bearing M_{sc} results exclusively from M_y while $M_x = 0$ Nm. Figure 9 shows the local pressure distribution in the lubricant gap for a straight gear (a) and a helical gear (b) with a helix angle $\beta = 7°$ in two different views. Here, the term "aggregate pressure" refers to the sum of hydrodynamic and asperity contact pressure. While the aggregate pressure in case (a) purely results from hydrodynamic fluid film, the edge loading in case (b) that is caused by the moment load due to the axial component of the mesh forces involves mixed friction, and therefore, a combination of hydrodynamic and solid contact pressure in the bearing. Moreover, the aggregate pressure significantly increases in case (b) and reaches a level that exceeds the strength level of common sliding bearing materials. In addition to these results, Figure 10 includes the characteristic of minimum film thickness, solid contact and hydrodynamic pressure as well as the alignment between planet and pin as a function of the helix angle. A significant rise of the pressure level exists for helix angles $\beta > 4°$. As expected, the load moment M_y causes an alignment of the bearing about x- and y-axis due to the cross-coupling effects in the fluid film.

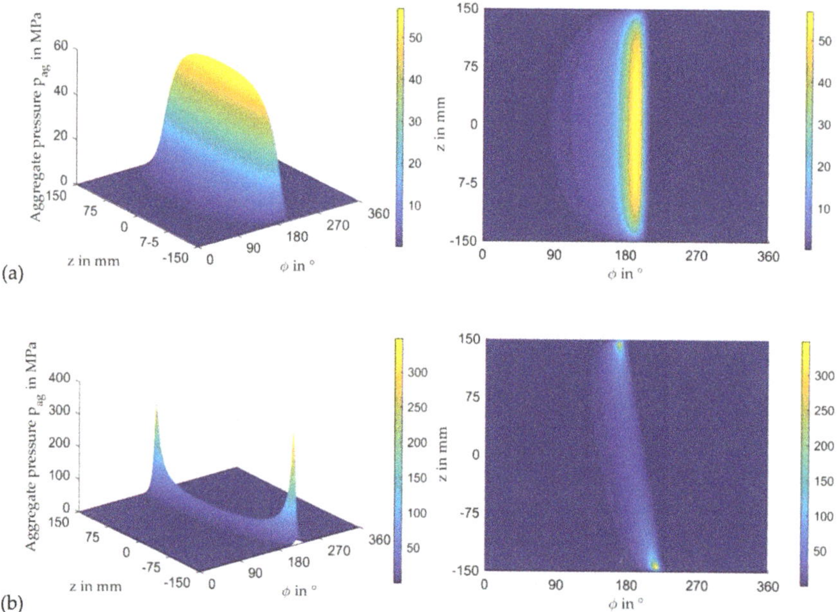

Figure 9. Pressure distributions @ nominal operating conditions (**a**) straight gear ($\beta = 0°$) and (**b**) helical gear ($\beta = 7°$) ($n_{pl} = 30$ rpm, $F_{sc} = 900$ kN, $T_{sup} = 60\,°C$, $p_{sup} = 0.2$ MPa, $M_{sc} = 27.6$ kNm (**b**)).

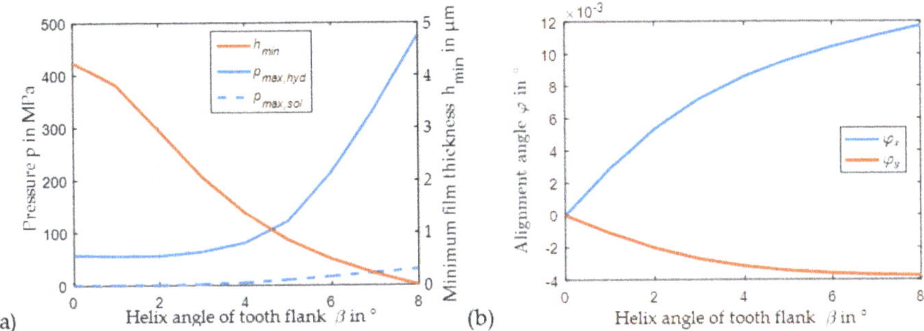

Figure 10. Impact of helix angle of helical gear on (**a**) maximum pressures, minimum film thickness, and (**b**) alignment angles between planet and pin (n_{pl} = 30 rpm, F_{sc} = 900 kN, T_{sup} = 60 °C, p_{sup} = 0.2 MPa).

Deformation of planet and pin potentially reduce the extremely high maximum pressures. However, the pressure level might remain in a critical range for technically required helix angles in the magnitude of $\beta \approx 7°$. Therefore, measures have to be defined to reduce maximum local load levels in the contact between planet and pin.

3.2. Impact of Radial Clearance on Load Carrying Capacity

Radial clearance is an important design parameter of hydrodynamic bearings. In particular, a suitable value of this parameter has to ensure two aspects. First, positive operating conditions for the entire nominal operating range of the bearing have to be guaranteed. Second, the bearing has to be robust against any disturbances from these conditions and tolerate shape deviations of the components due to the manufacturing process. Both aspects especially have to allow thermal changes of the radial clearance due to the transient operation of the machine.

Figure 11 exhibits the impact of the relative clearance on maximum pressure and temperature for different gear helix angles. While a comparably moderate influence of the relative clearance on maximum pressure exists for the lower helix angles of $\beta = 0°$ and $\beta = 3°$ in Figure 11a, a very significant change for the practical highly relevant helix angle $\beta = 7°$ is observable as maximum pressure rises progressively with increasing relative clearance. Moreover, the maximum temperature on the pin sliding surface increases with rising relative clearance in Figure 11b for the maximum helix angle due to contact intensity, whereas a commonly expected decrease in this value is present for $\beta = 0°$. Generally, maximum temperature is only slightly higher than the nominal supply temperature, and consequently, thermal effects are not of major importance in this study. Results indicate that relative clearance should be designed as low as possible in the investigated range for $\beta = 7°$.

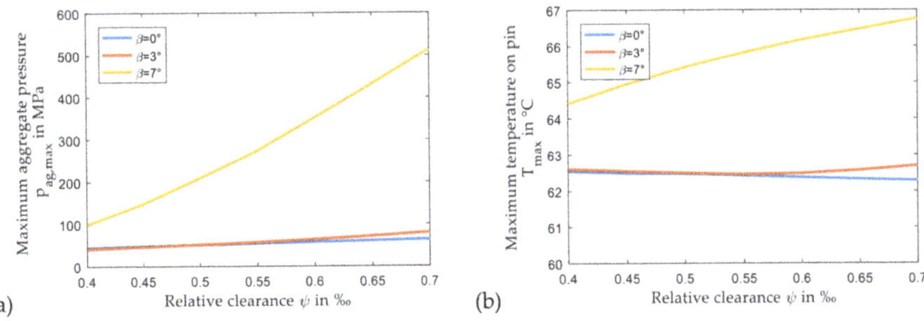

Figure 11. Maximum pressure (**a**) and maximum temperature on the pin sliding surface (**b**) for variable relative loads (n_{pl} = 30 rpm, F_{sc} = 900 kN, T_{sup} = 60 °C, p_{sup} = 0.2 MPa).

3.3. Impact of Axial Profiling

An axial crowning of contact partners with high edge loading is a common measure to reduce local maximum contact intensity. Examples for such applications are rolling element bearings or tooth flanks. To investigate the impact of axial crowning on the performance of the planet gear bearing, two different artificial crowning characteristics are defined according to Figure 12a and applied at nominal load conditions with a helix angle of $\beta \approx 7°$. The first one focuses on profile relief at the bearing edges where maximum pressures occur, according to Figure 9b. The second one exhibits continuous parabolic shaped characteristic across the bearing width. Though the maximum pressure in Figure 12b slightly decreases compared to Figure 9b, the first profile tends to reduce the load-carrying bearing width. However, the second profile provokes a significant reduction of the pressure level. According to Figure 12c,d, a maximum pressure of 77.6 MPa is reached which is approximately 37% higher than the one for the straight tooth flanks in Figure 9a and acceptable for several bearing materials. Figure 12d illustrates the continuous pressure level in axial direction accompanying a homogenous load distribution in the bearing. The main cause for the advantages of the second crowning profile is that edge loading results from the tilting movement of the planet relative to the pin about the axis in load direction. Therefore, the crowning profile involves a comparably high region of nearly constant film thickness at nominal load that starts at $\phi \approx 160°$ at z = 150 mm and ends at $\phi \approx 235°$ at z = −150 mm, as depicted in Figure 13.

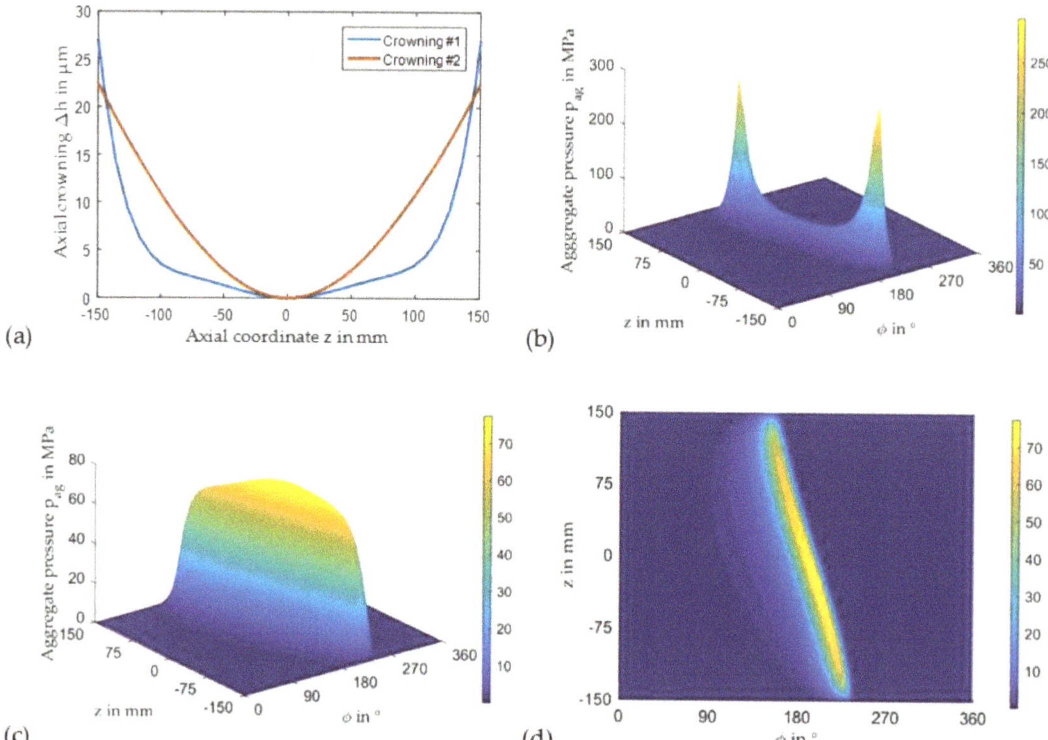

Figure 12. Impact of axial crowning (**a**) on the pressure distribution: (**b**) crowning #1, (**c,d**) crowning #2 (n_{pl} = 30 rpm, F_{sc} = 900 kN, T_{sup} = 60 °C, p_{sup} = 0.2 MPa, β = 7°, M_{sc} = 27.6 kNm).

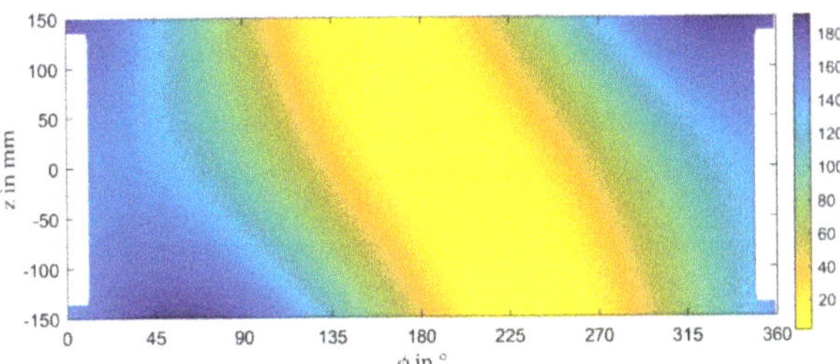

Figure 13. Film thickness @ nominal operating conditions with crowning #2 (n_{pl} = 30 rpm, F_{sc} = 900 kN, T_{sup} = 60 °C, p_{sup} = 0.2 MPa, β = 7°, M_{sc} = 27.6 kNm).

3.4. Operation at Part-Load and Over-Load Conditions

For the investigation of part-load and over-load conditions, the driving torque is varied, while planet speed remains constant. Assuming a homogenous load distribution in the gear mesh, the driving torque is proportional to the gear mesh forces causing the bearing load forces and moments. The subsequently presented figures include the relative input torque T_r. Figure 8 includes the corresponding force and moment loads.

3.4.1. Part-Load Conditions for Homogenous Load Distribution on Tooth Flank

In many applications, a geometrical optimization for nominal load conditions accompanies a degradation of the bearing performance in the part-load operating range. Figure 14 contrasts maximum pressure and minimum film thickness for the investigated planet gear bearing with crowning profile #2 and without crowning for relative loads ranging from T_r = 10% to 100%. Results clearly show that the crowning exhibits advantages in the entire investigated load range as maximum pressure decreases and minimum film thickness rises. Moreover, a strictly monotonous characteristic of these parameters can be observed, and operating conditions in the bearing become unfavorably with a rising relative load.

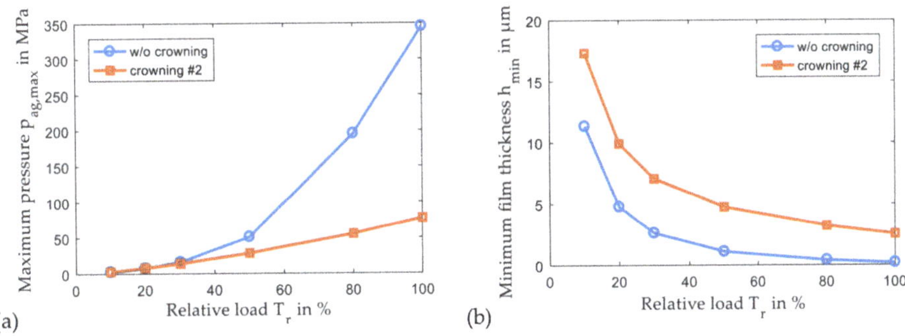

Figure 14. Maximum pressure (**a**) and minimum film thickness (**b**) for variable relative loads (n_{pl} = 30 rpm, T_{sup} = 60 °C, p_{sup} = 0.2 MPa, β = 7°).

3.4.2. Part-Load and Over-Load Conditions for Constant Tooth Flank Geometry

While the previously presented results concordantly assume a homogenous load distribution on the gear mesh, this ideal case is now supplemented by the practical case of a certain tooth flank geometry designed for the nominal load case. The optimization of the tooth flank geometry generally aims to match the previously studied ideal case for the corresponding operating conditions. Consequently, only very low eccentricities of the tangential and radial gear mesh forces to the middle plane of the bearing exist at nominal load conditions. This shape optimization accompanies the eccentricities of these forces for all other load cases. With increasing load, the resulting mesh forces move along the tooth flank from one side of the planet to the other one and across the center plane for the nominal load case. Therefore, a load-dependent derivation of the moments from the ideal case is present and depicted for an artificial tooth flank geometry in Figure 15. According to Figure 15a, a parabolic shape of the characteristic of moment M_y is assumed between relative loads $T_r = 0$ and 100%, and a linear one is present in the overload range. The moment M_x in Figure 15b remains unchanged as no shifts are considered between the meshing points of sun and planet gear and planet and ring gear, respectively. In opposite to the ideal case studied so far, the direction of the resulting moment load varies with driving torque. Figure 15c shows a rise of the resulting moment load due to the eccentricity of the tangential mesh forces depicted in Figure 15d. In particular, the moment loads in the part-load region are much higher, as the yellow curve in Figure 15c illustrates.

The modified bearing loads are applied to the bearing with crowning #2 and cause a significant modification of the characteristics previously shown in Figure 14. Figure 16 indicates a degradation of the characteristic parameters, in particular, in the part-load range. While the mesh forces, and therefore, the bearing force loads are identical for the practical and ideal case, their application point changes, and the bearing moment load rises. Consequently, the minimum film thickness for the practical case in Figure 16b decreases and is partly below the value predicted for the nominal load case. Moreover, monotonous characteristics of maximum pressure and minimum film thickness do not exist anymore.

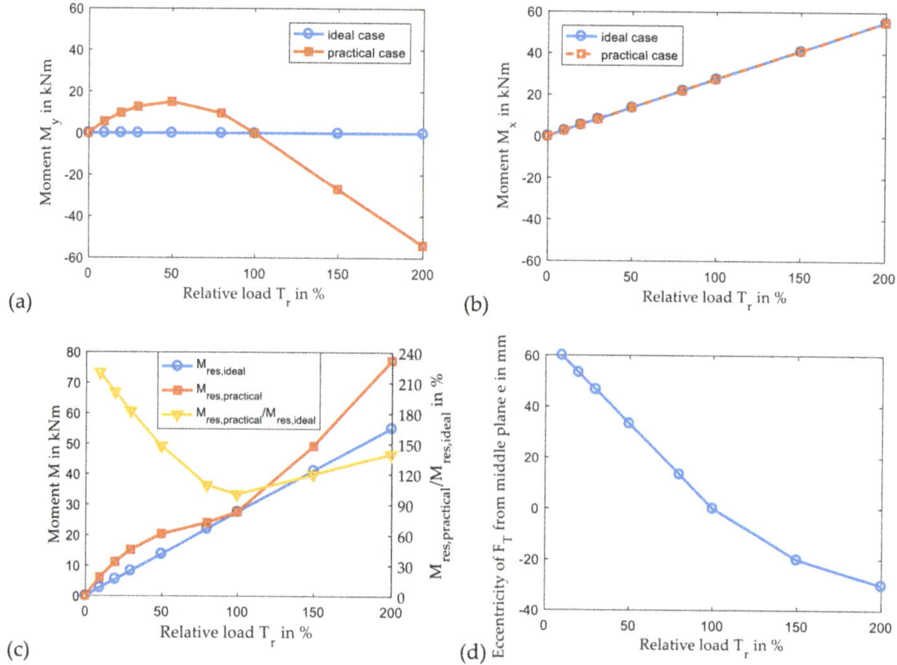

Figure 15. Load moments M_y (**a**), M_x (**b**), and M_{res} (**c**) for variable relative loads ($n_{pl} = 30$ rpm, $T_{sup} = 60\,°C$, $p_{sup} = 0.2$ MPa, $\beta = 7°$); eccentricity of tangential mesh force from the middle plane (**d**).

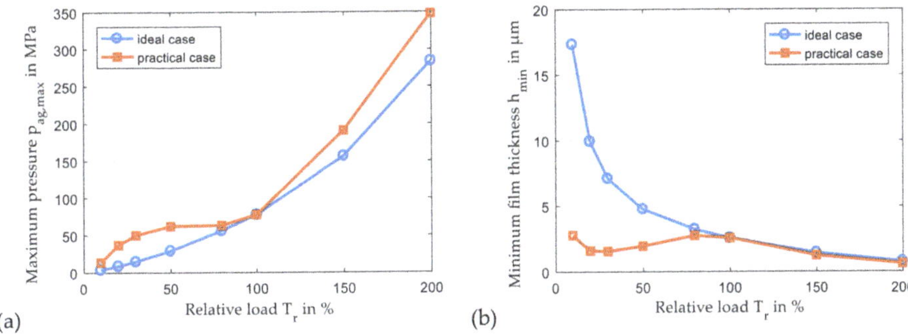

Figure 16. Maximum aggregate pressure (**a**) and minimum film thickness (**b**) for variable relative loads (n_{pl} = 30 rpm, T_{sup} = 60 °C, p_{sup} = 0.2 MPa, β = 7°).

3.5. Modification of the Lubricant Gap by Wear

According to the previous results, load conditions of planet gear bearings might incorporate operation in the mixed friction regime. Consequently, wear mechanisms can arise. To study their general impact on the conditions in the contact between planet and pin, solid contact forces F_{sol} are used to evaluate wear according to Archard's law [27].

$$V_w = \frac{k}{H} \cdot F_{sol} \cdot L = K \cdot F_{sol} \cdot L \tag{10}$$

The wear volume V_w is a function of the contact force F_{sol}, the sliding distance L and the wear coefficient K. It is assumed that the softer material is located on the stator side provoking a two-dimensional variable distribution of wear on the sliding surface of the bearing without axial crowning. In the concrete case, a wear coefficient of $K = 8.5 \cdot 10^{-9}$ mm^3/J and an operation time of t = 280 h at nominal load is assumed. These operating conditions cause significant wear according to Figure 17a. Figure 17b shows the characteristic of the maximum aggregate pressure and minimum film thickness during the run. Here, wear increases minimum film thicknesses and decreases maximum pressure monotonously in the first 160 h. After 160 h a nearly constant level of minimum film thickness close to pure hydrodynamic operation is reached that slightly reduces in the further run. The pressure level decreases monotonously throughout the entire investigated time range. Local maximum pressures tend to shift to the middle plane similar to the behavior observed in the results for crowning #1 in Figure 12b. Nevertheless, maximum pressures significantly reduce as wear accompanies regions of only slightly changed film thickness. To show this phenomenon more clearly, the run time is extended to 560 h. Figure 18b depicts an extension of the load-carrying region in the circumferential direction of the bearing due to wear. Therefore, maximum pressure reaches a level comparable to the one predicted for the unworn bearing with crowning #2 in Figure 12d. Though the pressure is more homogenous in the axial direction with the crowning, Figure 18a indicates that the load-carrying region in the circumferential direction, which is enclosed by the two red lines in both partial figures, is smaller than the one of the worn bearing.

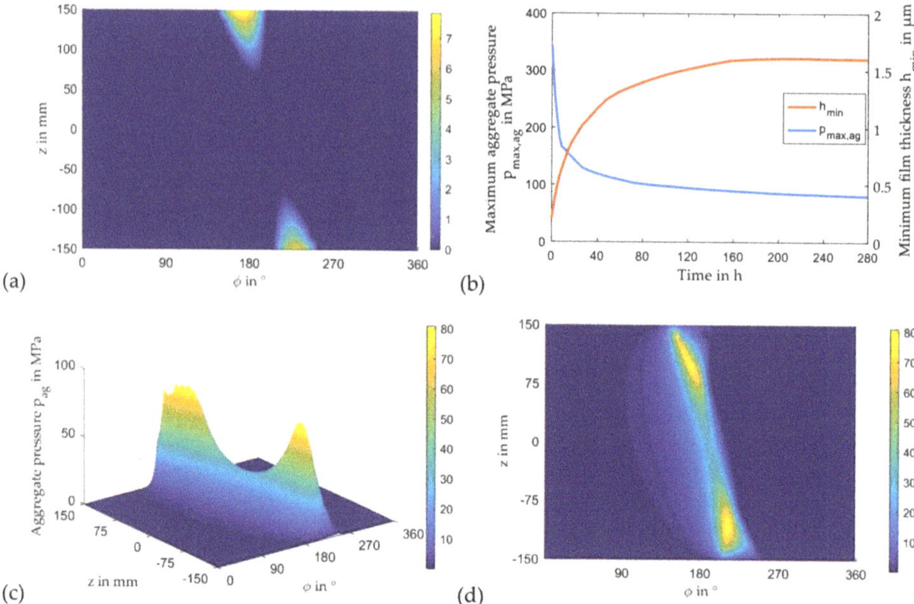

Figure 17. Wear after 280 h run (**a**), maximum aggregate pressure and minimum film thickness (**b**), and pressure distribution after 280 h run (**c**,**d**) (n_{pl} = 30 rpm, T_{sup} = 60 °C, p_{sup} = 0.2 MPa, β = 7°, t = 280 h).

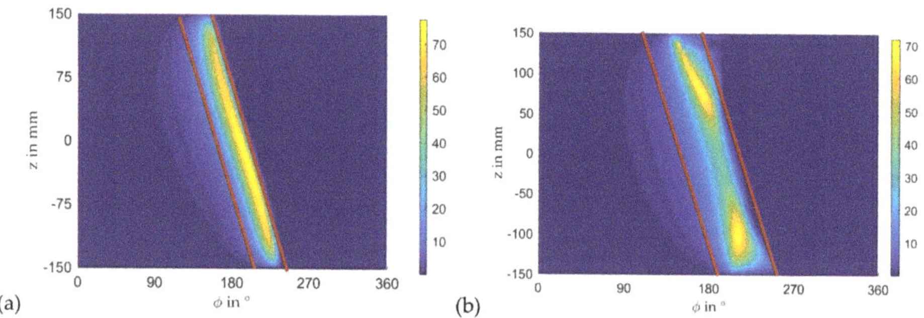

Figure 18. Comparison of pressure distribution with crowning #2 (**a**) and without crowning after 560 h (**b**) (n_{pl} = 30 rpm, T_{sup} = 60 °C, p_{sup} = 0.2 MPa, β = 7°).

4. Discussion and Conclusions

This paper focuses on specific operating conditions of sliding planet gear bearings for wind turbine gearbox applications as well as on the impact of particular design parameters on the properties in the contact between pin and planet. Special emphasis is given to the influence of the helix angle of the gear. A helix angle greater than zero incorporates additional moment loads that have to be restored by the bearing and accompany the edge loading phenomena and mixed friction processes in the lubricant film.

The level of edge loading is very sensitive to relative bearing clearance for higher helix angles. Results of a variation of relative clearance in a practical relevant range indicate an improvement of the load-carrying capacity with a reduction of this parameter. However, for practical applications, sufficiently high clearances are required that prevent problems during transient operation or undesirable influences of manufacturing tolerances. Moreover, the system has to be robust to sustain the deformation of its flexible components due to the high mechanical loads. As a second geometrical design parameter, axial crowning of the sliding surface is an appropriate measure to reduce maximum contact pressures. Contrary to the straight cylindrical pin and planet bore geometry, a homogenization of the contact pressure on the oblique line of maximum load exists. The shape of the crowning has to approach the contact surfaces of planet and pin in this region to reduce high deviations of the film thickness that exists due to misalignment between the contact partners.

Results for part-load conditions reveal critical characteristic operating parameters if the load-carrying behavior of the gear mesh contact is considered. In particular, the minimum film thickness is of the magnitude as at nominal operation or can be even lower. Though the intensity of mixed friction might be lower due to the lower mechanical load, issues may arise over the lifetime of the machine. Consequently, distinct characterizations of tooth flank geometry as well as part-load conditions have to be the input parameters of the bearing design process. Moreover, neglecting the additional load moments caused by the eccentricity of tangential forces in the gear mesh contact might lead to a completely different result assessment.

Wear processes in the bearing are able to adapt the shape of the lubricant gap to the specific requirements of the bearing at operation. While load-carrying capacity is improved by wear in the investigated case, the characteristic of the gap deviates from the one determined, applying a crowning to the bearing because wear particularly arises on the bearing edges. Therefore, the worn gap profile represents an improvement but not an optimization of the bearing geometry. If the occurrence of wear is tolerable by the tribologic system, it might be a measure to adjust the bearing itself to the operating conditions, involving leaving out manufacturing steps for the axial crowning. However, this option must be carefully evaluated because the danger of bearing damage arises if the intensity of the wear becomes too high.

According to these investigations, the design process of planet gear bearings for wind turbine applications involves several specific aspects that need to be studied in detail to meet the high requirements on the reliability of wind turbine applications. Additionally, the impact of deformation of the comparably flexible planet stage structure on the operating behavior of the bearing needs to be considered according to part 2 of these investigations [27]. Furthermore, future studies must provide details on the validation of these results.

Author Contributions: Conceptualization, methodology, T.H., H.D., E.R.; software, T.H.; investigation, writing, and visualization, T.H., H.D., E.R.; supervision and funding acquisition, H.S. All authors have read and agreed to the published version of the manuscript.

Funding: This research was funded by the German Federal Ministry of Economic Affairs and Energy. The financial support was assigned by the Industrial Research Association (AiF e. V.) in project number IGF 19425 N/1.

Acknowledgments: The authors thank the expert committees of the German research community FVA e. V. for the technical and scientific steering of this research project.

Conflicts of Interest: The authors declare no conflict of interest.

Nomenclature

c	lubricant specific heat
C_R	radial clearance
d_1	pitch diameter
e	Eccentricity between pin and planet
F_0, F_1, F_2	viscosity factors
F	force
h	film thickness
M	moment
n	rotor speed

p	pressure
R_q	root mean square surface roughness
T	temperature, torque
U	surface speed
u, v, w	flow velocities
x, y, z	Cartesian coordinates
X, Y	translational displacement relative to equilibrium position
Θ	lubricant density ratio
γ	attitude angle
η	lubricant dynamic viscosity
λ	lubricant conductivity
ρ	lubricant density
φ_x, φ_y	alignment angle about x-,y-axis
φ_X, φ_Y	alignment angle about x-,y-axis relative to equilibrium position
ϕ	angular coordinate
$\phi_x, \phi_z{}^p$	pressure flow factors
$\Phi_x{}^S$	shear flow factor

References

1. Ragheb, A.; Ragheb, M. Wind turbine gearbox technologies. In Proceedings of the 2010 1st International Nuclear & Renewable Energy Conference (INREC), Amman, Jordan, 21–24 March 2010; pp. 1–8.
2. Tavner, P.J.; Xiang, J.; Spinato, F. Reliability analysis for wind turbines. *Wind Energy* **2007**, *10*, 1–18. [CrossRef]
3. Ribrant, J.; Bertling, L. Survey of failures in wind power systems with focus on Swedish wind power plants during 1997–2005. In Proceedings of the 2007 IEEE Power Engineering Society General Meeting, Tampa, FL, USA, 24–28 June 2007; pp. 1–8.
4. Qiao, W.; Lu, D. A survey on wind turbine condition monitoring and fault diagnosis—Part I: Components and subsystems. *IEEE Trans. Ind. Electron.* **2015**, *62*, 6536–6545. [CrossRef]
5. Jones, A.; Harris, T.A. Analysis of a Rolling-Element Idler Gear Bearing Having a Deformable Outer-Race Structure. *J. Basic Eng.* **1963**, *85*, 273–278. [CrossRef]
6. Fingerle, A.; Hochrein, J.; Otto, M.; Stahl, K. Theoretical Study on the Influence of Planet Gear Rim Thickness and Bearing Clearance on Calculated Bearing Life. *J. Mech. Des.* **2019**, *142*, 031102. [CrossRef]
7. Bouyer, J.; Fillon, M. An Experimental Analysis of Misalignment Effects on Hydrodynamic Plain Journal Bearing Performances. *J. Tribol.* **2001**, *124*, 313–319. [CrossRef]
8. Sun, J.; Changlin, G. Hydrodynamic lubrication analysis of journal bearing considering misalignment caused by shaft deformation. *Tribol. Int.* **2004**, *37*, 841–848. [CrossRef]
9. Hili, M.A.; Bouaziz, S.; Maatar, M.; Fakhfakh, T.; Haddar, M. Hydrodynamic and Elastohydrodynamic Studies of a Cylindrical Journal Bearing. *J. Hydrodyn.* **2010**, *22*, 155–163. [CrossRef]
10. Hagemann, T.; Kukla, S.; Schwarze, H. Measurement and prediction of the static operating conditions of a large turbine tilting-pad bearing under high circumferential speeds and heavy loads. In Proceedings of the ASME Turbo Expo 2013, San Antonio, TX, USA, 3–7 June 2013.
11. Prölß, M. Berechnung Langsam Laufender und Hoch Belasteter Gleitlager in Planetengetrieben unter Mischreibung, Verschleiß und Deformationen. Ph.D. Thesis, Clausthal University of Technology, Clausthal-Zellerfeld, Germany, 2020.
12. Hagemann, T.; Schwarze, H. A Model for Oil Flow and Fluid Temperature Inlet Mixing in Hydrodynamic Journal Bearings. *J. Tribol.* **2018**, *141*, 021701. [CrossRef]
13. Muzakkir, S.M.; Hirani, H.; Thakre, G.D. Lubricant for Heavily Loaded Slow-Speed Journal Bearing. *Tribol. Trans.* **2013**, *56*, 1060–1068. [CrossRef]
14. Linjamaa, A.; Lehtovaara, A.; Larsson, R.; Kallio, M.; Söchting, S. Modelling and analysis of elastic and thermal deformations of a hybrid journal bearing. *Tribol. Int.* **2018**, *118*, 451–457. [CrossRef]
15. Xiang, G.; Han, Y.; Wang, J.; Wang, J.; Ni, X. Coupling transient mixed lubrication and wear for journal bearing modeling. *Tribol. Int.* **2019**, *138*, 1–15. [CrossRef]
16. Garabedian, N.T.; Gould, B.J.; Doll, G.L.; Burris, D.L. The Cause of Premature Wind Turbine Bearing Failures: Overloading or Underloading? *Tribol. Trans.* **2018**, *61*, 850–860. [CrossRef]
17. Hagemann, T.; Ding, H.; Radtke, E.; Schwarze, H. Operating behavior of sliding planet gear bearings in wind turbine gearbox applications—Part II: Impact of structure deformation. *Lubricants* **2021**, in press.
18. Patir, N.; Cheng, H.S. An Average Flow Model for Determining Effects of Three-Dimensional Roughness on Partial Hydrodynamic Lubrication. *J. Lubr. Technol.* **1978**, *100*, 12–17. [CrossRef]
19. Falz, E. *Grundzüge der Schmierungstechnik*; Springer: Berlin/Heidelberg, Germany, 1931.
20. Elrod, H.G. A Cavitation Algorithm. *J. Lubr. Technol.* **1981**, *103*, 350–354. [CrossRef]
21. Patankar, S.V. *Numerical Heat Transfer and Fluid Flow*; CRC Press: Boca Raton, FL, USA, 1980; ISBN 9781315275130.

22. Waltermann, H. Optimierte Thermo-Elasto-Hydrodynamische Berechnungsverfahren für Gleitlager. Ph.D. Thesis, Rheinisch-Westfälisch Technische Hochschule Aachen, Aachen, Germany, 1992.
23. Hagemann, T.; Zemella, P.; Pfau, B.; Schwarze, H. Experimental and theoretical investigations on transition of lubrication conditions for a five-pad tilting-pad journal bearing with eccentric pivot up to highest surface speeds. *Tribol. Int.* **2020**, *142*, 106008. [CrossRef]
24. Greenwood, J.A.; Tripp, J.H. The Contact of Two Nominally Flat Rough Surfaces. *Proc. Inst. Mech. Eng.* **1970**, *185*, 625–633. [CrossRef]
25. Prölß, M.; Schwarze, H.; Hagemann, T.; Zemella, P.; Winking, P. Theoretical and Experimental Investigations on Transient Run-Up Procedures of Journal Bearings Including Mixed Friction Conditions. *Lubricants* **2018**, *6*, 105. [CrossRef]
26. Hu, Y.; Cheng, H.S.; Arai, T.; Kobayashi, Y.; Aoyama, S. Numerical Simulation of Piston Ring in Mixed Lubrication—A Nonaxisymmetrical Analysis. *J. Tribol.* **1994**, *116*, 470–478. [CrossRef]
27. Archard, J.F. Contact and Rubbing of Flat Surfaces. *J. Appl. Phys.* **1953**, *24*, 981–988. [CrossRef]

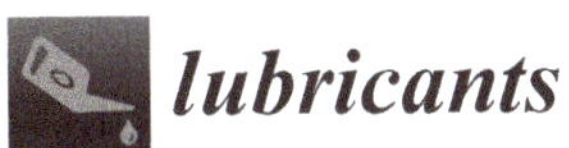 *lubricants*

Article

Operating Behavior of Sliding Planet Gear Bearings for Wind Turbine Gearbox Applications—Part II: Impact of Structure Deformation

Thomas Hagemann *, Huanhuan Ding, Esther Radtke and Hubert Schwarze

Institute of Tribology and Energy Conversion Machinery, Clausthal University of Technology, 38678 Clausthal-Zellerfeld, Germany; ding@itr.tu-clausthal.de (H.D.); radtke@itr.tu-clausthal.de (E.R.); schwarze@itr.tu-clausthal.de (H.S.)
* Correspondence: hagemann@itr.tu-clausthal.de; Tel.: +49-5323-72-2469

Abstract: The use of planetary gear stages intends to increase power density in drive trains of rotating machinery. Due to lightweight requirements on this type of machine elements, structures are comparably flexible while mechanical loads are high. This study investigates the impact of structure deformation on sliding planet gear bearings applied in the planetary stages of wind turbine gearboxes with helical gears. It focuses on three main objectives: (i) development of a procedure for the time-efficient thermo-elasto-hydrodynamic (TEHD) analysis of sliding planet gear bearing; (ii) understanding of the specific deformation characteristics of this application; (iii) investigation of the planet gear bearing's modified operating behavior, caused by the deformation of the sliding surfaces. Generally, results indicate an improvement of predicted operating conditions by consideration of structure deformation in the bearing analysis for this application. Peak load in the bearing decreases because the loaded proportion of the sliding surface increases. Moreover, tendencies of single design measures, determined for rigid geometries, keep valid but exhibit significantly different magnitudes under consideration of structure deformation. Results show that consideration of structure flexibility is essential for sliding planet gear bearing analysis if quantitative assertions on load distributions, wear phenomena, and interaction of the bearing with other components are required.

Keywords: planet gear bearing; journal bearing; structure deformation; mixed friction; wear; wind turbine gearbox

Citation: Hagemann, T.; Ding, H.; Radtke, E.; Schwarze, H. Operating Behavior of Sliding Planet Gear Bearings for Wind Turbine Gearbox Applications—Part II: Impact of Structure Deformation. *Lubricants* **2021**, *9*, 98. https://doi.org/10.3390/lubricants9100098

Received: 30 July 2021
Accepted: 21 September 2021
Published: 1 October 2021

Publisher's Note: MDPI stays neutral with regard to jurisdictional claims in published maps and institutional affiliations.

1. Introduction

The majority of journal bearing analyses, in practice and literature, neglect the impact of structure deformation of the bearing components. However, many studies show that this influencing factor can become one of the most important ones in analysis of bearing operating behavior if structures are sufficiently flexible. Different authors investigate the impact of misalignment between bearing and shaft caused by a load dependent bending deformation of the shaft [1–4]. Generally, misalignment reduces minimum film thickness and increases maximum film pressure. Consequently, it contributes to wear intensity in the bearing. Some of the most typical applications showing this behavior are crankshaft bearings of combustion engines [5,6]. Design measures can help to improve journal bearings' performance under misalignment. While Zhang et al. [7] report on optimum radial clearance for given misalignment angles, Bouyer and Fillon [8] observe improved operating behavior by application of local or global defects to the bush. In planet gear bearings with helical teeth, a misalignment between pin and planet exists that is induced by the load distribution in the gear mesh. In the first part of this study, the authors show an improvement of bearing operating behavior by an axial crowning, which can either be applied to the pin or the planet [9]. This measure represents a geometrical modification, similar to the idea of a global defect from Ref. [8], but specific in its shape for the load situation in planet gear bearings.

While misalignment caused by shaft deformation is predominantly influenced by global deformation of this bearing component, changes of the shape of the sliding surface also modify bearing operating behavior. Due to their highly flexible structure, tilting-pad journal bearings represent an example for an application that exhibits deformation of the sliding surface that significantly influences local film thickness and, therefore, characteristics of hydrodynamic operation [10–12]. Concordantly, pad deformation causes a reduction in minimum film thickness but with different characteristics for the shape of the structure deformation. While [10,11] identify and predict minimum film thickness in the center plane of the bearing due to an axial bending of the pad, ref. [12] predicts minimum film thickness at the bearing edges due to maximum radial deformation in the center plane. This contradiction shows that journal bearing structure deformation highly depends on assumed load and displacement boundary conditions, and generalization, as a bearing property, is limited. Lahmar et al. [13] study the impact of liner deformation on operating behavior of a compliant journal bearing. At same journal eccentricity, the region of pressure build-up spreads in circumferential direction, whereas peak pressure significantly decreases. Prölß [14] investigates the impact of deformation, on planet gear bearings with spatial fixed pins, and finds similar tendencies due to the high flexibility of the highly loaded structures. The complexity of structure models requires a weak iterative coupling between fluid and structure analysis in many applications. Here, the strategy to update investigated parameters during the iterative procedure strongly influences its convergence speed, as Profito et al. [15] show for a connecting-rod big-end bearing investigation.

Planetary gearboxes enable torque conversion at very high power densities due to the separation of the tooth forces to the single planets and the entire compact designs. Therefore, the level of specific mechanical loads on the components of the gearbox are extremely challenging and accompany significant structure deformation. In extension to the first part of this study [9], the subsequent investigations focus on the impact of deformation on predicted results and the comparison of general tendencies between rigid and flexible analysis of planet gear bearings for wind turbine gearbox applications. As thermal effects are not of major impact for this low speed application, according to the results in [9], only structure deformation induced by mechanical loads are considered in the analysis. The displacement boundary conditions of the planet stage components feature a character significantly differing from other applications. Consequently, a specific deformation behavior of the planet gear bearing sliding surface is expected.

2. Materials and Methods

2.1. Procedure for Consideration of Planet and Pin Deformation

High mechanical loads and, simultaneously, high structure flexibility require consideration of fluid structure interaction in planet gear bearing analyzes. The full or strong coupling of the fluid and structure problem in one system of equation is numerically complex, and its application in the literature is limited to less detailed structure models, focusing on point or line contacts of an elasto-hydrodynamic (EHD) problem [16,17]. For the weak coupling of separated fluid and structure analysis, generally, two different iterative procedures exist. The first one is to perform a co-simulation between the thermo-hydrodynamic (THD) bearing analysis and a structure mechanics software. This procedure involves the advantage of an arbitrary complexity of the structure model. Consideration of non-linear contacts or boundary conditions, varying due to the current operating conditions predicted in the iterative procedure, are possible. However, this analysis is time consuming as an analysis of full, unreduced structure model is conducted in each iterative step. Moreover, the investigated planet gear bearing features low absolute values and gradients of the film thickness in the highly loaded film region, requiring a strong under-relaxation to achieve reliable convergence. Therefore, low convergence speeds of the weak coupling between fluid and structure analysis is present here. The application of reduced stiffness matrices for the structures represents the second alternative. While this analysis is less flexible to varying boundary conditions and requires a linear structure model, it reduces computational self-times of the simulation, based on a weak coupling, as the structure deformation can be evaluated based on a priori determined information. Prölß [14] successfully used stiffness matrices in planet gear bearing analyses for wind turbine gearbox applications. The procedure will also be applied for further investigations in this study and is described subsequently.

According to the theory of Craig and Bampton [18], the reduced stiffness matrix $\overline{\overline{C}}_{red}$ characterizing the behavior of the master nodes on sliding surface can be determined according to Equations (1)–(4).

$$\left\{ \begin{array}{c} \overline{F}_M \\ \overline{F}_S \end{array} \right\} = \left[\begin{array}{cc} \overline{\overline{C}}_{MM} & \overline{\overline{C}}_{MS} \\ \overline{\overline{C}}_{SM} & \overline{\overline{C}}_{SS} \end{array} \right] \cdot \left\{ \begin{array}{c} \overline{h}_M \\ \overline{h}_S \end{array} \right\} \tag{1}$$

Herein, M and S are the indexes for master nodes and slave nodes and $\overline{\overline{C}}$ is the stiffness matrix of the structure. All nodes that are loaded by varying forces in the calculation procedure represent master nodes, while slave nodes are all other remaining ones. In the concrete case, the master nodes are all located on the sliding surface. The force vector $\overline{F}_M$ and $\overline{h}_M$ represent the film force $\overline{F}_{film}$ and the deformation of master nodes $\overline{h}_{def}$ on the sliding surface, respectively. Applying these parameters to Equation (1) gives

$$\overline{F}_{film} = \overline{\overline{C}}_{MM} \cdot \overline{h}_{def} + \overline{\overline{C}}_{MS} \cdot \overline{h}_S \tag{2}$$

$$\overline{F}_S = \overline{\overline{C}}_{SM} \cdot \overline{h}_{def} + \overline{\overline{C}}_{SS} \cdot \overline{h}_S \tag{3}$$

Combining Equations (2) and (3) provides

$$\overline{F}_{film} = \underbrace{\left(\overline{\overline{C}}_{MM} - \overline{\overline{C}}_{MS} \cdot \left[\overline{\overline{C}}_{SS} \right]^{-1} \cdot \overline{\overline{C}}_{SM} \right) \cdot \overline{h}_{def}}_{\overline{\overline{C}}_{red}} + \underbrace{\overline{\overline{C}}_{MS} \cdot \left[\overline{\overline{C}}_{SS} \right]^{-1} \cdot \overline{F}_S}_{\overline{F}_{red}} \tag{4}$$

The reduced stiffness matrix $\overline{\overline{C}}_{red}$ in Equation (4) contains the information about structure elasticity and boundary conditions. The reduced load vector $\overline{F}_{red}$ characterizes the impact of the mesh forces on sliding surface deformation. While the description of planet structure requires $\overline{F}_{red}$, the reduced force vector becomes equal zero vector for the pin.

The modification of stiffness matrix dimensions due to the reduction procedure is explained for the planet model. The sliding surfaces of planet and pin are discretized with 128 elements in circumferential and 16 elements in axial directions. Due to condensation for reduced stiffness matrix, the number of degrees of freedom for the spatial deformable 178,987 nodes of the unreduced planet model is reduced from $DOF = 178,987 \times 3 = 536,961$ to $DOF_{red} = 128 \times 16 \times 3 = 6144$ of master nodes on the sliding surface. If only the nodes on the sliding surface are defined as master nodes, the reduced description of the structure is specific for a load distribution on the tooth flank, as shown in Figure 1a. However, the nodes on the tooth flank can be considered as additional master nodes, enabling the analysis of arbitrary load distributions of the mesh forces. This paper studies the impact of structure deformation on bearing operating behavior under nominal load. Consequently, a definition of master nodes on the sliding surface is sufficient. In addition, the influence of the deformation behavior of the planet carrier on bearing properties is also considered. In Figure 2a, the linear contact type 'Bonded' is defined at the contact surfaces between carrier and pin to ensure linearity of the entire model. This contact type merges the two structures and prohibits sliding or separation of its contact surfaces. Therefore, the model consisting of planet and pin behaves as a homogenous structure.

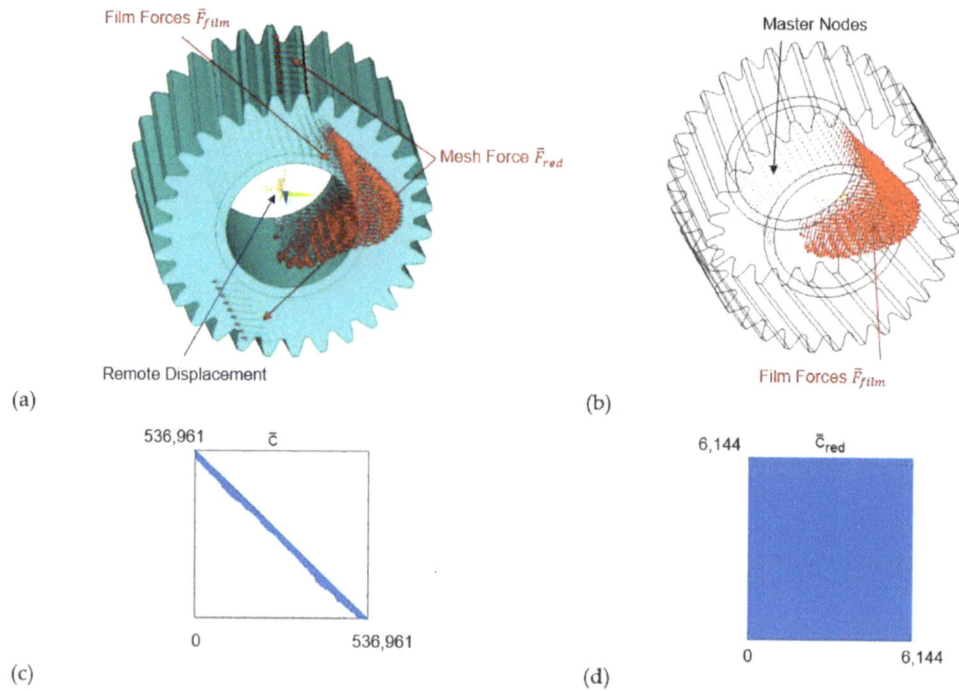

Figure 1. (**a**) Complete FE-Model and (**c**) complete stiffness matrices, (**b**) reduced FE-Model and (**d**) reduced stiffness matrices for planet.

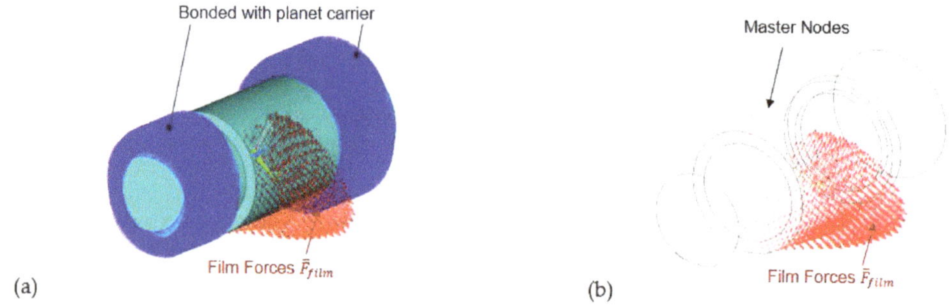

Figure 2. (**a**) Complete and (**b**) reduced FE-Model for pin.

Based on the reduced structure properties introduced in Equation (4), deformation $\bar{h}_{def}^{(i)}$ of the sliding surface is evaluated by Equation (5).

$$\bar{h}_{def}^{(i)} = \bar{\bar{C}}_{red}^{-1} \cdot \left(\bar{F}_{film}^{(i)} - k \cdot \bar{F}_{red} \right) \tag{5}$$

Herein, $\bar{\bar{C}}_{red}$ and $\bar{F}_{red}$ are the reduced stiffness matrix and load vector, and $\bar{F}_{film}^{(i)}$ is the film force representing the sum of hydrodynamic and asperity contact forces. For the planet, the force vector $\bar{F}_{red}$ models the influence of the mesh forces on the sliding surface deformation, while this component does not exist for the pin. Assuming a constant shape of the load distribution on the tooth flank, $\bar{F}_{red}$ can be scaled linearly by k to consider different levels of mesh forces.

Figure 3 shows the algorithm for iterative calculation of the mechanical equilibrium for the planet gear bearing considering structure deformation. After the start of the iteration, the planet bearing code calculates the pressure field p and the planet position for the undeformed gap contour according to the procedure explained in part I of this study [9]. Here, an equilibrium between the film forces and moments and the outer forces and moments, generated by the mesh load, exists. Predicted film forces $\bar{F}_{film}^{(i)}$ are applied, on the nodes of the sliding surface of pin and planet, to analyze resulting deformation $\Delta \bar{h}_{def}^{(i)}$. This deformation is numerically damped by $\xi^{(i)}$ and considered in the gap function of the next fluid film analysis, according to Equation (6). These iterative steps repeat for the highly loaded system until maximum local difference of deformation, between two subsequent iterations, is less than 1 μm, and the maximum local difference of pressure remains below 0.2 MPa.

$$h_{def}^{(i+1)} = h_{def}^{(i)} + \xi^{(i)} \cdot \Delta h_{def}^{(i)}; \quad i = i+1 \tag{6}$$

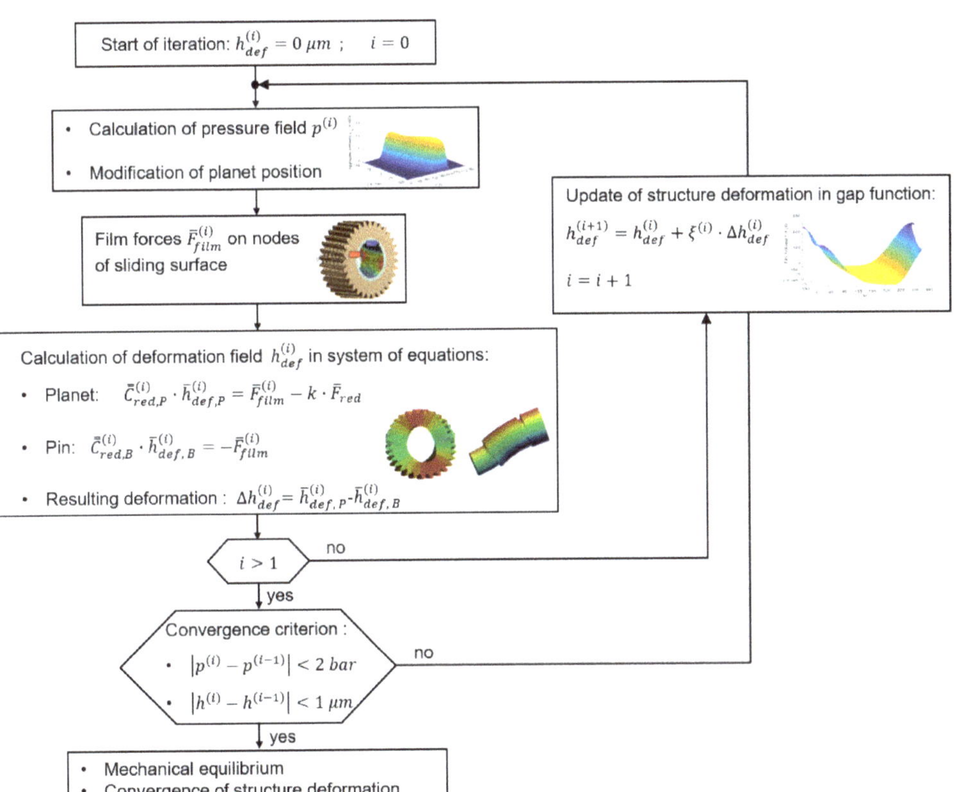

Figure 3. Flowchart for consideration of deformation.

The factor $\xi^{(i)}$ varies between 0.01 and 0.2 in this study and is modified by a heuristic algorithm based on the convergence behavior of the calculation procedure.

2.2. Investigated Gear Set: Planet Gear Bearing

Table 1 includes the basic analysis parameters of the investigated planet gear bearing that equals the one studied in part I [9]. The values are in typical magnitude of planet gear bearings applied in 3 MW wind turbine gearboxes. Figure 4 depicts a pin with the sliding planet gear bearing. The oil supply pocket is located in the opposite direction of the bearing load.

Table 1. Planet gear bearing parameters.

Parameter	Value
Geometrical Properties	
Number of pads, -	1
Nominal diameter, mm	250
Pitch circle diameter, mm	499
Helix angle, degrees	7
Bearing width, mm	300
Angular span of lube oil pocket, degrees	20.5
Width of lube oil pocket, mm	260
Radial clearance, μm	138
Pad sliding surface preload, -	0.0
Static Analysis Parameters	
Nominal rotational speed, rpm	30
Nominal bearing load, kN	900
Nominal bearing moment, kNm	27.6
Lubricant supply temperature, °C	60
Lube oil supply pressure, MPa	0.2
Lubricant Properties	
Lubricant	ISO VG 320
Lubricant density kg/m^3	865 @ 40 °C
Lubricant specific heat capacity kJ/(kg·K)	2.0 @ 20 °C
Lubricant thermal conductivity, W/(m·K)	0.13

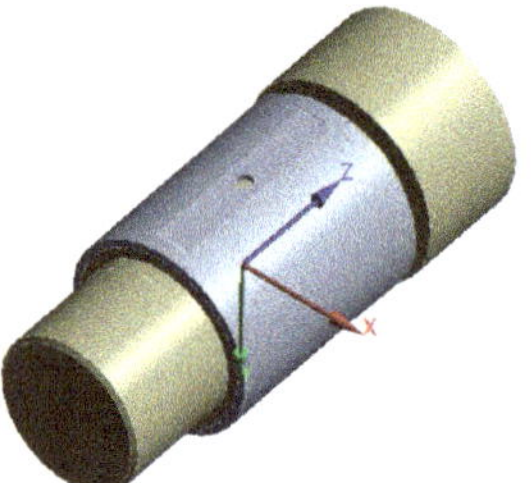

Figure 4. CAD model of the pin with sliding planet gear bearing.

2.3. Investigated Gear Set: CAD Model and Material Properties

Figure 5 shows the CAD model of the investigated planetary stage. It consists of a carrier, five pins, and five planets with a helix angle $\beta = 7°$. Due to the periodicity of the system, one-fifth of the model in Figure 6 is utilized to reduce the required number of nodes in the subsequently derived structure analysis. Table 2 includes material properties of the planetary stage solid body components.

2.4. FEM Approximation for Structure Analysis: Meshing

Figure 7a shows that the planet is divided into two parts. The inner structure (part 1) features a cylinder geometry that can be well approximated by hexahedral meshes. The mesh of the outer structure (part 2) can be optionally set to a tetrahedral mesh. Both sub-structures are joint via a bonded contact to behave as a homogenous structure. The same method of mesh generation is applied to the pin shown in Figure 7b. Additionally, pin and carrier are merged by a bonded contact.

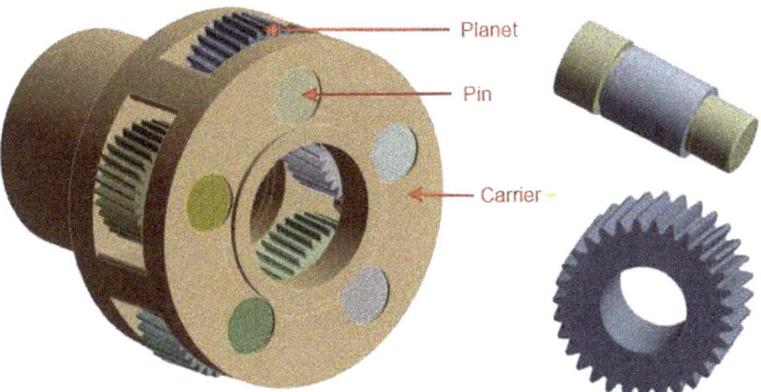

Figure 5. CAD model of planet gear stage and its components.

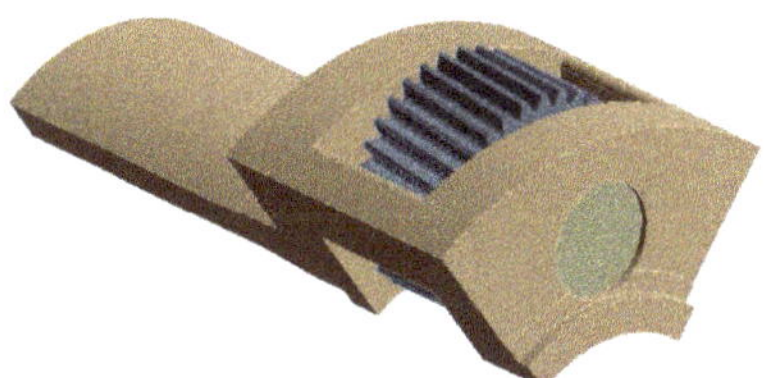

Figure 6. One-fifth of the CAD model of planet gear stage.

Table 2. Material properties.

	Planet	Pin	Carrier
Parameter		Value	
Young's Modulus, MPa	210,000	210,000	176,000
Poisson's Ratio, -	0.3	0.3	0.275
Coefficient of Thermal Expansion, 10^{-6}/K	12	11.1	12.5
Isotropic Thermal Conductivity, W/(m·K)	39.8	42.6	31.1
Tensile Yield Strength, MPa	850	1000	420

The position of the nodes on the sliding surface in FEM model matches the one of the film analysis to prevent interpolation uncertainties. For the following investigations, the sliding surface is discretized with 128 in circumferential and 32 elements in axial directions by using the hexahedral element type. The entire unreduced FE-model features 338,241 nodes. To prove independency of the results, due to level of discretization, the number of nodes is varied homogenously in all three space directions. Figure 8 shows predicted radial deformation of the sliding surface at the origin of the angular coordinate for the same load distribution on the sliding surface of the planet. Results indicate that maximum relative deviation of radial deformation, between two levels of investigated mesh density, is below 0.2% starting with the discretization chosen by the authors. Therefore, the impact of structure discretization on predicted results is expected to be below the remaining uncertainties of the entire simulation procedure.

2.5. FEM Approximation for Structure Analysis: Boundary Conditions and Mesh Forces

At the planet, a stationary mechanical equilibrium between the external mesh forces, the gravity forces, and the film forces exists. However, numerical differences between these forces remain and require a 'Remote Displacement' boundary condition, with 0 degrees of freedom, to prevent rigid body movement. In Figure 9a, this boundary condition is applied on the internal surface of part 2 of the planet. Figure 9b shows the bonded contact model for the connection between the planet carrier and planetary pins. Additionally, tapered rolling element bearings support both sides of the planet carrier and limit movement in radial and axial directions. The inner surface of the rotor side hub of carrier is fixed to provoke the twist deformation due to the applied driving torque. The two cross sections of the one-fifth model of the carrier feature cyclic periodicity boundary conditions.

Figure 7. Hexahedral mesh of sliding surface for planet (**a**) and pin with carrier (**b**).

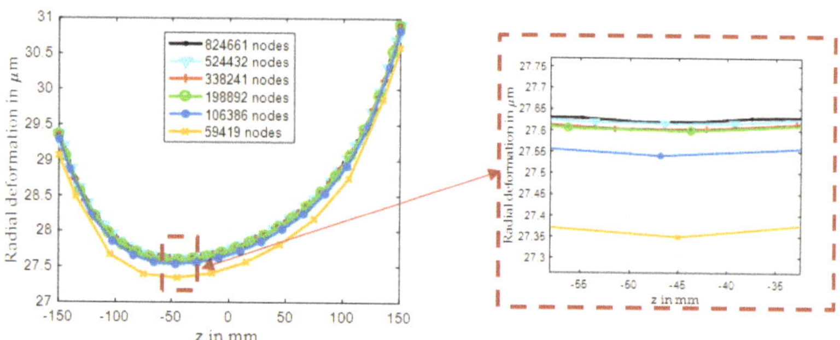

Figure 8. Radial deformation at 0° degree across the bearing width for different mesh densities of the structure model.

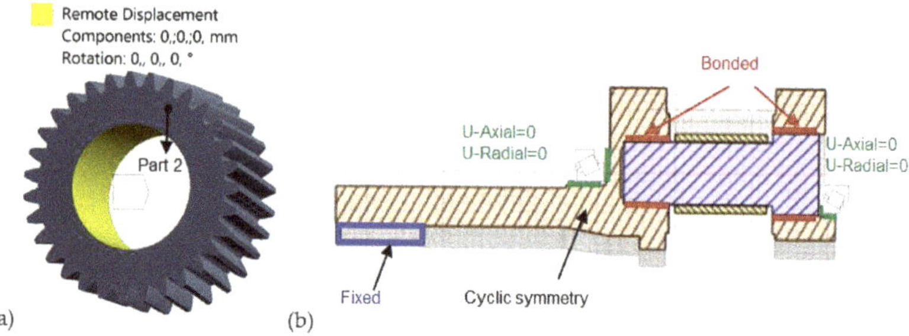

Figure 9. Boundary condition for planet (**a**) and pin with carrier (**b**).

Figure 10 shows the homogenous load distribution applied at 4 nodes on each tooth flank, and the total force components on each side of the tooth flank are included in the Table 3.

Figure 10. Homogenous load distribution on each tooth flank.

Table 3. Mesh load on the each tooth flank at nominal load conditions.

Parameter	Ring Gear Side	Sun Gear Side
Radial force F_r, kN	−164	164
Tangential force F_t, kN	−450	−450
Axial force F_{ax}, kN	55.3	−55.3

3. Results

3.1. Mesh and Bearing Loads

The relations between mesh forces and the investigated static bearing loads are comprehensively described in part I of this study. Considering the helix angle of the helical gear of 7°, according to Table 1, the nominal load case corresponds to a bearing force of F_{sc} = 900 kN and a moment load of M_{sc} = 27.6 kNm. This load case exists for a relative input torque of T_r = 100%. According to the explanations in part I, bearing force and moment loads vary linearly with the relative input torque.

3.2. Verification of the Calculation Procedure

The procedure for consideration of planet and pin deformation, using a reduced structure model and its implementation in the planet bearing code, is verified by a comparison to predicted deformation results of the full, unreduced structure model. This study applies ANSYS for structure mechanical analyzes. Figure 11 shows the results of the deformation analysis with the full structure model in ANSYS and with the reduced model in the bearing code. In this case, the same exemplary pressure distribution is applied to both structures. Results at the center and the bearing end show very good agreement with a maximum deviation of 1.7 μm. Furthermore, the distribution of deviations that is not shown here features a harmonic shape and, therefore, indicates differences of predicted rigid body displacements without significant impact on gap contour and hydrodynamics. Consequently, these results verify the application of the reduced structure model, according to the procedure explained in Section 2.1.

3.3. Deformation Behavior of Components

The general component deformation behavior, which is already introduced in Figure 11, is analyzed in more detail in this section. Figure 12 shows the bearing clearance, together with the deformed contour of pin and planet, at the bearing edge ($z = -150$ mm) and in the center of the bearing ($z = -4.7$ mm) under nominal operating conditions, according to Table 1. In addition to these deformed components, the resultant gap contour, modifying the shape of the lubricant gap, is depicted.

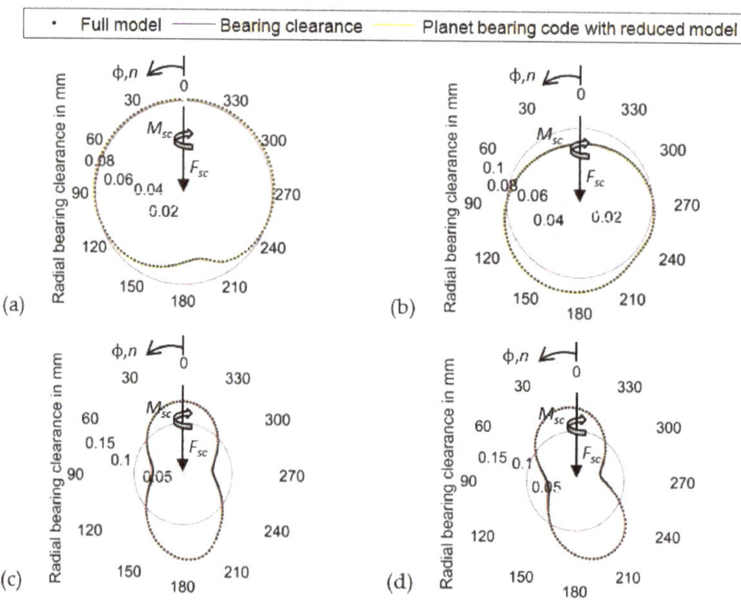

Figure 11. Comparison of gap contour in the center of bearing $z = -4.7$ mm for (**a**) pin and (**c**) planet, and at the bearing edge $z = -150$ mm for (**b**) pin and (**d**) planet @ nominal operating conditions ($n_{pl} = 30$ rpm, $F_{sc} = 900$ kN, $T_{sup} = 60$ °C, $p_{sup} = 0.2$ MPa, $M_{bear} = 27.6$ kNm, $\beta = 7°$).

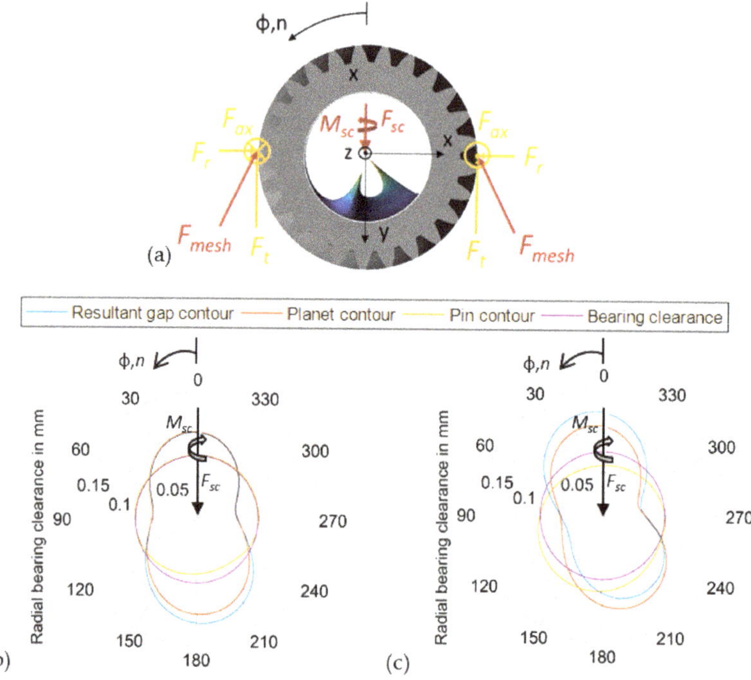

Figure 12. Load conditions (**a**) and gap contour in the center of bearing $z = -4.7$ mm (**b**) and at the bearing edge, $z = 150$ mm (**c**) @ nominal operating conditions ($n_{pl} = 30$ rpm, $F_{sc} = 900$ kN, $T_{sup} = 60$ °C, $p_{sup} = 0.2$ MPa, $M_{bear} = 27.6$ kNm, $\beta = 7°$).

Figure 12 indicates three major characteristics of the deformation distribution for the two components, due to the load conditions in Figure 12a. First, the pin shows an indentation due to the pressure load that can be clearly seen by the comparison of the bearing clearance and the deformed pin contour in Figure 12a. Second, the pressure load provokes a twist of the components that is well recognizable for the planet, due to its higher flexibility. Maximum radial deformation in load direction is shifted from approximately 185°, in the center of the bearing, to 205° at its front end. Third, tangential and film forces pull the planet while the orthogonally acting radial forces push the planet. Consequently, the planet exhibits an oval shape that also dominates the resulting gap contour.

In addition, Figure 13 shows the variation of major and minor radial clearances at the front ($z = 150$ mm) and rear ($z = -150$ mm) of the planet gear bearing over the entire load range. The results show that the major and minor radial clearances exhibit nearly linear increasing and decreasing trends with rising load, respectively. Moreover, ovalization of the two lateral bearing ends is different. Here, the rear end shows a higher level due to larger maximum and lower minimum clearance. Figure 14 visualizes this property by the change of the total radial deformation, which represents the sum of radial planet and pin deformation, over the bearing width. Here, the major axis on the lateral bearing ends are shown, additionally, to express the previously described twist of the load zone in a three-dimensional view.

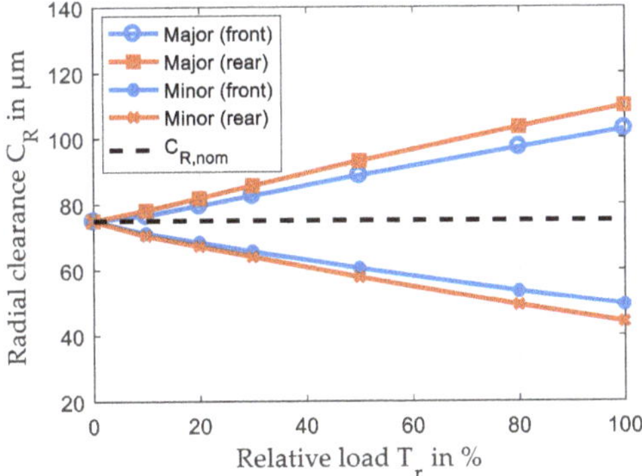

Figure 13. Radial clearance of elastic calculation for variable relative loads ($n_{pl} = 30$ rpm, $F_{sc} = 0$–900 kN, $T_{sup} = 60\,°C$, $p_{sup} = 0.2$ MPa, $M_{bear} = 0$–27.6 kNm, $\beta = 7°$).

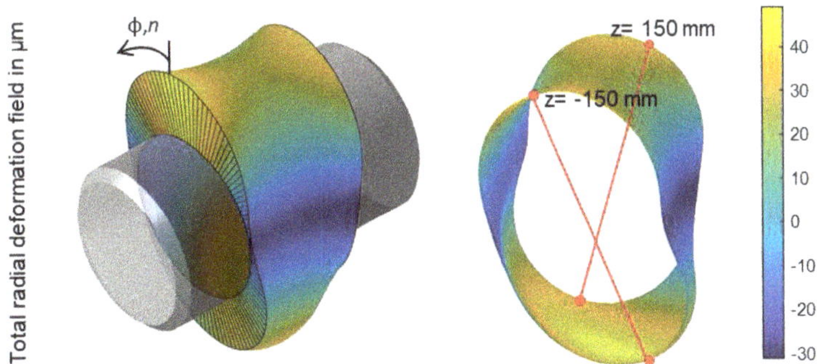

Figure 14. Change of ovalization of the total radial deformation field over the bearing width @ nominal operating conditions ($n_{pl} = 30$ rpm, $F_{sc} = 900$ kN, $T_{sup} = 60\,°C$, $p_{sup} = 0.2$ MPa, $M_{bear} = 27.6$ kNm, $\beta = 7°$).

Figure 15 depicts radial deformation of the full three-dimensional structure model for further analyses of deformation behavior and its major characteristics. Figure 15a visualizes the planet deformation with its significant ovalization that is previously explained. Figure 15b describes pin deformation. The global pin deformation is dominated by the twist of the two carrier cheeks, caused by the transmitted torque in the planetary stage. Therefore, the pin that is connected on both ends, by a crimp to the cheeks, exhibits an s-shape deformation clearly observable in Figure 15b.

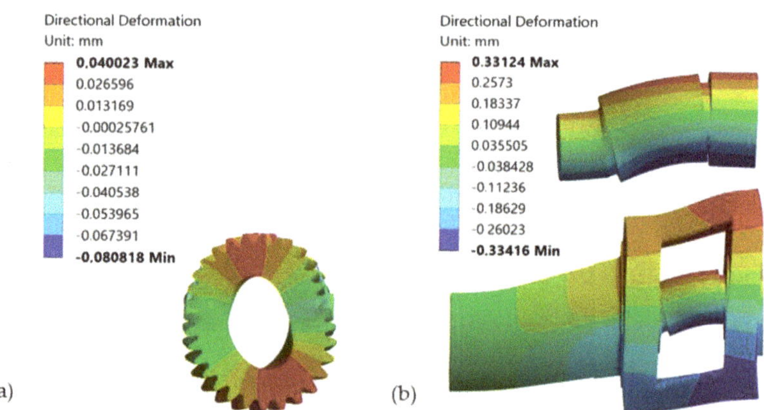

(a)   (b)

Figure 15. Radial deformations of components of planet (**a**) and pin and carrier (**b**) in ANSYS @ nominal operating conditions (n_{pl} = 30 rpm, F_{sc} = 900 kN, T_{sup} = 60 °C, p_{sup} = 0.2 MPa, M_{bear} = 27.6 kNm, β = 7°).

Figure 16 shows the comparison of the film thickness predicted by the rigid and the flexible calculation. Compared to the rigidly predicted film thickness in Figure 16a,b, the elastic calculation provides a nearly parallel film thickness with low gradients in circumferential direction of the loaded zone in Figure 16c,d. The red marks, in the side view in Figure 16a,c, as well as the top views in Figure 16b,d, give a deeper impact on this difference. This behavior is typical for highly loaded contacts and incorporate a decrease in the maximum pressure level.

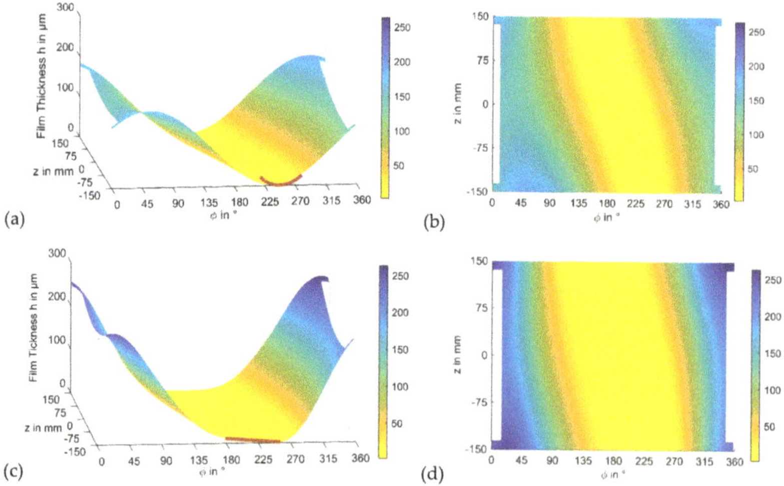

(a)   (b)   (c)   (d)

Figure 16. Film thickness of rigid (**a**,**b**) and flexible calculation (**c**,**d**) @ nominal operating conditions with crowning #2 (n_{pl} = 30 rpm, F_{sc} = 900 kN, T_{sup} = 60 °C, p_{sup} = 0.2 MPa, M_{bear} = 27.6 kNm, β = 7°).

3.4. Impact of Axial Profiling Model on Bearing Considering Elastic Deformation of Components

The first part [9] of this study shows that an axial crowning #2, exhibiting a continuous parabolic shape among the bearing width, is suitable to reduce maximum pressure and increase minimum film thickness. In the next step, this measure is investigated considering elastic deformation. Figure 17 illustrates that comparable maximum pressures exist with, and without, crowning at nominal operation. This result is in contrast to the rigid one presented in [9], where essential differences are determined. In accordance with the rigid results, the crowning #2 provides an advantage. However, the maximum pressure of 23.6 MPa with crowning, in Figure 17b, is only 7.1% lower than the one without axial crowning. Again, the crowning provides a more homogeneous pressure profile in lateral direction.

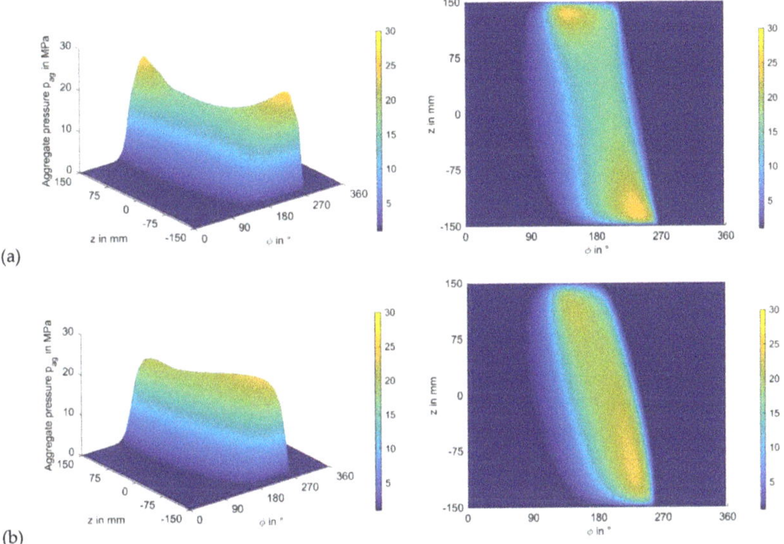

Figure 17. Pressure distributions of elastic calculation at nominal operating conditions without crowning (**a**) and with crowning #2 (**b**) (n_{pl} = 30 rpm, F_{sc} = 900 kN, T_{sup} = 60 °C, p_{sup} = 0.2 MPa, M_{bear} = 27.6 kNm, β = 7°).

Figure 18 shows maximum pressure and minimum film thickness for rigid and flexible analyses for the entire speed range. The crowning #2 provides advantage in the entire load range with increasing minimum film thickness and decreasing maximum pressure. While the reduction in maximum pressure in Figure 18a is comparably low, as explained before, the rise of minimum film thickness Figure 18b is significant and underlines the importance of this measure for safe operation.

While results predicted under consideration of crowning show nearly pure hydrodynamic operation with maximum asperity contact pressures of p_c < 0.04 MPa, significant deviations between the results for flexible and rigid geometries exist for constant clearance in axial directions. As shown in Figure 18b, minimum film thickness increases under consideration of deformation. Consequently, the intensity of mixed friction is reduced, and maximum solid contact pressure decreases, according to Figure 19a. Assuming a boundary coefficient of friction of μ = 0.03, to derive solid contact shear stress from solid contact pressure, a notable impact of asperity contacts on maximum temperature and frictional power loss can be observed, starting with a relative load of T_r = 50%, in Figure 19b.

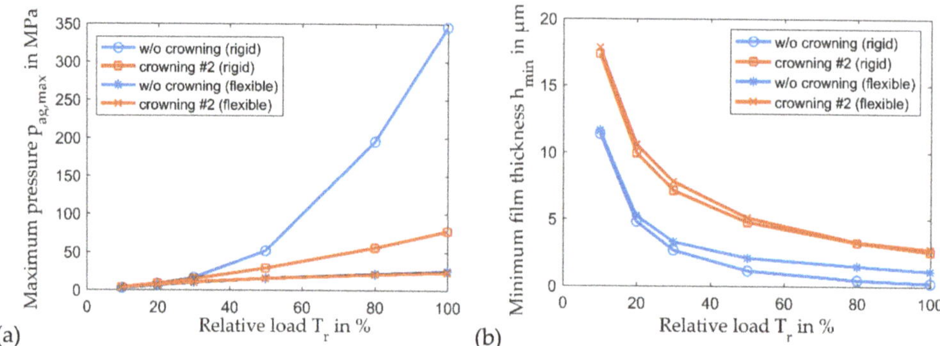

Figure 18. Comparison of maximum pressure (**a**) and minimum film thickness (**b**) between rigid and elastic calculation for variable relative loads (n_{pl} = 30 rpm, F_{sc} = 0–900 kN, T_{sup} = 60 °C, p_{sup} = 0.2 MPa, M_{bear} = 0–27.6 kNm, β = 7°).

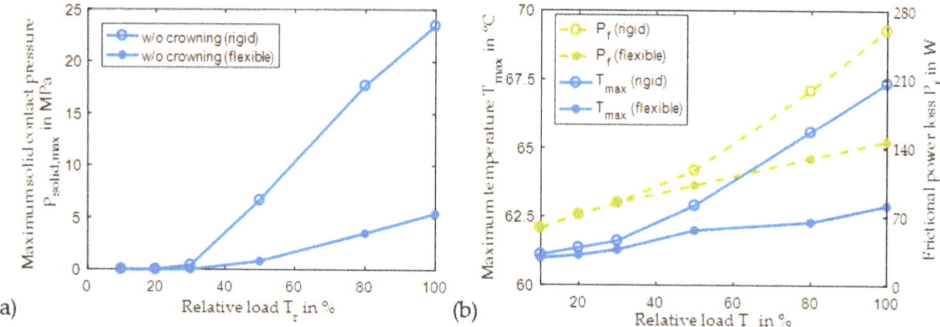

Figure 19. Comparison of maximum solid contact pressure (**a**) and temperature and frictional power (**b**) between rigid and elastic calculation for variable relative loads without axial crowning (n_{pl} = 30 rpm, F_{sc} = 0–900 kN, T_{sup} = 60 °C, p_{sup} = 0.2 MPa, M_{bear} = 0–27.6 kNm, β = 7°).

3.5. Modification of the Lubricant Gap by Wear Considering Elastic Deformation of Components

To analyze the impact of flexibility on predicted wear distributions, rigid and flexible simulations are conducted for 560 h of operation. According to part I of this study, wear is evaluated according to Archard's law with a wear coefficient of $K = 8.5 \cdot 10^{-9}$ mm^3/J.

$$V_w = \frac{k}{H} \cdot F_{sol} \cdot L = K \cdot F_{sol} \cdot L \tag{7}$$

It is assumed that the softer material is located on the stator side, provoking a two dimensional variable distribution of wear on the sliding surface of the bearing that does not feature the axial crowning.

Figure 20 includes the wear distributions for the nominal load case. In comparison to the wear distribution predicted for rigid geometries, the flexible analysis shows three major differences. First, the entire wear level decreases and exhibits a reduction in wear height by approximately 40%. Second, the extent of the areas featuring wear significantly reduces in lateral direction and concentrates more on the bearing ends. Third, the angular span of the area with wear in a circumferential direction rises, due to the characteristic of the deformed lubricant gap, that shows a wider region of low film thickness, according to Figure 16c,d.

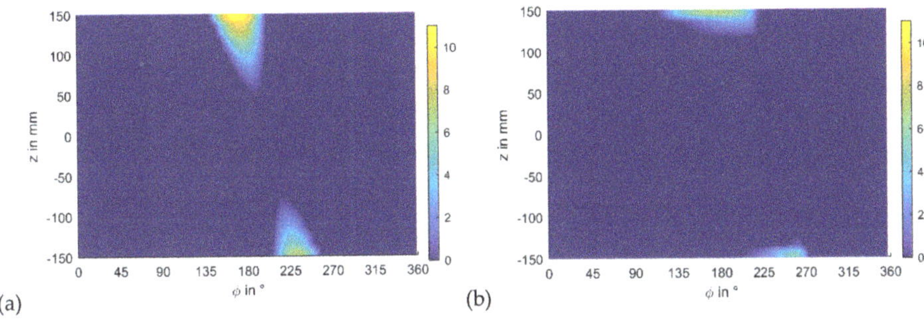

Figure 20. Wear after 560 h run with rigid geometries (**a**) and under consideration of elastic deformations (**b**) at nominal operating conditions (n_{pl} = 30 rpm, F_{sc} = 900 kN, T_{sup} = 60 °C, p_{sup} = 0.2 MPa, M_{bear} = 27.6 kNm, β = 7°, t = 560 h, no axial crowning).

Maximum pressure, determined for the flexible planet gear bearing in Figure 21b, is only 26.7 MPa and is, therefore, approximately 38% of the one predicted for the rigid bearing that is reported, already, in part I [9] and shown in Figure 21a. However, in contrast to the rigid case, wear has no significant impact on maximum pressure, as film thickness modification concentrates on the lateral bearing end and has little influence on hydrodynamics. A comparison of Figure 17b or Figure 21b underlines this statement, as there is no qualitative difference of the pressure distributions. Additionally, Figure 21 illustrates the extent of the load region in circumferential direction, caused by elastic deformation and already described for film thickness, in Figure 16. Here, the two red lines enclosing the load region in the top view of each pressure distribution show a wider range of contact pressures in circumferential direction and a more homogenous distribution in lateral direction for the flexible case in Figure 21b than for the rigid case in Figure 21a.

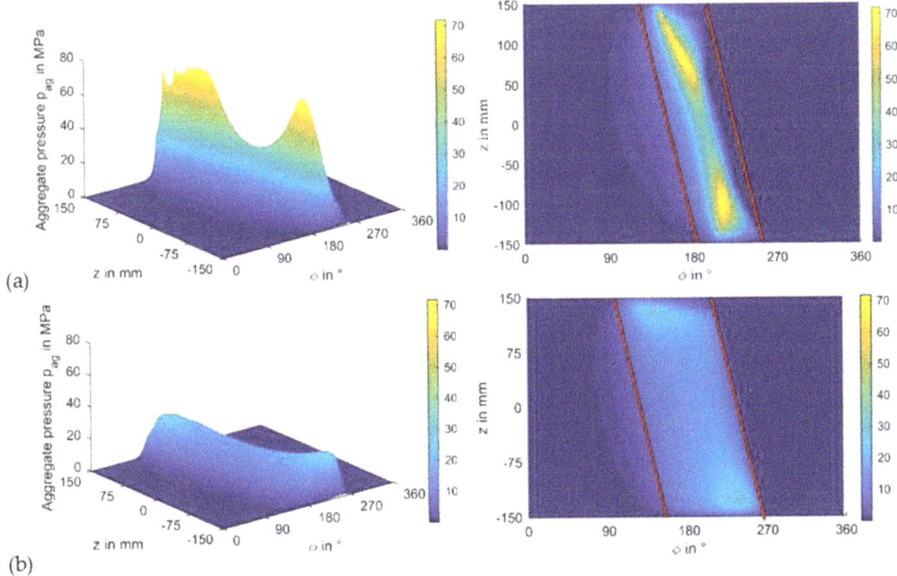

Figure 21. Pressure distribution after 560 h with rigid geometries (**a**) and under consideration of elastic deformations (**b**) @ nominal operating conditions (n_p = 30 rpm, F_{sc} = 900 kN, T_{sup} = 60 °C, p_{sup} = 0.2 MPa, M_{bear} = 27.6 kNm, β = 7°, t = 560 h, no axial crowning).

Figure 22 shows the film thickness of rigid and elastic calculations for the worn bearing after 560 h. As expected, results are qualitatively comparable to the ones in Figure 16 with the exception of the extension of the loaded area, due to wear in the rigid case. The minimum film thicknesses of Figures 16 and 22 are summarized in Table 4. By comparing rigid and elastic calculations, the elastic deformation of components have little impact on predicted minimum film thickness, whereas the characteristic of the lubricant gap significantly changes if the impact of deformation is considered.

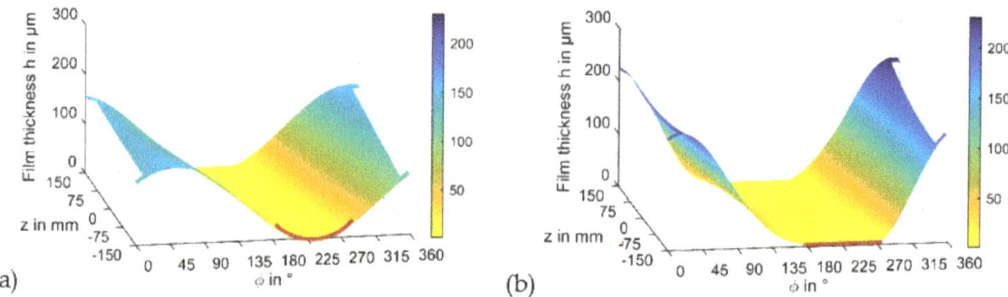

Figure 22. Film thickness of rigid calculation (**a**) and under consideration of elastic deformations (**b**) at nominal operating conditions without crowning after 560 h (n_{pl} = 30 rpm, F_{sc} = 900 kN, T_{sup} = 60 °C, p_{sup} = 0.2 MPa, M_{bear} = 27.6 kNm, β = 7°, t = 560 h, no axial crowning).

Table 4. Minimum film thickness of rigid and elastic calculation at nominal operating conditions with crowning #2 or without crowning after 560 h.

Parameter	Rigid Calculation	Elastic Calculation
Minimum film thickness with crowning #2, µm	2.6	2.748
Minimum film thickness with wear after 560 h, µm	1.55	1.75

Figure 23 shows the variation of maximum pressure and minimum film thickness over the entire load range for rigid and elastic calculations, considering wear after 560 h. The comparison with the unworn results from Figure 16 indicates that the wear process has a great impact on predicted maximum film pressure for rigid geometries by decreasing its value by almost 70%, while it modifies the flexible results less significantly, and maximum pressure remains nearly constant. However, for the same operating conditions, minimum film thickness increases from the theoretical value of 0.21 µm to 1.55 µm for the rigid analysis. Concordantly, a significant rise from 1.11 µm to 1.75 µm is predicted under consideration of flexible geometries. As this modification concentrates on the bearing edge, it has little impact on maximum pressure.

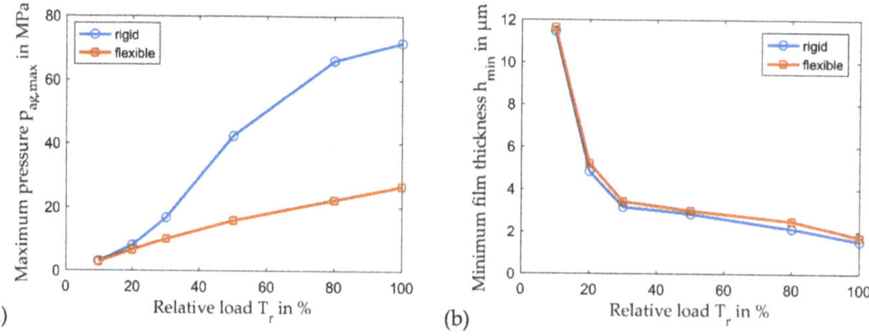

Figure 23. Comparison of maximum pressure (**a**) and minimum film thickness (**b**) between rigid and elastic calculation for variable relative loads, without crowning, after 560 h (n_{pl} = 30 rpm, F_{sc} = 900 kN, T_{sup} = 60 °C, p_{sup} = 0.2 MPa, M_{bear} = 27.6 kNm, β = 7°, t = 560 h, no axial crowning).

4. Discussion and Conclusions

This paper introduces a time-efficient procedure for sliding planet gear bearing analysis, based on reduced structure models. The numerical procedure is verified by comparisons to co-simulation results, determined with an unreduced structure model of a planetary gear stage, with dimensions that are typical for wind turbine gearbox applications. Further structure analyses with this model provide an impact on deformation behavior of the bearing components caused by the specific load conditions of the helical gear. Under load, the planet exhibits a characteristic ovalization of its structure that twists in lateral direction. The pin deforms to an s-shape by the twist of the carrier cheeks it is connected to. Moreover, both planet gear bearing components show a deformation that provokes lower film thickness gradients in the loaded region. The modification of the lubricant gap, generated by this behavior, leads to an enlargement of the loaded area and a reduction in local load maxima. Consequently, consideration of structure flexibility is essential for quantitative sliding planet gear bearing analyses.

General tendencies of the impact of an axial crowning and wear on operating behavior match the ones predicted, based on the presumption of rigid bearing geometries, in part I [9] of this study. However, the magnitude of their influence changes significantly, as structure deformation modifies the shape of the lubricant gap in wide ranges. For the investigated cases of this study, consideration of structure flexibility provides lower maximum film pressures and higher minimum film thickness. This improvement of predicted operating behavior reduces the intensity of asperity contacts and of the wear derived from it. Therefore, results indicate that consideration of structure deformation is also important to predict wear-lifetime. Moreover, the modified operating behavior has to be taken into account to characterize the bearing as a component of the entire gear system in the design procedure. Here, the impact of the planet gear bearing properties on the gear mesh design represents an example. In summary, part I of this study indicates that different measures exist to optimize planet gear bearings, while part II additionally shows that a quantitative judgement about these measures requires a consideration of structure deformation, due to its high impact on predicted operating behavior.

The entire study uses a THD model that has been validated comprehensively for journal bearing applications. It is adapted to the planet gear bearing case, and the procedure for the consideration of structure deformation is verified by a full model structure analysis in a co-simulation. However, validation of the entire procedure is not possible yet.

Author Contributions: Conceptualization, methodology, T.H., H.D. and E.R.; software, T.H. and H.D.; investigation, writing, and visualization, T.H., H.D. and E.R.; supervision and funding acquisition, H.S. All authors have read and agreed to the published version of the manuscript.

Funding: This research was not supported by external funding.

Conflicts of Interest: The authors declare no conflict of interest.

References

1. Jang, J.Y.; Khonsari, M.M. On the characteristics of misaligned journal bearings. *Lubricants* **2015**, *3*, 27–53. [CrossRef]
2. Jang, J.Y.; Khonsari, M.M. Performance and characterization of dynamically-loaded engine bearings with provision for misalignment. *Tribol. Int.* **2019**, *130*, 387–399. [CrossRef]
3. Sun, J.; Gui, C.; Li, Z.; Li, Z. Influence of journal misalignment caused by shaft deformation under rotational load on performance of journal bearing. *Proc. Inst. Mech. Eng. Part J J. Eng. Tribol.* **2005**, *219*, 275–283. [CrossRef]
4. Sun, J.; Gui, C.; Li, Z. An Experimental Study of Journal Bearing Lubrication Effected by Journal Misalignment as a Result of Shaft Deformation Under Load. *ASME J. Tribol.* **2005**, *127*, 813–819. [CrossRef]
5. Sander, D.E.; Allmaier, H.; Priebsch, H.H.; Reich, F.M.; Witt, M.; Skiadas, A.; Knaus, O. Edge loading and running-in wear in dynamically loaded journal bearings. *Tribol. Int.* **2015**, *92*, 395–403. [CrossRef]
6. Lahmar, M.; Frihi, D.; Nicolas, D. The Effect of Misalignment on Performance Characteristics of Engine Main Crankshaft Bearings. *Eur. J. Mech. A* **2002**, *21*, 703–714. [CrossRef]
7. Zhang, X.; Yin, Z.; Dong, Q. An experimental study of axial misalignment effect on seizure load of journal bearings. *Tribol. Int.* **2019**, *131*, 476–487. [CrossRef]

8. Bouyer, J.; Fillon, M. Improvement of the THD performance of a misaligned plain journal bearing. *J. Trib.* **2003**, *125*, 334–342. [CrossRef]
9. Hagemann, T.; Ding, H.; Radtke, E.; Schwarze, H. Operating behavior of sliding planet gear bearings in turbine gearbox applications—Part I: Basic relations. *Lubricants* **2021**, in press.
10. Desbordes, H.; Fillon, M.; Frene, J.; Chan Hew Wai, C. The Effects of Three-Dimensional Pad Deformations on Tilting-Pad Journal Bearings under Dynamic Loading. *J. Tribol.* **1995**, *117*, 379–384. [CrossRef]
11. Hopf, G. Experimentelle Untersuchungen an Großen Radialgleitlagern für Turbomaschinen. Ph.D. Thesis, Ruhr University Bochum, Bochum, Germany, 1989.
12. Hagemann, T.; Kukla, S.; Schwarze, H. Measurement and prediction of the static operating conditions of a large turbine tilting-pad bearing under high circumferential speeds and heavy loads. In Proceedings of the ASME Turbo Expo 2013, San Antonio, TX, USA, 3–7 June 2013. [CrossRef]
13. Lahmar, M.; Ellagoune, S.; Bou-Saïd, B. Elastohydrodynamic lubrication analysis of a compliant journal bearing considering static and dynamic deformations of the bearing liner. *Tribol. Trans.* **2010**, *53*, 349–368. [CrossRef]
14. Prölß, M. Berechnung Langsam Laufender und Hoch Belasteter Gleitlager in Planetengetrieben unter Mischreibung, Verschleiß und Deformationen. Ph.D. Thesis, Clausthal University of Technology, Clausthal-Zellerfeld, Germany, 2020.
15. Profito, F.J.; Zachariadis, D.C.; Dini, D. Partitioned Fluid-Structure Interaction Techniques Applied to the Mixed-Elastohydrodynamic Solution of Dynamically Loaded Connecting-Rod Big-End Bearings. *Tribol. Int.* **2019**, *140*, 105767. [CrossRef]
16. Habchi, W.; Eyheramendy, D.; Vergne, P.; Morales-Espejel, G. A full-system approach of the elastohydrodynamic line/point contact problem. *J. Tribol.* **2008**, *130*, 021501. [CrossRef]
17. Oh, K.P. The numerical solution of dynamically loaded elastohydrodynamic contact as a nonlinear complementarity problem. *J. Tribol.* **1984**, *106*, 88–95. [CrossRef]
18. Craig, R.; Roy, R.; Bampton, M.C.C. Coupling of substructures for dynamic analyses. *AIAA J.* **1968**, *6*, 1313–1319. [CrossRef]

MDPI
St. Alban-Anlage 66
4052 Basel
Switzerland
Tel. +41 61 683 77 34
Fax +41 61 302 89 18
www.mdpi.com

Lubricants Editorial Office
E-mail: lubricants@mdpi.com
www.mdpi.com/journal/lubricants